Intelligent Innovations in Autonomous Business Process Management

Houssem Chemingui
https://orcid.org/0000-0002-3351-250X
Brest Business School, France

IGI Global
Scientific Publishing
Publishing Tomorrow's Research Today

Vice President of Editorial Melissa Wagner
Director of Acquisitions Mikaela Felty
Director of Book Development Jocelynn Hessler
Production Manager Mike Brehm
Cover Design Jose Rosado

Published in the United States of America by
IGI Global Scientific Publishing
701 East Chocolate Avenue
Hershey, PA, 17033, USA
Tel: 717-533-8845 | Fax: 717-533-7115
Website: https://www.igi-global.com E-mail: cust@igi-global.com

Library of Congress Cataloging-in-Publication Data

LCCN: 2026016576 (CIP Data Pending)
ISBN13: 979-8-2600-0050-2
Isbn13Softcover: 979-8-2600-0051-9
EISBN13: 979-8-2600-0052-6
British Cataloguing in Publication Data
A Cataloguing in Publication record for this book is available from the British Library.

Table of Contents

Detailed Table of Contents

agentic AI introduces autonomous, goal-oriented agents capable of reasoning, planning, and coordinating complex organizational workflows. The chapter analyzes how these systems contribute to cognitive management by augmenting decision-making processes and enhancing organizational adaptability. Drawing on theoretical perspectives such as dynamic capabilities, absorptive capacity, and distributed cognition, it proposes a conceptual framework explaining how agentic AI supports different stages of the BPM lifecycle, including process design, execution, monitoring, and optimization. The chapter also discusses emerging governance challenges related to transparency, accountability, and human oversight. Overall, it highlights the shift from AI as an operational assistant toward AI as a cognitive partner in organizational management.

This chapter examines how large language models are reshaping business process management from process documentation to intelligent decision support. It shows that the relevance of LLMs in BPM lies not only in text generation but also in their ability to connect dispersed procedural knowledge, formal process representations, analytical interpretation, and managerial judgment. The chapter adopts a continuum-based perspective covering process documentation, process analysis, execution support, and decision support. It argues that LLMs can help formalize fragmented process knowledge, interpret textual evidence on bottlenecks and deviations, support users during process execution, and improve decision preparation through contextualized explanations and recommendations. At the same time, their contribution remains dependent on contextual grounding, data quality, governance mechanisms, and human oversight. The chapter concludes that the future of BPM will likely rely on hybrid sociotechnical arrangements in which LLMs augment rather than replace human actors.

The incorporation of Large Language Models (LLMs) into Decision Support Systems (DSS) signifies a major transformation in the domain of Business Intelligence

Process Sentience has been introduced as a higher order of Autonomous Business Process Management whereby workflows become more dynamic, instead of being rule-based automation, and cognitively coordinated and governance-conscious systems. The response to real-time cues and risk response is achieved through dynamic reconfiguration of processes using structured Cognitive-Orchestration Framework of sensing, reasoning, learning, and adaptive action. Evidence of cross-domain proves that there are measurable improvements in resilience, efficiency, accuracy, and compliance. Autonomy gradients guarantee calibrated autonomy aided by transparency, auditability and human monitoring, making BPM a strategic ability in intelligent these days firms.

The emergence of Autonomous Business Process Management (ABPM) has significantly transformed the way organizations manage their operations and data. This paper explores the integration of governance, risk management, and ethical considerations in ABPM, emphasizing the challenges and solutions associated with data management. Governance frameworks play a crucial role in ensuring that ABPM systems operate within legal, regulatory, and organizational boundaries, while robust risk management strategies are necessary to mitigate potential threats associated with automation and data privacy concerns. Furthermore, ethical considerations, particularly in data handling and decision-making algorithms, must be at the forefront of ABPM implementation to ensure transparency, fairness, and accountability. This study examines best practices in managing data in autonomous environments, proposing an integrated approach that aligns technological advancements with ethical and governance standards to foster trust and sustainability in business processes.

Preface

The rapid evolution of Artificial Intelligence (AI), Large Language Models (LLMs), intelligent automation, and autonomous decision-making systems is fundamentally reshaping how organizations design, manage, and optimize business processes. Traditional Business Process Management (BPM), once centered on process standardization and operational efficiency, is now entering a transformative era characterized by cognitive automation, adaptive intelligence, real-time analytics, and human–AI collaboration. This transformation has given rise to Autonomous Business Process Management (ABPM), an emerging paradigm that integrates AI-driven intelligence into organizational workflows to create adaptive, resilient, sustainable, and self-optimizing enterprise systems.

This book, *Autonomous Business Process Management and Intelligent Enterprise Transformation*, explores the theoretical foundations, technological innovations, organizational implications, and strategic applications of AI-driven process management in the context of Industry 5.0 and intelligent enterprises. The contributions gathered in this volume collectively investigate how organizations can leverage AI technologies—including generative AI, agentic AI, machine learning, predictive analytics, and intelligent workflow systems—to redesign processes, support decision-making, enhance sustainability, and foster organizational resilience.

The relevance of this topic has never been greater. Organizations across industries are facing unprecedented levels of complexity, uncertainty, and digital disruption. The rise of generative AI and autonomous systems has accelerated the transition from static process architectures toward intelligent ecosystems capable of continuous learning and adaptation. Enterprises are increasingly expected to manage unstructured data, automate knowledge-intensive tasks, ensure ethical governance of AI systems, and balance technological innovation with human-centered values. In this context, ABPM emerges not merely as a technological evolution, but as a strategic transformation framework that redefines organizational intelligence itself.

The subject matter of this book sits at the intersection of Business Process Management, Artificial Intelligence, Intelligent Information Systems, Organizational

Transformation, Data Governance, and Industry 5.0. It contributes to ongoing academic and professional discussions surrounding digital transformation, intelligent enterprises, AI governance, sustainability, and cognitive automation. By combining conceptual discussions, theoretical models, applied frameworks, and sector-specific analyses, the book provides a multidimensional perspective on how AI technologies are redefining business processes and organizational decision-making.

This volume is intended for a diverse audience. Researchers and academics in the fields of Information Systems, Business Process Management, Artificial Intelligence, Innovation Management, and Organizational Studies will find theoretical insights and emerging research directions relevant to the advancement of the field. Graduate students and doctoral candidates may use this book as a reference for understanding the evolution of intelligent process management and identifying future research opportunities. Practitioners, consultants, digital transformation managers, and enterprise architects will benefit from the applied perspectives on AI-enabled workflows, autonomous systems, sustainability, governance, and intelligent decision support. Policymakers and innovation strategists may also find value in the discussions surrounding ethical governance, regulatory considerations, and organizational adaptation in AI-driven environments.

The chapters included in this volume collectively address key dimensions of Autonomous Business Process Management and intelligent enterprise transformation.

CHAPTER OVERVIEW

Chapter 1: Foundations and Theoretical Frameworks of Autonomous Business Process Management (ABPM): AI-Driven Process Modeling and Intelligent Workflow Design

Autonomous Business Process Management (ABPM) represents an emerging paradigm that integrates artificial intelligence with traditional Business Process Management to enable self-optimizing and intelligent organizational workflows. This chapter explores the foundational theories and conceptual frameworks underlying ABPM, emphasizing how AI-driven process modeling, machine learning, and advanced analytics transform conventional workflow design into adaptive and autonomous systems. It examines the evolution of BPM toward intelligent automation, highlighting technologies such as predictive analytics, process mining, and intelligent decision-support mechanisms that allow processes to monitor, analyze, and optimize themselves in real time. The chapter also discusses theoretical perspectives that support autonomous process orchestration, including systems theory, digital transformation frameworks, and intelligent workflow architectures.

Chapter 2: From Automatic Assistant to Cognitive Partner: The Role of Agentic AI in Cognitive Management of Business Processes

This chapter examines the growing role of Agentic Artificial Intelligence (AI) in transforming Business Process Management (BPM) within contemporary organizations. Unlike traditional AI systems designed mainly for task automation, agentic AI introduces autonomous, goal-oriented agents capable of reasoning, planning, and coordinating complex organizational workflows. The chapter analyzes how these systems contribute to cognitive management by augmenting decision-making processes and enhancing organizational adaptability. Drawing on theoretical perspectives such as dynamic capabilities, absorptive capacity, and distributed cognition, it proposes a conceptual framework explaining how agentic AI supports different stages of the BPM lifecycle, including process design, execution, monitoring, and optimization. The chapter also discusses emerging governance challenges related to transparency, accountability, and human oversight. Overall, it highlights the shift from AI as an operational assistant toward AI as a cognitive partner in organizational management.

Chapter 3: Large Language Models in Business Process Management: From Process Documentation to Intelligent Decision Support

This chapter examines how large language models are reshaping business process management from process documentation to intelligent decision support. It shows that the relevance of LLMs in BPM lies not only in text generation but also in their ability to connect dispersed procedural knowledge, formal process representations, analytical interpretation, and managerial judgment. The chapter adopts a continuum-based perspective covering process documentation, process analysis, execution support, and decision support. It argues that LLMs can help formalize fragmented process knowledge, interpret textual evidence on bottlenecks and deviations, support users during process execution, and improve decision preparation through contextualized explanations and recommendations. At the same time, their contribution remains dependent on contextual grounding, data quality, governance mechanisms, and human oversight. The chapter concludes that the future of BPM will likely rely on hybrid sociotechnical arrangements in which LLMs augment rather than replace human actors.

Chapter 4: Intelligent Business Decisions with Large Language Models

The incorporation of Large Language Models (LLMs) into Decision Support Systems (DSS) signifies a major transformation in the domain of Business Intelligence (BI). Traditionally, DSS relied primarily on structured datasets and rule-driven algorithms to generate insights and assist decision-makers. With the emergence of LLMs, organizations now have the capability to examine unstructured information, identify complex patterns in textual data, and produce responses that resemble human communication. This chapter examines the paradigm created by integrating the advanced functionalities of LLMs with DSS, thereby strengthening their ability to deliver meaningful insights, support effective decision-making, and enhance organizational performance. It highlights real-time natural language interaction, improved contextual comprehension, and the capacity to process information from varied data sources. In addition, the chapter discusses implementation challenges, including issues related to data privacy, model interpretability, and seamless integration with existing organizational workflows.

Chapter 5: Role of AI in design thinking and innovation

In this study, the author focuses on the application of AI (Artificial Intelligence) to the human-centric and iterative design thinking approach. The research deploys AI tools such as machine learning, natural language processing, generative design, and predictive analytics during the five steps of the design thinking process namely; empathize, define, ideate, prototype and test. It emphasizes the issue that AI can introduce creativity, empathy, and innovativeness to the global trades that would otherwise have been difficult to accomplish. By combining AI with design thinking, it can be possible to finish repetitive work with no human intervention, which is why it is effortless to have instant decisions, as well as, being creative to a significantly higher degree. The application of generative AI technologies offers rapid concept creation, and the analytics fuelled by AI enhances the discovery of the requirements of the stakeholders and testing of the solution.

Chapter 6: Autonomous Business Process Management for Organizational Resilience: Designing Self-Adaptive Process Systems Under Disruption

This chapter examines autonomous business process management as a resilience-oriented process logic for organizations operating under disruption, uncertainty, and rapid change. It argues that conventional BPM, while effective in stable environments, remains limited when processes require runtime adaptation, contextual interpretation, and coordinated response to unexpected events. In response, the chapter conceptualizes ABPM as a socio-technical and governance-bounded approach that integrates sensing, interpretation, decision-making, reconfiguration, and learning within self-adaptive process systems. It further explains that ABPM should not be viewed as unrestricted automation but as bounded autonomy supported by digital infrastructures, human oversight, and resilience-oriented design principles. The chapter concludes that ABPM can strengthen continuity, responsiveness, recovery, and organizational renewal, while its effectiveness depends on governance, observability, knowledge integration, and adaptive process design.

Chapter 7: Autonomy Gradient Architecture-Designing Self-Evolving and Governance-Aware Business Process Systems in the Age of Intelligent Enterprises

The chapter introduces the Autonomy Gradient Architecture (AGA) that is a governance aware platform applied in the creation of self-evolving business process systems in intelligent business firms. Beyond binary automation conceptualizations, AGA conceptualizes autonomy as a continuum with structured components, that is, a continuum between assisted decision support and adaptive orchestration and self-learning processes. Embedded governance layers that bring transparency, accountability, and regulatory alignment are integrated within the structure, using analytics, orchestration, sensing, feedback and governance. Evidence and data in the financial services and healthcare cases demonstrate that resilience, efficiency and accuracy were improved and in measurable ways. The chapter entails autonomy and co-development of the governmental strategies as a strategic foundation of sustainable competitive advantage in Industry 5.0 settings.

Chapter 8: Process Sentience-A Governance-Embedded Cognitive-Orchestration Framework for Autonomous Business Process Management

Process Sentience has been introduced as a higher order of Autonomous Business Process Management whereby workflows become more dynamic, instead of being rule-based automation, and cognitively coordinated and governance-conscious systems. The response to real-time cues and risk response is achieved through dynamic reconfiguration of processes using structured Cognitive-Orchestration Framework of sensing, reasoning, learning, and adaptive action. Evidence of cross-domain proves that there are measurable improvements in resilience, efficiency, accuracy, and compliance. Autonomy gradients guarantee calibrated autonomy aided by transparency, auditability and human monitoring, making BPM a strategic ability in intelligent these days firms.

Chapter 9: "Governance, Risk Management, and Ethical Considerations in ABPM Data Management,"

The emergence of Autonomous Business Process Management (ABPM) has significantly transformed the way organizations manage their operations and data. This paper explores the integration of governance, risk management, and ethical considerations in ABPM, emphasizing the challenges and solutions associated with data management. Governance frameworks play a crucial role in ensuring that ABPM systems operate within legal, regulatory, and organizational boundaries, while robust risk management strategies are necessary to mitigate potential threats associated with automation and data privacy concerns. Furthermore, ethical considerations, particularly in data handling and decision-making algorithms, must be at the forefront of ABPM implementation to ensure transparency, fairness, and accountability. This study examines best practices in managing data in autonomous environments, proposing an integrated approach that aligns technological advancements with ethical and governance standards to foster trust and sustainability in business processes.

Chapter 10: Sustainable and Green Autonomous BPM: Designing Process Intelligence for Environmental Performance

This chapter develops the concept of Sustainable and Green Autonomous BPM as a higher-order process capability through which environmental objectives are embedded into the sensing, analysis, adaptation, execution, and governance of business processes. It argues that environmental performance should not be treated as a peripheral reporting outcome but as an internal dimension of process intelligence. Building on BPM, Green BPM, self-adaptive process systems, and process-level environmental measurement, the chapter explains how environmentally intelligent process systems can detect sustainability-relevant conditions, interpret their operational significance, and adjust execution within defined governance boundaries. It also highlights the role of process mining, environmental metrics, enabling analytics, human oversight, and staged implementation. The chapter concludes that the future of BPM will increasingly depend on its ability to integrate ecological intelligence into adaptive and accountable process management.

Collectively, these chapters demonstrate that Autonomous Business Process Management is not simply an extension of traditional BPM, but rather a paradigm shift toward intelligent, adaptive, and human-aware organizational ecosystems. The book contributes to the advancement of knowledge by integrating interdisciplinary perspectives from AI, information systems, organizational management, sustainability, governance, and digital transformation.

This volume also contributes to the broader transition from automation-centric thinking toward cognitive enterprise transformation. While earlier generations of BPM focused primarily on efficiency and optimization, the emerging ABPM paradigm emphasizes intelligence, adaptability, collaboration, resilience, and ethical responsibility. The chapters included in this book reflect this transition by exploring both technological capabilities and organizational implications.

We hope this book stimulates future research, encourages interdisciplinary collaboration, and supports practitioners and scholars in understanding the evolving role of AI within business process ecosystems. As organizations continue to navigate the challenges and opportunities of intelligent transformation, the insights presented in this volume aim to contribute to the responsible development of adaptive, sustainable, and human-centered enterprise systems.

Ultimately, this book positions Autonomous Business Process Management as a critical pillar of the intelligent enterprise of the future. By bridging theoretical foundations with practical applications, it offers a comprehensive perspective on how AI-driven technologies are redefining organizational processes, managerial decision-making, and enterprise transformation in the era of Industry 5.0.

Acknowledgment

The editor would like to express their sincere gratitude to all authors who contributed to this book. Their valuable research, dedication, and scholarly contributions have made this volume possible and have significantly enriched the discussions surrounding Autonomous Business Process Management, Artificial Intelligence, and Intelligent Enterprise Transformation.

We also extend our appreciation to the reviewers for their time, expertise, and constructive feedback, which greatly contributed to improving the quality and academic rigor of the chapters included in this book.

Special thanks are due to the publishing and editorial teams for their continuous support, professionalism, and guidance throughout the preparation and publication process.

Finally, we would like to acknowledge the academic and professional communities whose ongoing research and innovation continue to advance the fields of Business Process Management, Artificial Intelligence, Industry 5.0, and digital transformation.

We hope this book will contribute to future research and inspire continued exploration in the development of intelligent, adaptive, and human-centered enterprise systems.

Chapter 1
Foundations and Theoretical Frameworks of Autonomous Business Process Management (ABPM):
AI-Driven Process Modeling and Intelligent Workflow Design

Amit Verma
http://orcid.org/0000-0002-9283-7873
Kalicharan PG College, Lucknow, India

Y. Srinivasa Rao
http://orcid.org/0000-0001-9483-0707
Vignan's Foundation for Science, Technology, and Research, Valdamudi, India

Nidhi Oswal
http://orcid.org/0000-0001-5045-7226
Liwa University, UAE

Punit Pathak
Auro University, India

ABSTRACT

Autonomous Business Process Management (ABPM) represents an emerging paradigm that integrates artificial intelligence with traditional Business Process

DOI: 10.4018/979-8-2600-0050-2.ch001

Management to enable self-optimizing and intelligent organizational workflows. This chapter explores the foundational theories and conceptual frameworks underlying ABPM, emphasizing how AI-driven process modeling, machine learning, and advanced analytics transform conventional workflow design into adaptive and autonomous systems. It examines the evolution of BPM toward intelligent automation, highlighting technologies such as predictive analytics, process mining, and intelligent decision-support mechanisms that allow processes to monitor, analyze, and optimize themselves in real time. The chapter also discusses theoretical perspectives that support autonomous process orchestration, including systems theory, digital transformation frameworks, and intelligent workflow architectures.

1. INTRODUCTION

Instead, more companies are implementing tools that enable them to coordinate digital processes and workflows so they can complete this work faster. In this context, Business Process Management (BPM) has emerged as a key area for analyzing, modeling and optimizing how work is conducted across any organization. A business process is a series of linked tasks performed in an organization to achieve a specific goal. Examples of B2B tasks include processing customer orders, doing financial transactions, delivering services and managing internal processes. If organized and structured, businesses can make a seamless run and give their customers more value. BPM allows companies to monitor their activities and identify opportunities for improvements. It also ensures that every step progresses the business toward its objective (Wu et al. 2019).

Business process management systems (BPMS) were mainly used for the tracking of processes and automation of tedious tasks. Organizations introduced software systems to log activities, assign tasks and ensure the process was followed. These technologies contributed to minimizing human errors, automating routine tasks and providing uniformity in operations. However, most of these systems were rule-based with predetermined steps to take. If something unexpected occurred, human managers would usually have to intervene and change the method by hand. Unlike tightly bound rules-based automation, these systems know better. They scan operational data for patterns and recommend ways to improve processes (Davenport 1993).

We welcome the continuous, diligent assessment and enhancement of automated process systems. These systems analyze historical data and monitor the execution of tasks over time to identify issues and recommend remedies. This makes it easier for companies to address issues quickly. For example, when monitoring transaction activity for anomalies, banks can identify potential risks. Digital process systems can help healthcare companies coordinate scheduling of patients and resources. Similarly, technology that can track demand trends may aid supply chain operations and provide the information managers need to make rapid logistical decisions. Having good stock management practice strengthens organizations. Disruptions are quite common in an industry as changing in nature as business, and causes can range from alterations in the market to operational hurdles and abrupt drops in demand. Set of technologies that MONITORS the performance of a process and enables management to detect deviations from specifications rapidly enough to attribute them to specific causes. This may make companies more resilient as things change. Computer-human Partnerships. Many proponents of these benefits are widely used and have often to discuss them, but we should always note that technology can never fully replace human judgment. There will continue to be a need for people to monitor the processes, detecting intelligence insights and determining whether any decisions are consistent with the morals and values of the business. So, if technology is to serve well, it must be able to do the things that people can do (Mendling 2008).

1.1 Research Questions

This chapter is guided by the following research questions:

RQ1. How has Business Process Management evolved from rule-based workflow automation to AI-enabled autonomous process management?
RQ2. What theoretical foundations explain the development of Autonomous Business Process Management in AI-driven organizational environments?
RQ3. How do AI-driven process modeling, intelligent workflow design, and autonomous decision mechanisms transform business process management?
RQ4. What role do process intelligence, feedback loops, and continuous optimization play in autonomous BPM systems?
RQ5. How can organizations govern autonomous BPM systems through transparency, accountability, fairness, compliance, and human–AI collaboration?

1.2 Methodology and Research Design

The research design of this chapter is conceptual synthesis that builds on a structured literature review of studies in Business Process Management, Autonomous Business Process Management, process mining, AI-based process modeling, intelligent workflow design, workflow orchestration and work-process intelligence as well as human-AI governance. The core statistics and primary data are not included in the chapter. Instead, it develops a conceptual understanding of the AI technology that has theoretical bases that help to model business processes, design, decision support and optimization.

Theme synthesis in literary studies. The study traced the evolution of process management from classical BPM, the study of workflow automation, business process reengineering (BPR) and enterprise integration. Second, AI product and process modeling based on a systematic synthesis of the process mining, machine learning, predictive analytics and natural language processing (NLP) literature. Third, we investigated different constructs, e.g., adaptive workflows, decision engines, autonomous agents, process intelligence and feedback mechanisms within the context of the ABPM operational logic. Fourthly, responsible performance was based on a combination of governance, human oversight, transparency, fairness and accountability literature.

The conceptual framework consists of five iterative dimensions, namely data acquisition, AI analytics, workflow orchestration, autonomous process execution, and continuous monitoring. Conceptually, the justification is based on theoretical coherence, internal consistency, business relevancy and fit with digital transformation and intelligent process automation trends.

2. EVOLUTION OF BUSINESS PROCESS MANAGEMENT

Like everything else in a firm, business process management (BPM) has changed over time. In the past, businesses put a lot of effort into making sure that the exchanges that happen naturally in the course of everyday work were formalized and that the culture of the workplace was preserved even when trends came along. However, the growth of digital technology and the huge amounts of data that are now available have made organizational settings more complicated and less stable in the last several decades. So, BPM's evolution may be seen as a journey from automating tasks to creating smart process systems (Rauch et al. 2025).

2.1 Early workflow automation and process digitization

During the first growth phase of BPM, the main goal was to digitize everyday administrative tasks. Before the digital revolution, most business operations were written down on paper. Several things, including getting permission to buy something, taking time off, and getting money, needed tangible papers to be sent from one department to another. It was hard to trace, took a lot of work, and was easy to make mistakes. Companies started using process automation technology to get around these problems. These allowed employees to fill out electronic forms, which were subsequently sent along predetermined approval pathways. Using digital procedures is a game changer because it lets companies cut down on the amount of paperwork they must do by hand, which speeds up processing and makes it easier to make sure the data is correct. Information about the third dimension—specifically, the office level that was most important in this situation, sharing positive and bad news for Biden—was no longer physically carried from one office to another. Instead, it moved across digital networks. But early process automation systems were rather rigid. They were usually based on strict rules that outlined how clean rooms should work and how each step of the process should go. One of these would be to take action to change the system in reaction to an unanticipated event. Managers or system administrators must add new requirements to the process configuration (Aalst & Ter 2004).

Still, process automation marked an important step in the digitization of enterprises. It set the strategy for coordinating work on discrete and structured digital processes, along with the technology infrastructure for what would become known as BPM.

2.2 Enterprise Integration & Data-Driven Process Optimization

As organizations grew more experienced with this technology, it became clear that while automation offered promising improvements to the organization, it alone was not sufficient to achieve performance gains. In the 90's, Many Organizations started rebuilding the structure of their operational processes through initiatives such as Business Process Reengineering BPR. So instead of just taking existing processes and putting them in a digital frame, BPR suggested that organizations need to redesign their processes to counteract any time wasted thereby increasing productivity. During this period, the popularity of Enterprise Resource Planning (ERP) technologies increased. Powerful ERP platforms are designed to integrate various business functions, including finance, purchasing, production, and human resources, into a single system (Krogstie 2016).

This provided management with a better overall view of the organization's operations and harmonized the flow of information between different departments. This era encompassed some major developments such as data-driven process analysis methodologies. As businesses moved to more reliance on digital entities, large amounts of operational data were generated in the form of system events and transaction logs. Citation: The data sources enabled researchers to study the real world of how militants operate within organizational settings. Managers could then review the event logs and have a better representation of what was happening inside holistic systems by using process mining techniques. This enabled companies to detect delays, inefficiencies and divergences from desired processes. Instead of depending only on theoretical process diagrams and theory, managers could observe how processes were executed in practice and use that information to optimize performance.

These advancements resulted in BPM transitioning into a discipline that encompassed continuous monitoring and process analytical optimization. Processes have changed from being static entities to systems that could be analyzed and improved by empirical data, when combined with the appropriate other processes (Chen 2012).

2.3 Business Process Management with AI And Autonomous

In its latest iteration, BPM has expanded to include artificial intelligence and advanced analytic technologies as part of a large expression of process management systems. Markets of the twenty-first century are characterized by instability, rapid technological development and complex operational demands. Process systems must be both adaptive and facilitate effective decision making to enable operations in such environments. Artificial intelligence provides process systems with the ability to analyze vast data sets, identify patterns within those data sets and create predictive insights. It is in the context of BPM that these capabilities will enable organizations to prepare for disruptive forces, better deploy their resources, and ultimately improve service delivery. Customer service operations, for example, can apply predictive analytics to estimate the expected levels of demand and adjust their supply chain processes based on projected changes in demand or transportation conditions (Burgueño et al. 2021).

This was the beginning of Autonomous Business Process Management powered by these technologies. In this model, process systems can monitor operational data in real time and adapt workflows as needed. For example, intelligent software agents can recommend or carry out adjustments for reallocating resources, changing the sequence of tasks and initiating alternative procedures when disruptions occur.

Table 1 outlines the various levels of BPM, demonstrating how process management has grown from simple rule-based automation to intelligent self-adaptive systems. This transformation reflects a larger shift in corporate technology, with digital systems now playing an increasingly important role in both operational execution and analytical decision-making. BPM maturity is marked by a clear transition from the digital management of jobs to the smart and flexible orchestration of processes Automation of processes from the beginning brought more uniformity, but still was tied to strict rules. Cross-functional collaboration was enhanced by enterprise integration and process reengineering. Intellectual transparency was achieved through event logs and process mining with data-oriented BPM. Then you have AI-augmented and autonomous BPM that further pushes out these capabilities with predictive insights, adaptive routing, AI-enabled intelligent decision-making support and continuous process on top of that. Such a development requires a theory that describes ABPM as an IT- and organization-related transformation.

Table 1. Evolution of Business Process Management

BPM Stage	Key Technology	Characteristics
Workflow Automation	Rule-based systems	Static process flows
Process Reengineering	ERP systems	Integrated enterprise processes
Data-driven BPM	Process mining	Analytics-based optimization
AI-augmented BPM	Machine learning	Predictive insights
Autonomous BPM	Intelligent agents	Self-adaptive workflows

3. CONCEPTUAL FOUNDATIONS OF AUTONOMOUS BUSINESS PROCESS MANAGEMENT

In this section we describe the structure of the theory of Autonomous Business Process Management. ABPM is grounded in systems thinking, organizational learning theory, cybernetic control, data-driven decisions, and Human–AI cooperation. Formal theoretical foundations explain why autonomous process system should be seen not only as an automation technology, but rather as adaptive socio-technical systems. The central thesis is that ABPM delivers value through an integrated process management framework that brings together data, AI analytics, workflow orchestration, human oversight and feedback loops.

3.1 Systems Thinking and Process Interconnectivity

Systems theory is considered as the theoretical biosphere for self-led BPM ecosystems which are complex systems of entangled actuators and sensors in constant feedback. This view is that organization operates through networks of relationships, information flows and shared resources rather than discrete department or task. From a systems perspective, a business process is not just a sequence of independent actions; there seems to be an inherent purpose for which it exists as part of the larger mechanism that is the organization. For example, fulfilling a customer order may involve collaboration between inventory systems, finance departments, transportation departments and sales teams. This means that changes in one phase of the process can affect others. However, traditional methods of process management have never ceased to focus on the efficiency or effectiveness of each individual process at the expense of a higher-level view (Leno et al. 2020).

Autonomous BPM: Moving up a notch! It studies the interaction of processes in an organization. Intelligent process systems aggregate data from the multitude of sources and formats, accurately present and evaluate it for the purposes of establishing relationships between operational activities. This helps the company decrease operational fragmentation, boost inter-departmental collaboration and minimize waste. In complex systems, if a system has an aim, then the aim is likely more about its ability to adapt than any one piece which it already has. All businesses need to keep pace with the changing consumer needs, technological innovation, and competition in the market.

This flexibility is enabled through autonomous BPM systems, which allow the process to be adjusted on the fly as conditions change. Historically, processes tended to follow structured and non-adaptive systems; these next-generation systems consider data outside of traditional operating silos and suggest subtle adjustments to optimize the entire system (Zhu et al. 2011).

3.2 Adaptive Learning and Data-Driven Organizational Learning

An additional pillar of autonomous BPM is organizational learning theory, which underlines the importance of knowledge learning in organizations. This concept suggests that businesses enhance their performance by looking back on past events and using what they learned to make better decisions going forward. For instance, firms could collect information on successful practices and waste over a long period. People at mature organizations relied more on human experience and managerial judgement. Managers reviewed past results, shared experiences and adjusted operating procedures on the fly. This has always been important, but new opportunities evolved with data analysis on current digital learning systems (Casati & Shan 2001).

Autonomous BPM systems closely relate to organizational learning, analyzing historical process data for patterns of operational behavior. Machine learning algorithms can process massive datasets created by corporate systems and identify trends or patterns that human analysts might overlook. For instance, such systems may spot repeat bottlenecks in specific stages along the process path or predict customer demand patterns and potential threats within operational procedures. Intelligent process systems help organizations make more informed choices related to process structure and management by transforming raw data into actionable insights.

3.3 Cybernetic Control and Feedback Mechanisms

But in organizational operational controls, feedback is critical. Time tracking is used by managers to monitor operational performance service time, resource usage, error rates, etc. During troubled times, managers make changes to enhance performance. Similar feedback mechanisms are applied in autonomous BPM systems that employ digital observation technology. These systems integrate streaming data from corporate apps, workflow services and back-end databases. This data is then analyzed to gain insights into the operations' performance. When detected, inefficient or aberrant patterns can trigger the system to automatically take corrective action. It might suggest moving resources, rearranging process priorities or changing the order in which work gets done to maximize performance. The above feedback loops stabilize the system as it attempts to self-improve operational performance on a recursive basis (Reijers & Aalst 2005).

The conceptual architecture of such a system is shown in **Figure 1** where the various data sources (sensor and camera records) interact with analytical systems (scanner, loader, converter), workflow coordination mechanisms (process engine) and performance monitoring functions in an autonomous BPM framework. ABPM is built on the theory of systems thinking, adaptive learning and cybernetic feedback. And this is where systems thinking helps explain the interconnectivity of the processes that cross over the departments and functions. Process systems learn through Organizational Learning, which integrates data and real-world expertise. Cybernetic Control describes the nature of feedback loops for observation, correction and adaptation. These notions point out that ABPM is not just plain automation but a smart process management approach which is adaptive during execution, reacting to emerging situations and fostering continuous organizational growth.

Figure 1. Conceptual Architecture of Autonomous BPM

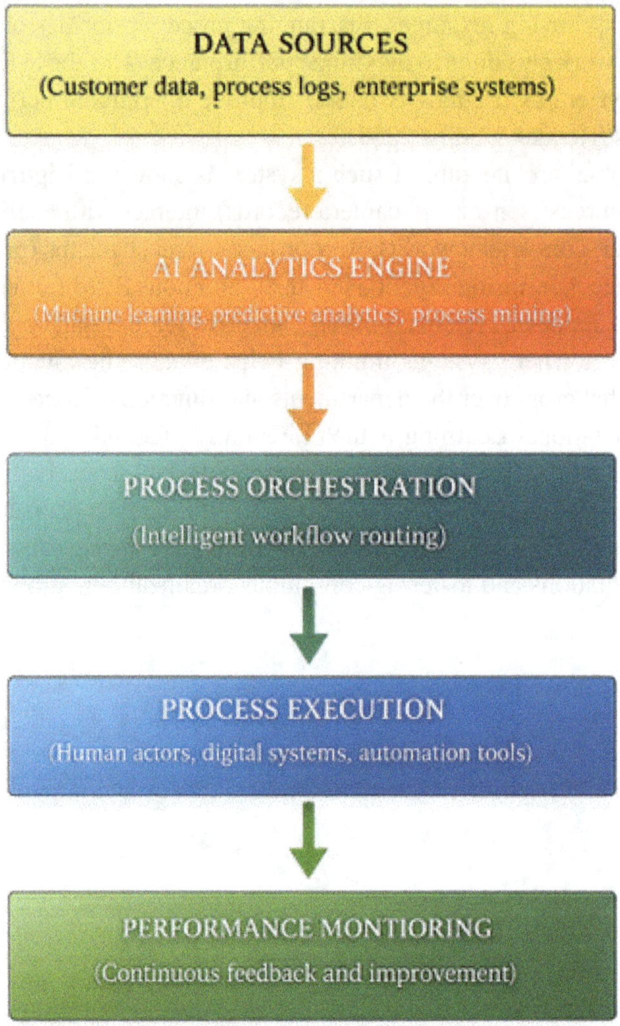

In short, you must embrace the philosophical principles underpinning Autonomous BPM — effective process management requires automating more than just technology. It involves the construction of adaptive decision-making feedback loops, the ongoing learning from operational experience and treating organizations as correlated systems. By integrating these theoretical insights, autonomous BPM systems help enterprises cut through complexity, attain coordination and sustain performance in rapidly changing business contexts.

4. ARCHITECTURE OF AUTONOMOUS BPM SYSTEMS

ABPM shows that autonomous process management requires multilevel interconnection. The data layer captures operational evidence, the intelligence layer converts raw data to analytic insights, the orchestration layer manages workflows, the execution layer executes process actions, and the monitoring layer evaluates results. The results help suggest that the effectiveness of ABPM depends more on the synergies between these layers than on any one technology. This building block thus gives the operational basis for intelligent and adaptive process management.

4.1 Data Collection and the Data Management Layer

The data layer aggregates most of these sources and converts it to the format understandable by upper-level systems. Typical data sources include enterprise resource planning systems, customer relationship management solutions, supply chain databases and digital work management platforms. All these systems monitor occurrence scenarios triggered by real business processes: order taking, service delivery or provisioning, resource allocation or financial transaction. You will appreciate that this information is fundamental, as it presents the plating required to comprehend practically how business processes work. Data issues are a hindrance to identify wasteful or unfit business processes Elsey tayo on Unsplashed Ultimately, where the process fails in any re-enforcing loop will be where all other results differ from expectations. The data layer is essentially the knowledge base upon which all other autonomous BPM architectures are founded (Bandara et al. 2005).

4.2 Intelligence and Analytical Processing Layer This

After the operational data is collected, it needs to be analyzed to extract insights. This is the work of a lawyer of architecture responsible for intelligence. This intelligence layer leverages algorithms and analytical tools to analyze process data for determining areas of potential anomalies, trends, and patterns. Predictive analytics techniques and machine learning models are commonly used in this layer. These tools then analyze historical data, extracting insights from past operational results.

Slowly they learn to recognize patterns that develop as processes are carried out. Analytic systems, for example, may not only identify when service delays are more common across certain phases or operational scenarios (Huser 2012).

Also, the intelligence layer enables anomaly detection. To justify these intelligent systems, need at least a lot of data to have an idea of what normal is and then define every deviation anomaly as defects in operation or behavior predictably, so potentially dangerous. This could help detect unusual transaction volume or system behavior, for example through classifying a normative operating point versus abnormal one several times (Njima et al. 2024). Beyond identifying issues, the intelligence layer can also offer predictions about upcoming operational states. Forecasting models can predict customer (e.g., food or news delivery) demand, resources needed, and lead time to service. These predictions provide insight to organizations about what upcoming changes are most likely and how they can appropriately allocate resources.

4.3 Workflow Orchestration & Coordination Layer

Though the insight from data analysis helps, organizations need to interpret these insights into coordinated implementation of operational processes. The orchestration layer of autonomous BPM architecture takes ownership of this responsibility. Orchestration layer — the layer that determines how different departments and digital systems interact as part of the workflow. Digital workflows that live in this layer define and determine how work is divided, what resources to allocate where, and how decisions are channeled through the organization. When a customer service request is submitted, the workflow system decides what department should address the request, what information is needed to process it and how the request should flow through the system (Ter et al. 2010). Static rules are not the only way to orchestrate in autonomous BPM systems. Instead, it may draw on insights produced by the intelligence layer. This layer enables organizations to adhere to an execution structure by separating concerns for process executions, thereby enabling flexibility around workflow management.

4.4 Process Execution Environment

Execution layer operational environment in which business activities are performed. This layer contains the man worker, software applications and automation systems to accomplish process tasks. This is the domain in which organizational workflows manifest as actual operational results. Human actors are still an important component of the execution layer. Employees exercise judgment, engage with customers, and perform work that involves creativity or contextual knowledge. Simultaneously, automated software agents and enterprise applications execute common routines like data processing, transaction validation, and system updates (Dumas et al. 2013).

While the automation provided by the execution layer streamlines many steps, it still relies on human talent at various stages to account for more complex decisions and other areas that require flexibility.

4.5 Continuous Monitoring & Feedback Mechanisms

A critical feature of autonomous BPM systems is that they can instrument and track the performance of a process over time. Monitoring systems monitor indicators like processing time, resource utilization, service quality and error rate. Organizations can utilize this data to determine if processes are working and deliver the results they were intended to. Performance monitoring yields feedback loops that help drive continuous improvement as well. Systems can alert managers or start analytic processes to discover potential optimizations when performance metrics indicate inefficiency or other unexpected behavior. Some of these can be done automatically in the orchestration layer (Li et al. 2014).

A general overview of this architecture is depicted in **Figure 2**, showing the data flow from analytical systems to a decision maker and/or workflow environment, returning to the monitoring step for evaluation. And **Table 2** showing core components of Autonomous BPM Architecture

Figure 2. AI-Driven BPM Architecture

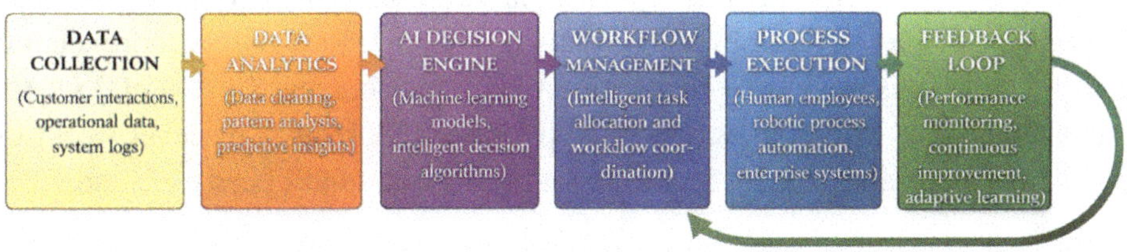

Table 2. Core Components of Autonomous BPM Architecture

Component	Function
Data Layer	Collects operational information
Intelligence Layer	Analyzes process data
Orchestration Layer	Coordinates workflows
Execution Layer	Executes operational tasks
Monitoring Layer	Tracks process performance

The architecture of autonomous BPM systems shows how contemporary businesses use a single technology platform to handle data, use analytical information, coordinate workflows, and carry out operations. By putting these parts together, companies may better handle complicated procedures and change how they do business as circumstances change.

5. AI-DRIVEN PROCESS MODELING

Process has always been a critical aspect of Business Process Management because modeling helps organizations to visualize how different activities will flow and work is transferred between departments. In the past, organizations depended on business analysts and process experts to use graphical modeling tools to document workflows. These models were made by interviewing employees and observing procedures and using that to create diagrams. Although this method allowed organizations to visualize their processes, it also came with limitations. They could require a lot of time to put together and often reflect the idealistic picture rather than the actual layout of how work is done within business (Ong et al. 2025).

Employees often informally adjust such procedures to suit workplace realities; as a result, operational practice may diverge from documented workflow. The digital systems and enterprise software have led to large amounts of operational data being accrued for different processes. This has opened new avenue to create more accurate and better process models by integrating these data sources. Such digital records are leveraged by AI-based process modeling to autonomously discover workflow patterns and generate precise replicas of how processes have been executed.

5.1 Traditional Process Modeling: Shortcomings

Business analysts examined the existing procedures and transformed them into graphical diagrams of the process. Additionally, they were employed to direct operational activities and design workflow systems. Even though this method provided a precise representation of organizational operations, it frequently lacked an understanding of the complexities of genuine systems. Occasionally, employees modify processes to accommodate unforeseen events without formally documenting the changes. This suggests that the which process model is likely an idealized view of labor rather than a day-to-day reality (Lübke et al. 2021).

Manual modeling does not work either, since processes are dynamic and not static. The change of organization operating techniques is often forced by two factors — global economic developments and rapid change in consumer demand forecasting and technology innovation. Manually updating process models can be tedious every time a change happens, resulting in stale or outdated documentation. Sourcing and interpreting the data became challenging, thus needing innovative process models that are not dependent on human interpretation.

5.2 The Business Process Discovery Powered by AI

AI based process modeling uses automation and analyzes operational data to decide how processes are executed in an organization. Instead of plotting out each stage of a process by hand, AI systems analyze historical event logs generated by corporate applications and automatically reconstruct the events from them. Process mining is a powerful tool for that. Enterprise systems produce a data trail of transactions, such as time stamps, unique task identifiers and resource consumption. Process mining algorithms analyze this data to gain insights about how activities are performed and what types of processes are interconnected.

Organizations can do this by monitoring event logs. Finally, process mining discovers bottlenecks, delays and deviations during workflow execution. These insights allow managers to identify and improve operational efficiency and adapt processes accordingly. Machine learning approaches go even further by looking for patterns in process data, as described earlier (Singh et al., 2026).

5.3 NLP and Knowledge Systems-Based Intelligent Modeling

An especially valuable application of natural language processing (NLP) is translating proposed operations (found in manuals and procedures) into structured commentaries. Most of business organizations have documentation about their habits, regulations, and standards of operation. These documents would then have to be analyzed manually by analysts in traditional process modeling. It can automatically read activity, decision point, and task sequence text descriptions with the help of natural language processing (NLP) (Reichert & Weber 2012).

This ability allows organizations to quickly turn existing paper assets into digital assets and words on a page into executable, graphical workflow designs. This makes it far easier and quicker to model your processes. Process models edit an essential use of this type of data is for the development AI-based process models, where they structure information about interconnections between events (861), resources (860) and organizational roles. Graph databases can serve as a data structure to represent the organizational knowledge of activities in the structured BPM systems previously defined, thereby allowing any way of querying their relationships. It is an evolution in modeling that can predict will perform processes during a variety of conditions. Predictive models are built using historical data to predict processing times, pinpoint potential bottlenecks, and estimate the resource requirements. The current characteristics of the primary AI technologies that could support process modeling extraction are detailed in this table (**Table 3**). This table illustrates the diversity of software analytics tools of this nature and their potential to enhance business processes and provide insights, (Indulska et al. 2009). Process modeling is research done with AI platforms – a process model is dynamic, predictive and evidence-backed. Traditional modeling was largely based on expert opinions, and often depicted fictive processes. AI assisted modelling uses event logs, process mining, machine learning and natural language processing (NLP) and knowledge graphs to know what the real process behaviour is. It enhances the model's accuracy and allows for the detection of bottlenecks and prediction of performance—this is data-driven restructuring. The key gap is therefore the need to combine skills in technological modelling with knowledge of governance and human judgement to ensure that the implementation of the new technology is done properly.

Table 3. AI Technologies Used in Process Modeling

Technology	Role
Process Mining	Discover workflows from event logs
Machine Learning	Predict process outcomes
Natural Language Processing	Extract processes from documents
Knowledge Graphs	Represent organizational knowledge
Predictive Analytics	Forecast process performance

AI in Process Modeling is a game-changing technology covering Business Process Management. Organization Process may be represented as models with process mining, machine learning, natural language processing and predictive analytics can represent reality more accurately. These technologies allow organizations to constantly monitor their workflows, detect problems in the early stages, and adapt processes while they grow.

6. INTELLIGENT WORKFLOW DESIGN

Workflow design I the Business Process Management domain that defines how work should flow through an enterprise and specify the distribution between workers and digital technologies. Traditional BPM systems outlined workflows in advance according to rigid rules and structured diagrams. For the sake of dynamism and both fluidized enterprises, daily activity in what have been (at least) a highly modularized work processes: based on sequencing. But with enterprises moving towards dynamic data-focused strategies, flexible workflows also grew in demand Mishra et al. (2026).

This automation of processes is only the tip of the iceberg when it comes down to intelligent design for workflow, which takes this cup a notch higher. In contrast, intelligent workflows are analytical and can respond to business rules — which they interpret rather than simply react to — that enables adjustments of processes with respect to multiple business needs. These systems combine contextual information — levels of workload and available resources, as well as priorities set by clients — to decide next steps in dealing with tasks. This means that processes can adjust dynamically according to what is to be done at any given time, rather than being carried out in a fixed order.

6.1 From Static Workflow Automation to Intelligent Process Routes

Traditional workflow systems were intended to automate routine processes. When a process was modeled, the workflow engine performed the tasks in the same order every time that process was invoked. The service desk would pass a request from the client to support, and it supported would pass it to finance to ensure payment. This approach worked for day-to-day business functions, but it fell short in more complex scenarios. In a real-world business, the scenario is often more attached to "dynamic case management": Resources go unavailable at times, the consumer demand goes up and down, and some of the departments have much more workload than others. In some contexts, there is a relatively narrow workflow framework that allows them to come to a standstill or even stop a work process because functions cannot be assigned to other employees (Smirnov 2012).

Intelligent Workflow Solutions mitigate this drawback through adaptive process routing apparatus. Intelligent workflow engines do not assign work based on rigid rules; it first evaluates real-time operating conditions and decides the most efficient way to complete a task. is an automated system that can detect when service teams are overwhelmed with work and automatically forward incoming tickets to other available teams. The advantage of being able to dynamically reroute is in operational effectiveness for companies when faced with varying operating circumstances. These data chunks make intelligent processes promote adaptive dynamics and transparent organizational scaffolding.

6.2 AI-Enhanced Decision Making in Work Process Management

The ability to create smart workflows through AI. Decision engines driven by AI analyze operational data to determine how the work will assign and create execution strategies. These decision systems rely on a wide array of elements: mission assignment, resource availability, process urgency and importance levels assigned to consumers (Rozinat & Aalst 2008).

For example, before a task is assigned to a person or department, the system could check previous case breakouts, resource status and estimated response times when it receives a service request. This allows the relevant resource to manage each of these processes in real-time. Machine learning models also enable better workflows by detecting patterns in historical data associated with these processes. It can also look for patterns to avoid future logjams and suggest alternate ways of routing through the process if various channels cannot be executed as execution becomes more chronically delayed or perceivably inefficient (via predictive modeling). The system adjusts it automatically as more production data is collected.

It will be critical to work through the elements of AI that decide whether you do or do not meet the criteria for a certain role. This will not mean that urgent client complaints get priority over usual inquiries or that verification checks may be added for transactions above a certain value. These intelligent prioritization techniques empower organizations to allocate resources more effectively and deliver better service quality in the long-term (Panwar et al., 2024).

6.3 Dynamic Monitoring and Continuous Workflow Enhancement

For instance, monitoring systems can find that a particular step of the workflow consistently takes longer than expected and some jobs require more resources than anticipated. Informed by this data, the system can suggest improvements to process architecture. Under some circumstances, workflow paths may be modified to eliminate nonessential stages or redistribute accountability among divisions. Automation will, with use, be progressively applied across the solution scope to remove manual work where necessary (Aalst .2013)

As shown in **Figure 3**, intelligent workflows continuous improvement cycle consists of analysis framework that incorporates process design, execution framework and monitor framework with feedback from optimization to adopt the change in adaptive workflow environment as the system learns from its operating experience and identifies areas for improvement, workflows are adjusted through several iterations. Intelligent Workflow design research indicates that workflow systems are evolving from rigid routing to more flexible process coordination. Intelligent processes respond in real time to workload, resource availability, job priority, customer urgency and operational risk. This means that companies can achieve greater efficiencies than static rule based systems in dealing with complexity. But adaptive process design has different decision boundaries, transparency and review requirements to prevent ungoverned automation. Thus, the smart design of workflow should be regarded as a technological competence and a governance challenge.

Figure 3. Intelligent Workflow Lifecycle

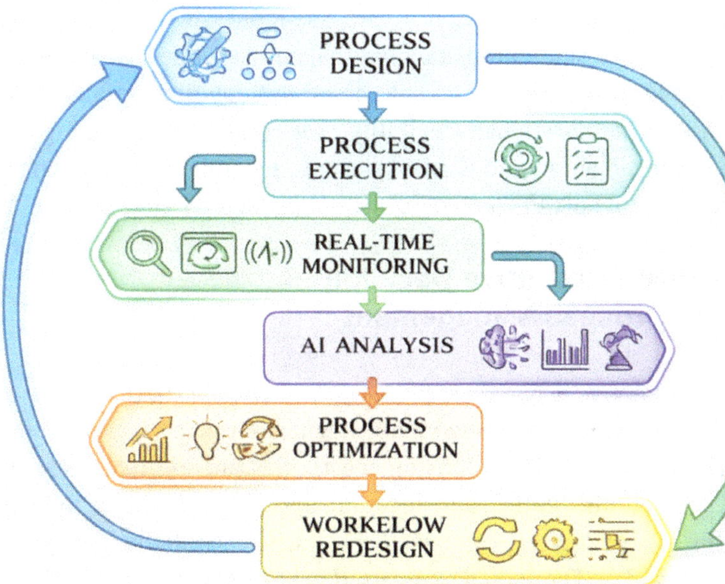

Modern process management systems have been greatly benefited by smartly designed workflows. Using AI, Organizations can develop workflows that evolve based on the working circumstances by integrating AI-based decision structures, real-time monitoring techniques and adaptive workflow routing approaches among others. With these features firms can enhance the quality of their services, remain efficient as well as continuously adapt their operational processes over time.

7. AUTONOMOUS DECISION-MAKING IN BUSINESS PROCESSES

Decision making is one of the most important parts of many organizational processes. From authorizing a financial transaction to assigning a service request, most operational activities require some form of judgment. Such decisions have traditionally been made by managers or employees who evaluated the information at their disposal and decided on a course of action. And although human judgment will always be crucial, most modern companies want to assist or automate certain kinds of decisions using digital tools (Fouda, 2026).

They are autonomous Business Process Management (ABPM) systems where intelligent decision mechanisms analyze various large data volumes and provide recommendations in real time. In traditional workflow systems, decisions were made using static rules. As an example, if a purchase request was over a certain dollar amount, the system would automatically escalate the request to another manager for approval. While those kinds of rule-based systems helped automate routine decisions, they didn't have much flexibility over complex or changing conditions. Even more advanced autonomous BPM systems are embedding analytical models that consider multiple parameters to decide on the best possible step.

7.1 Data-Driven Executive Decision Support for Process Management

Decision autonomy is based on data, and it forms the backbone of a sizeable portion of it. Until now, corporate systems generate operational data for how the system works, transaction details, customer activities and resources used. Intelligent BPM does not store this in the same location, allowing for entrees to be managed, providing for more extensive options. This is where predictive analytics models start coming into play. These models are used to process past data and detect trends/patterns affecting operational performance. Predictive models use historical process performance to predict future results, which might be vulnerable to operational harmonic disruptive, demand conditions or schedule disruptions. If managers can understand how to better use the data they have, they can make more informative decisions (Aziz & Nuryana, 2025).

For example, predictive analytics models might analyze data points such as a client's credit history, income level and repayment histories to approve their loan application within the financial services sector. For example, retailers might use predictive algorithms to project product demand and control inventory levels. Clearly, by inserting the predictive insight directly into the workflow system organizations have been able to make decisions faster and mitigate uncertainty in their operational planning. There are other examples.

7.2 AI-Enabled Decision Engines and Automated Process Actions

Plus, they have specialized decision engines that evaluate operational conditions and recommend actions accordingly. These are decision engines powered by analytical algorithms, which analyze many variables at the same time. Rather than using just one rule, it could also consider a multitude of other factors including operational constraints, client priorities, resource availability and demand levels. For example, the decision engine might consider how urgent a service request is, whether there are any qualified individuals available and what the expected response time would be. This analysis is used by the system to allocate work to the most productive resource available (Kumar, 2025).

This automation of decision-making also allows firms to bypass red tape and clear service delivery logjams. Also, there are software agents that take decisions on their own. These are agents that implement the actions based on objectives and analyses, considering workflows. These may include notations on systems, investigations of particular transaction details and follow up authentication processes if an intelligent agent perceives unusual behavior. Setting these decision paths though machines would give companies the tempo to regularize them and ultimately increase efficiency. From traditional capability in digitization and automation, allowing hymns to expedite the achievement of things once reliant on undetermined human judgement (Sutar et al., 2026).

7.3 Automating Just Enough, Not Everything

Afterward, it is entirely automated. While this has great benefits, it does not completely spare the need for humans to oversee things. Many organizational decisions are subject to moral, legal or contextual issues that a computerized system cannot grasp. This brings us to the area where human oversight remains a key aspect of autonomous BPM systems. managers and process administrators, for example, must oversee automated decision-making according to regulations that define organization-specific rules. Usually, these decision engines provide recommendations instead of choices. Those proposals must be evaluated by humans, whose specialists will decide which to undertake (Sarvate et al., 2025).

Striking the right balance between blind automation and strict reliance on humans would allow companies to derive technological efficiencies while still being accountable and transparent. Quicker at sorting through massive data sets than humans, automated systems, however, don't replace the need for human oversight to ensure decisions align with the organization's values and policies. In **Table 4,** we have summarized the newly introduced strategies and tools for decision automation that proved compatible with a new adapted BPM context of decisions facilitating their operationalization. The ability to make autonomous decisions is a watershed moment in ABPM, enabling process systems to evolve from supporting execution to supporting decisions and even to partial decision automation. Anomaly detection improves the quality. Predictive analytics, intelligent agents and decision engines strengthen others, making them uniform, speedy and agile during operations. But not all decisions need to be completely automated. Critical decisions on ethics, legal requirements, consumer or other obligations, financial risk, or impact on employees require human assessment. Therefore, ABPM should adopt a "human-in-the-loop" model, in which AI is used to aid decision making, but humans are ultimately responsible for the decision.

Table 4. Decision Automation Techniques in ABPM

Technique	Application
Predictive analytics	Demand forecasting
Intelligent agents	Automated task execution
Decision engines	Operational planning
Data analytics	Performance evaluation

Finally, autonomous decision-making is a game-changing innovation in the field of business process management. Intelligent decision engines, predictive analytics tools, and automated software agents can be integrated into workflow systems to improve operational efficiency and decision-making. Simultaneously, the incorporation of appropriate human oversight guarantees that automated determinations are consistent with the organization's ethical objectives and expectations.

8. PROCESS INTELLIGENCE AND CONTINUOUS OPTIMIZATION

As a result, this long-term shift from workflow automation to operational efficiency doesn't seem sound anymore — especially since the complexity of businesses tends to increase over time. Moreover, organizations need to understand how their processes behave over the long run, and these behaviors can be improved. This is where the notion of process intelligence arises, which involves applying analytic techniques and performance monitoring systems that help in measuring actual processes performance. Operational data analysis points to potential opportunities for increasing both efficiency and quality of service, which discloses process behavior (Leymann & Roller, 2001).

It is based on autonomous business process management solutions. These systems continuously mine execution and general performance data.] Some data is generated when an Online process task is executed. These data can be records of task accomplishment, resource usage logs, timestamps, or user interaction logs. That is, they tell us a timeline of how collective activity has changed through the ages. By leveraging this data, businesses can gain a more holistic view of their operations and spot bottlenecks. Instead of periodic evaluations or checking analytics manually, autonomous BPM provides continuous visibility of process performance (Kurniati et al., 2016).

8.1 Process data capture and analysis

Data Analytics: If the data is secured, analytical tools use it. Analytics systems, for example, might expose some locations in a workflow that take longer than expected or specific jobs that repeatedly require driving because of failures. And observing such trends helps the firms stamp out operational bottlenecks as well as growth opportunities (Aalst, 2016).

Hence, analytics systems are useful for comparing performance even across departments or operational units. When one team does a job, repeatedly and more quickly than other teams, managers can investigate why and reproduce good practices across the organization. This analytic method is used by process intelligence to extract knowledge from operational data.

8.2 Assessing Performance of Processes via Key Indicators

To be able to improve a business process, you need to evaluate its effectiveness. It would be helpful for companies to have concrete and quantitative metrics. Autonomous BPM systems analyze a wide range of performance metrics to determine the effectiveness of processes. This framework allows the administrators to measure service quality and performance. The process cycle time is a metric that measures the average time taken to complete one cycle within a process. It is among the most used metrics. However, if the process takes longer than expected, it could indicate that there are bottlenecks or constraints in certain areas (Brocke & Rosemann, 2015).

Process intelligence systems also track error rates, an important metric. This metric tracks how often process failures or operational errors happen. High error rates are often indicative of flaws in workflow design or process control mechanisms. Another key metric to evaluate the performance of a process is customer satisfaction. Even if this isn't measured directly in the context of workflow systems, it describes how well processes are delivering value to customers. Organizations frequently analyze customer feedback against operational data to determine if service processes are delivering on expectations. **Table 5:** Performance indicators used in autonomous BPM environments As shown in Table 5, which presents the performance indicators used to evaluate workflow effectiveness (Mishra et al., 2024).

Table 5. Performance Metrics in Autonomous BPM

Metric	Description
Cycle Time	Time required to complete processes
Resource Utilization	Efficiency of resource usage
Error Rate	Frequency of process failures
Customer Satisfaction	Service quality perception

8.3 Continuous Optimization Through Intelligent Feedback

The real value of intelligence comes from enabling continuous improvements. Continuous optimization: Continuous optimization is about constantly improving workflows, utilizing insights derived from analysis and feedback on performance. Rather than treating process design as a one-off activity, organizations continuously monitor processes and adapt them whenever they identify areas for improvement. Autonomous BPM systems enable such an improvement cycle by combining monitoring tools with analytical algorithms. These systems track performance metrics, identifying potential areas where changes can improve operational efficiency (Handbook, 2010).

For example, if certain analytics reveal that a certain step in the workflow is creating significant delays during the process for each input while processing them to achieve improved delivery times, it may recognize this and recommend that you redistribute your activities or implement additional automation. Sometimes, AI-enabled analytics can even recommend content connections for you. Such guidance could take the form of editing job orders, paring back tasks so they run in parallel, or creating processes to eliminate unnecessary steps. The managers can then review those comments and make process improvements for better performance. This is the basis of intelligent process management rooted in data collection, data analysis and data augmentation. Also, the feedback cycle is illustrated in **Figure 4** as firms keep collecting operational data, monitoring performance, and improving practices (Mendling et al., 2010). ABPM's ability to continuously improve comes from process intelligence. Companies can move from sporadic improvement to evolving optimizations by collecting process data, analyzing performance indicators, offering recommendations and feeding insights into workflow redesign. This produces a process environment that benefits itself. However, continuous optimization should not be only about efficiency. The challenge also needs to consider service quality, equity, regulatory compliance etc. not just staff burden/customer experience and organisational resilience. This is a good test to ensure that ABPM drives business success, yet remains accountable and human centered.

Figure 4. Process Intelligence Cycle

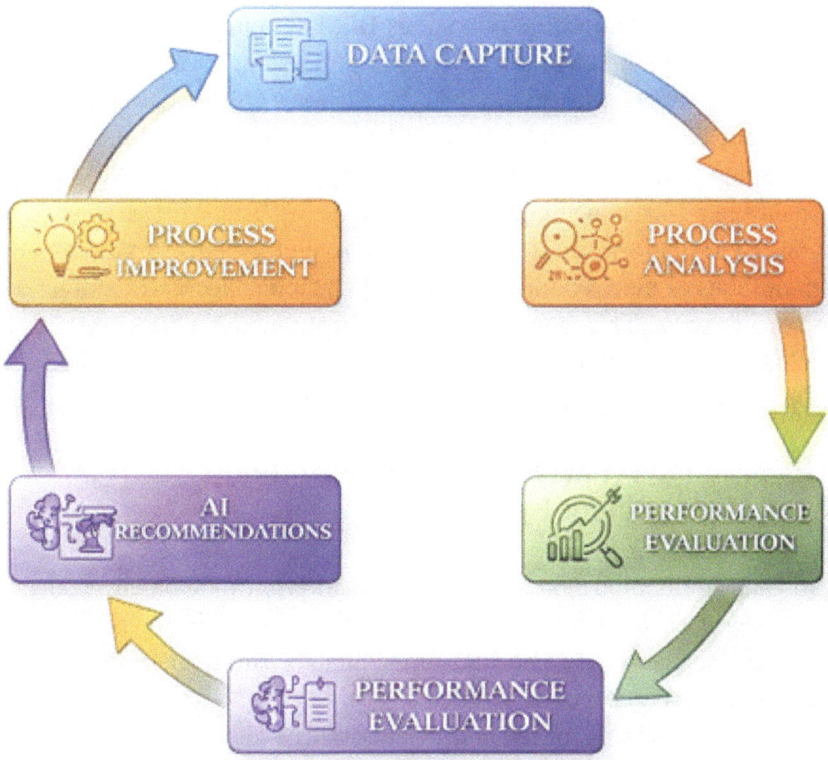

As companies start to look beyond automating workflows, process intelligence allows them to build a richer understanding of their businesses. No matter the diversity type organization monitoring continuously and assessing their performance and recommendations are performed on bases of analysis through data-driven approach which integrate into the system autonomous BPM which help to boost organizational productivity and ensure high standards service delivery. Such continuous evolution of business processes over time to adapt themselves in response to operational variables dynamic factors and market conditions.

9. GOVERNANCE AND HUMAN–AI COLLABORATION

Governance is the set of rules, regulations, and structures that ensure that organizational systems operate responsibly while simultaneously achieving their strategic objectives. Although autonomous BPM systems can perform numerous tasks independently, they will always require professional oversight to ensure that their operations are consistent with the organization's objectives and principles.

In modern businesses, technology and human decision-making are not two different things; they are aspects of a single operating system that work together. This could seem like something from an episode of Black Mirror, where I let autonomous BPM platforms make all the decisions and systems sort through dozens or even millions of pieces of data. However, human managers will still have some say in how we work. This is why governance in autonomous BPM systems has to be a collaboration between thousands of years of human knowledge and decades of analytical tools (Zur Muehlen & Indulska, 2009).

9.1 Governance Frameworks for Autonomous Process Management Systems

As organizations build more advanced automation into their process management systems, they need to ensure that technologies are operating within established governance frameworks. In general, autonomous systems make decisions from data-based insights and analytical models. These systems can process information more quickly than humans and recognize patterns faster than humans. However, they need to be monitored by ensuring their findings match enterprise-level policies and applicable outside regulations. It helps establish the rules under which inheritance is implicitly done. These frameworks delineate the monitoring process, the allocation of responsibilities, and the mitigation of risks associated with adverse events. In the absence of adequate governance structures, organizational governance issues, including accountability, transparency, and regulatory compliance, become apparent (Zur Muehlen & Indulska, 2009).

Automated decisions may influence financial transactions or patient services in specific industries, including banking and healthcare. In those circumstances, it is crucial that automated decision-making be audited and elucidated when applicable. This serves to preserve organizational control over automated processes while simultaneously providing them with a technological advantage.

9.2 Human–AI Collaboration in Decision-Making

Autonomous BPM contexts are predicated on a collaboration between human decision-makers and analytical systems that will increasingly be aided by AI. AI algorithms do not replace human judgment; they are decision-support tools that help us understand a flood of data.AI-enabled tools can draw insights from operational data, recognize trends, and create recommendations that guide managers toward data-driven decision-making. As such, an analytical system might identify trends in customer demand, highlight inefficiencies in workflow performance, or predict the effects of possible operational adjustments. Understanding this helps managers in comparing different options to find the best strategy to move forward (Davenport & Short, 1989).

Someone would still need to analyze the recommendations and decide when to act on them. Other organizational decisions depend on understanding the environment, deeper strategic imperatives, or ethical reasoning, all of which cannot be captured in analytical models. It requires human judgment to make sure those technological insights are translated properly. These AI Analysis components yield their predictability on all unique dimensions, providing context in which human intelligence creates an equally powerful decision-making process around organizational performance. In this context, the decisions that encompass all strategic and tactical requirements to schedule and execute analytics-driven operations in accordance with an autonomous BPM environment are based on real-time analytical insights (**Figure 5**). This prevents man exploitation while ensuring compliance under transparent oversight (Hundogan et al., 2023).

Figure 5. Human–AI Collaborative Governance Model

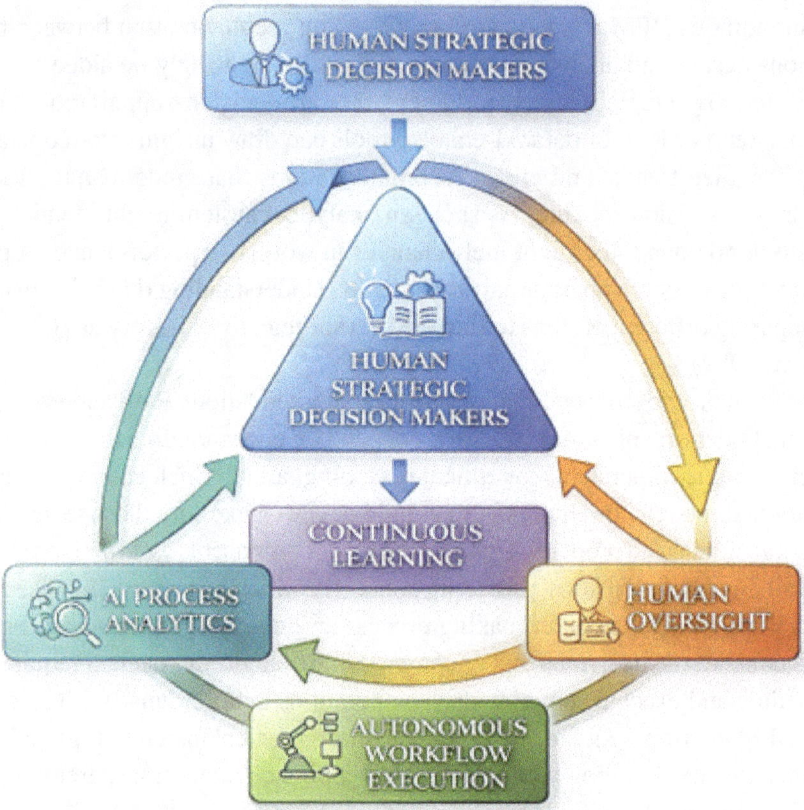

9.3 Ensuring Transparency, Fairness, and Accountability

Another key aspect of governance is accountability. Nonetheless, the outcomes of automated systems must always have a clear definition of responsibility when participating in decision-making. Human supervisors and process managers should be able to review actions taken by the system and intervene when necessary. Governance is also dependent on fairness. There are instances where analytical models may also depend on historical data which can potentially have hidden biases. Such biases, if not rigorously scrutinized, can lead to automated decisions perpetuating inequitable outcomes for members of the population. Governance frameworks thus encompass processes for reviewing analytical models and making sure that decision systems function without prejudice in an equitable manner (Aalst et al., 2012).

Third, organizations will need to ensure that automated systems meet legal and regulatory requirements. Various industries are bound by rigorous regulations governing issues of data privacy, financial transactions, or service delivery. Governance mechanisms are put in place to enforce compliance with these requirements and constraints imposed by regulatory bodies on autonomous BPM systems. **Table 6** summarizes the fundamental governance principles modulating autonomous BPM settings in terms of different governance aspects covering responsible process management (Aziz & Nuryana, 2025).

Table 6. Governance Framework for Autonomous BPM

Governance Element	Purpose
Transparency	Explain algorithmic decisions
Accountability	Human supervision
Fairness	Prevent algorithmic bias
Compliance	Follow regulations

Ultimately, the responsible implementation of autonomous BPM systems is dependent on human–AI collaboration and governance. Human expertise is still necessary to interpret insights, supervise automated decisions, and ensure ethical and regulatory conformance, even though analytical technologies enhance the speed and precision of process management.

10. CONCLUSION

ABPM transforms how companies plan, manage, and optimize operational processes. Traditional business process management (BPM) systems were primarily focused on making processes more formal and automating steps that needed to be performed repeatedly. Yet the process of managing them gets complex due to evolving corporate world. Autonomous BPM is about merging the traditional BPM halos, autonomous production operations and AI. Processes can evolve on the flight based on operational data so that performance can be optimized as conditions and goals change as humans are required. The biggest advantage of next-gen autonomous BPM lies in the ability to define and execute processes powered by analytics. Technology such as AI-based process modeling, intelligent workflow routing and predictive decision support enables companies to get a more complete picture of how their processes are really working. Besides being compliant, these systems analyze both historical and real-time data to spot trends, uncover inefficiencies and deliver recommendations for improvement. This allows businesses to address operational issues and maximize their operations in a timely manner.

Autonomous BPM helps organizations run with more agility. World 2:0 organizations must adapt their business models continuously to thrive at the dynamic pace of change driven by shifting customer expectations, technology evolution and competitive threats. These transformations of enterprises can be easily catered by autonomous BPM systems as real time monitoring and process improvement is possible. Data analysis for future-oriented organizations may reshuffle processes, redirect resources and develop new business models that help new operations to be considered more effective. However, the introduction of these autonomous process systems should come with analysis of key governance issues. These systems can make it easier for individuals to take decisions and enable them to work better; but humans are still required to verify that they work consistently and deliver the desired value to the organization. The more we prioritize transparency, accountability and impartiality, the easier it will be at times to understand, question and verify automated decisions. Hence, the need for an effective governance process is vital to ensure that autonomous BPM systems are deployed responsibly.

The future of BPM settings will still be influenced by the interplay between humans and AI, where people use an AI platform to help them better do their job. While tech allows individuals to better grasp concepts, it still requires human administrators to strategically interpret and implement such insights. By combining technology and human insight, organizations have an opportunity to enjoy the benefits of both. Emerging technologies — including generative AI, digital siblings and sophisticated analytics — will further enhance autonomous BPM. Under these benefits, companies will be better able to differentiate across process designs, enhance processes and test multiple operating scenarios. With the digital transformation greatly changing how businesses run, intelligent process automation will be an ever more decisive contributor to making processes smarter, faster and more adaptable. Those organizations that effectively leverage these technologies in their process management strategies will be well positioned to manage complexity, enhance performance and deliver lasting value. As an epilogue, it features an integrated ABPM system with five core pillars: data, AI analytics, orchestration, autonomous execution and continuous monitoring. The data collection permits us to unfold the reality of the real system. AI analytics turn process data into predictive insights by analyzing trends, detecting anomalies, identifying bottlenecks, and identifying hazards and improvement opportunities. Process orchestration takes the insights and translates them into synchronized process routing, resource allocation, and task prioritization. Autonomous execution is the capability of digital systems or intelligent agents to take over long-standing, routine or relatively low risk processes. This is done through

Continuous monitoring which monitors (as the name suggests) process performance and feeds this data back into the system for even more improvements.

You need to tell it to run the framework in a staged approach manner. The first step is process diagnosis, where the enterprise identifies high-volume, repetitive or data-intensive processes that are candidates for ABPM. The second step is data preparedness, generating event logs and enterprise system for process documentation with data quality standards. Step 3 Artificial Intelligence- AI modeling leverages process mining, machine learning, predictive analytics and natural language processing to understand workflow behavior. Regulated Process Automation (Intelligent routing, Decision Support Systems) Step five is human-AI governance, which builds transparency, accountability, fairness and compliance and escalation protocols. The last stage is continuous optimization, and this is where the workflow designs are refined through feedback loops by monitoring process metrics.

The framework can be assessed by means of measurable KPIs like reduced cycle times, lower error rates, better resource utilization, higher customer satisfaction, less process deviation and manual interventions in decision making processes to operational uncertainties like pandemic. This helps ensure ABPM is measured on the governance, decision reliability & organization value generation besides automation efficiency.

REFERENCES

Aalst, W. (2013). Aalst, W.M.P.: Business process management: A comprehensive survey. *ISRN Software Engineering*, 1–37. https://doi.org/DOI: 10.1155/2013/507984

Aalst, W., Adriansyah, A., & Dongen, B. (2012). Replaying history on process models for conformance checking and performance analysis. *Wiley Interdisciplinary Reviews. Data Mining and Knowledge Discovery*, *2*(2), 182–192. DOI: 10.1002/widm.1045

Aalst, W., & Ter, A. (2004). YAWL: Yet another workflow language. *Information Systems*, *30*(4), 245–275. DOI: 10.1016/j.is.2004.02.002

Aziz, A., & Nuryana, I. K. (2025). Business process analysis and modeling using the business process model notation (BPMN) method (Case study: Simo Jaya Printing). [JEISBI]. *Journal of Emerging Information Systems and Business Intelligence*, *6*(3). Advance online publication. DOI: 10.26740/jeisbi.v6i3.70359

Bandara, W., Gable, G. G., & Rosemann, M. (2005). Factors and measures of business process modelling: Model building through a multiple case study. *European Journal of Information Systems*, *14*(4), 347–360. DOI: 10.1057/palgrave.ejis.3000546

Brocke, J., & Rosemann, M. (2015). *Strategic alignment, governance, people and culture.*

Burgueño, L., Izquierdo, J. L. C., & Planas, E. (2021). An empirical study on the impact of introducing a modeling tool in a requirement engineering course. In *2021 ACM/IEEE International Conference on Model Driven Engineering Languages and Systems Companion (MODELS-C)* (pp. 712–720). DOI: 10.1109/MODELS-C53483.2021.00115

Casati, F., & Shan, M.-C. (2001). Dynamic and adaptive composition of e-services. *Information Systems*, *26*(3), 143–163. DOI: 10.1016/S0306-4379(01)00014-X

Chen, Y.-K. (2012). Challenges and opportunities of internet of things. In *17th Asia and South Pacific Design Automation Conference* (pp. 383–388). DOI: 10.1109/ASPDAC.2012.6164978

Davenport, T. (1993). *Reengineering work through information technology.*

Davenport, T., & Short, J. (1989). The new industrial engineering: Information technology and business process redesign. *Sloan Management Review*, •••, 31.

Dumas, M., La Rosa, M., Mendling, J., & Reijers, H. (2013). *Fundamentals of business process management.* Springer., DOI: 10.1007/978-3-642-33143-5

Fouda, E. (2026). *Data science in action*. DOI: 10.1007/979-8-8688-2013-7_1

Hundogan, O., Lu, X., Du, Y., & Reijers, H. A. (2023). CREATED: Generating viable counterfactual sequences for predictive process analytics. In *International Conference on Advanced Information Systems Engineering*. DOI: 10.1007/978-3-031-34560-9_32

Huser, V. (2012). Process mining: Discovery, conformance and enhancement of business processes. *Journal of Biomedical Informatics, 42*. Advance online publication. DOI: 10.1016/j.jbi.2012.06.007

Indulska, M., Recker, J., Rosemann, M., & Green, P. (2009). Business process modeling: Current issues and future challenges. https://doi.org/DOI: 10.1007/978-3-642-02144-2_39

Krogstie, J. (2016). *Business process modeling in practice.*

Kumar, B. R. (2025). *Handbook of brand management.*

Kurniati, A. P., Johnson, O., Hogg, D., & Hall, G. (2016). Process mining in oncology: A literature review. In *2016 6th International Conference on Information Communication and Management (ICICM)* (pp. 291–297). DOI: 10.1109/INFOCOMAN.2016.7784260

Leno, V., Dumas, M., Maggi, F. M., La Rosa, M., & Polyvyanyy, A. (2020). Automated discovery of declarative process models with correlated data conditions. *Information Systems, 89*, 101482. DOI: 10.1016/j.is.2019.101482

Leymann, F., & Roller, D. (2001). *Production workflow – Concepts and techniques.*

Li, H., Mohamed El-Amine, H., & Mohamed, H. (2014). Merging several business process variants. In *The 26th Chinese Control and Decision Conference (2014 CCDC)* (pp. 5218–5223). DOI: 10.1109/CCDC.2014.6853112

Lübke, D., Ahrens, M., & Schneider, K. (2021). Influence of diagram layout and scrolling on understandability of BPMN processes: An eye tracking experiment with BPMN diagrams. *Information Technology and Management, 22*(2), 1–33. DOI: 10.1007/s10799-021-00327-7

Mendling, J., Reijers, H. A., & Aalst, W. (2010). Seven process modeling guidelines (7PMG). *Information and Software Technology, 52*, 127–136. *Handbook on business process management 1: Introduction, methods, and information systems.* (2010).DOI: 10.1016/j.infsof.2009.08.004

Mishra, S. S., Panwar, D., Chouhan, R. P. S., & Shanmugapriya, V. (2024). The Internet of Things and its impact on EV charging stations. In *2024 2nd International Conference on Advancements and Key Challenges in Green Energy and Computing (AKGEC)* (pp. 1–6). IEEE. https://doi.org/DOI: 10.1109/AKGEC62572.2024.10868079

Mishra, S. S., Sharma, S., Sharma, Y., & Singh, M. (2026). Artificial intelligence for climate action and sustainable transitions: Conceptual frameworks, governance models, and responsible innovation pathways. In *Responsible AI for climate action: Governance, justice, and sustainable transitions*. IGI Global., DOI: 10.4018/979-8-2600-2020-3.ch005

Njima, C., Ghedira, C., Gamha, Y., & Romdhane, L. (2024). Web service composition in mobile environment: A survey of techniques. *IEEE Transactions on Services Computing*, *17*(2), 1–16. DOI: 10.1109/TSC.2024.3361655

Ong, H. J., Lim, B. G., Tan, R., & Ikeda, K. (2025). Causal discovery in additive noise models using beam search. *Artificial Life and Robotics*, *31*(1), 317–326. Advance online publication. DOI: 10.1007/s10015-025-01095-1

Panwar, D., Mishra, S. S., & Chouhan, R. P. S. (2024). A detailed examination of solar photovoltaic technology, design, efficiency improvement and applications. In *2024 International Conference on Communication, Computing and Energy Efficient Technologies (I3CEET)* (pp. 1709–1714). IEEE. https://doi.org/DOI: 10.1109/I3CEET61722.2024.10994009

Rauch, H., Henry, C., Mühlhaus, M., Kurz, F., Hellekes, J., & Merkle, N. (2025). Large-scale mapping of urban parking from aerial images: A case study in Berlin, Germany. *International Archives of the Photogrammetry, Remote Sensing and Spatial Information Sciences*, XLVIII-4/W16-2025, 83–89. DOI: 10.5194/isprs-archives-XLVIII-4-W16-2025-83-2025

Reichert, M., & Weber, B. (2012). Enabling flexibility in process-aware information systems. DOI: 10.1007/978-3-642-30409-5_11

Reijers, H., & Aalst, W. (2005). The effectiveness of workflow management systems: Predictions and lessons learned. *International Journal of Information Management*, *25*(5), 458–472. DOI: 10.1016/j.ijinfomgt.2005.06.008

Rozinat, A., & Aalst, W. (2008). Conformance checking of processes based on monitoring real behavior. *Information Systems*, *33*(1), 64–95. DOI: 10.1016/j.is.2007.07.001

Sarvate, D., Mishra, S. S., Shanmugapriya, V., & Panwar, D. (2025). Integrating blockchain and federated learning for enhanced security and privacy in smart cities. In Singh, K. K., Singh, A., & Mahesh, T. R. (Eds.), *Integration of federated learning and blockchain for smart cities*. Wiley., DOI: 10.1002/9781394167760.ch10

Singh, K. S., Shanmugapriya, V., Mishra, S. S., & Singh, M. (2026). Artificial intelligence-driven flow optimization for enhanced efficiency in renewable energy systems. In Awasthi, M. K., & Gupta, R. (Eds.), *Artificial intelligence and computational modeling in heat transfer and fluid dynamics*. Wiley., DOI: 10.1002/9781394433605.ch10

Smirnov, M. (2012). *Business process model abstraction.*

Sutar, D., Metri, R. A., Naik, R., Mishra, S., Thakkar, J., & Kolhe, M. (2026). Energy-efficient automation of industrial surface pre-treatment processes using PLC and SCADA. In *Proceedings of the Second International Conference on New Power System and Power Electronics (NPSPE 2025)* (Vol. 14129, Article 141292C). SPIE. https://doi.org/DOI: 10.1117/12.3110247

Ter, A., Aalst, W., Adams, M., & Russell, N. (2010). *Modern business process automation: YAWL and its support environment*. DOI: 10.1007/978-3-642-03121-2

van der Aalst, W. M. (2016). *Data science in action.*

Wu, Z., Pan, S., Chen, F., Long, G., Zhang, C., & Yu, P. S. (2019). A comprehensive survey on graph neural networks. *IEEE Transactions on Neural Networks and Learning Systems*, *32*(1), 4–24. DOI: 10.1109/TNNLS.2020.2978386 PMID: 32217482

Zhu, Z., Song, Z., & Palazoglu, A. (2011). Transition process modeling and monitoring based on dynamic ensemble clustering and multiclass support vector data description. *Industrial & Engineering Chemistry Research*, *50*(24), 13969–13983. DOI: 10.1021/ie201792r

Zur Muehlen, M., & Indulska, M. (2009). Modeling languages for business processes and business rules: A representational analysis. *Information Systems*, *35*(4), 379–390. DOI: 10.1016/j.is.2009.02.006

Chapter 2
From Automatic Assistant to Cognitive Partner:
The Role of Agentic AI in Cognitive Management of Business Processes

Brahim Sabiri
http://orcid.org/0009-0006-6314-3393
Hassan 2 University of Casablanca, Morocco

Othman Boulitama
http://orcid.org/0000-0003-3541-1582
Hassan 2 University of Casablanca, Morocco

Nissrine Mouchtakir
http://orcid.org/0009-0006-8816-8079
University Canada West, Canada

Karim Sabri
Hassan 2 University of Casablanca, Morocco

ABSTRACT

This chapter examines the growing role of Agentic Artificial Intelligence (AI) in transforming Business Process Management (BPM) within contemporary organizations. Unlike traditional AI systems designed mainly for task automation, agentic AI introduces autonomous, goal-oriented agents capable of reasoning, planning, and coordinating complex organizational workflows. The chapter analyzes how these systems contribute to cognitive management by augmenting decision-making processes and enhancing organizational adaptability. Drawing on theoretical

DOI: 10.4018/979-8-2600-0050-2.ch002

perspectives such as dynamic capabilities, absorptive capacity, and distributed cognition, it proposes a conceptual framework explaining how agentic AI supports different stages of the BPM lifecycle, including process design, execution, monitoring, and optimization. The chapter also discusses emerging governance challenges related to transparency, accountability, and human oversight. Overall, it highlights the shift from AI as an operational assistant toward AI as a cognitive partner in organizational management.

INTRODUCTION

The rapid evolution of artificial intelligence within organizations has profoundly transformed the way business processes are designed, executed, and managed (Nosova et al., 2022; Verma, 2025). Initially, AI agents were primarily designed as assistants operating according to predefined rules in order to improve organizational practices (Murire, 2024), positively influencing both individual performance (Benhmama & Sabiri, 2025) and collective performance (Ousghir et al., 2024). However, recent technological advances have led to the emergence of a new generation of artificial intelligence known as Agentic AI, characterized by a high level of technical sophistication. This form of intelligence is distinguished by its ability to pursue objectives, reason adaptively, and act autonomously in complex environments, thereby offering significant decision-making and operational potential (Sapkota et al., 2026).

According to Bandi et al. (2025), Agentic AI does not simply consist of a juxtaposition of artificial intelligence agents, but rather represents an integrated and orchestrated system capable of coordinating multiple autonomous agents to achieve complex organizational objectives. Nayebi (2025) explains that these agents are characterized by their autonomy, enabling them to make decisions without constant supervision, their responsiveness to real-time changes, and their proactivity in initiating actions aimed at performance optimization. Furthermore, their social capabilities allow them to communicate, collaborate, and coordinate effectively with other agents, systems, or human actors, thereby ensuring the simultaneous, coherent, and adaptive execution of multiple tasks.

At the organizational level, Agentic AI can therefore be compared to a highly qualified human assistant capable of managing several activities simultaneously, prioritizing actions, and effectively supporting decision-making. Pastakia et al. (2025) note that traditional AI agent systems generally operate according to a human-in-the-loop logic, whereas Agentic AI systems move toward an approach in which human intervention evolves from direct execution to orchestration. Similarly, Al-Mahdawi (2025) highlights that these new systems are capable of planning, coordinating, and executing complex multi-step processes with minimal supervision, while closely collaborating with humans and other agents.

The technological evolution of AI is particularly visible in the field of Business Process Management. A study conducted by the American consulting firm McKinsey highlights a paradox between the massive adoption of generative AI and its relatively limited economic impact. The study concludes that Agentic AI now represents a potential solution to overcome this paradox by enabling the automation of complex Business Process Management activities. However, such a transformation requires a redesign of organizational processes, the development of new technological architectures, and the evolution of governance mechanisms and human–AI collaboration models (Sukharevsky, Kerr, et al., 2025).

Today, organizations increasingly view AI as a cognitive partner rather than merely an automated assistant, reflecting a profound transformation in the human–AI relationship (Alshammari & Alshammari, 2026). Brynjolfsson and McAfee (2017) argue that the value of AI lies in its ability to complement human intelligence rather than replace it. This evolution implies a redefinition of managerial roles, which are now increasingly oriented toward more complex and strategic activities. In this context, Gavetti and Levinthal (2000) demonstrate that organizations rely on cognitive representations to guide their decisions in uncertain environments, highlighting the fundamental role of cognitive processes in contemporary management. Consequently, the concept of cognitive management is emerging as a central analytical framework capable of adapting to the evolution of artificial intelligence.

At the individual level, the emergence of Agentic AI profoundly redefines the role of managers by automating standardized operational and analytical activities through advanced data-processing and automated reasoning capabilities (Korzynski et al., 2025). A study conducted by MIT Sloan Management Review and the Boston Consulting Group, based on a sample of 2102 executives, shows that 76% of executive managers now consider Agentic AI as a "collaborator" rather than a simple tool, reflecting a major shift in managerial perceptions (Ransbotham et al., 2025).

At the organizational level, this transformation cannot be fully understood without referring to the absorptive capacity theory. Introduced by Cohen and Levinthal (1990), this theory defines absorptive capacity as the ability of an organization to recognize the value of new external knowledge, assimilate it, and exploit it effectively. In the context of Agentic AI, Hosseini and Seilani (2025) emphasize that organizations are increasingly required to integrate knowledge generated and continuously updated by intelligent systems into their decision-making processes in order to improve efficiency, adaptability, and overall organizational performance.

Within this framework, this chapter aims to analyze how Agentic AI transforms the management of business processes by evolving artificial intelligence from a simple automated assistant into a genuine cognitive partner. It proposes an integrated perspective on the transformations currently affecting business process management in contemporary organizations. The chapter is structured into several complementary sections. The first section presents the theoretical foundations of organizational change. The second section provides a deeper understanding of Agentic AI, highlighting its evolution toward autonomous action. The third section explores the concept of cognitive management and its theoretical foundations. The fourth section examines the contribution of Agentic AI to the transformation of process management. The fifth section analyzes the transition from AI as an assistant to AI as a cognitive partner. The sixth section highlights the main governance challenges and issues associated with this transformation. Finally, the chapter concludes with a concluding section.

1. THEORETICAL BACKGROUND

The interaction between an organization and its environment constitutes a strategic and dynamic determinant of organizational adaptation (Hrebiniak & Joyce, 1985; Miles et al., 1974; Parnell et al., 1996). The seminal work of Emery and Trist (1965) demonstrated that organizations progressively evolve from relatively simple and stable environments toward increasingly complex and even turbulent contexts. This transition is accompanied by growing uncertainty, unpredictability, and interdependence among actors, which compels organizations to adapt their modes of operation. In this context, firms are required to develop advanced analytical capabilities in order to better identify opportunities and threats while assessing their strengths and weaknesses in a constantly evolving environment.

Building on this perspective, Terreberry (1968) emphasizes that the contemporary organizational environment is characterized by increasing complexity resulting from the multiplication and dispersion of stakeholders. This configuration makes decision-making processes more difficult due to the diversity of interests, expectations, and sources of information involved. Uncertainty thus becomes a structural characteristic of the environment, challenging traditional management approaches based on predictability and control. As a result, conventional strategies appear insufficient to effectively address the demands of a dynamic and unstable environment.

This growing complexity has led to the development of structural contingency theory, which offers a more nuanced interpretation of the relationship between organizations and their environment (Luthans & Stewart, 1977; Mintzberg, 2009). According to this theory, there is no universally optimal organizational model. Instead, organizational effectiveness depends on the alignment between the internal characteristics of the firm (e.g., structure, resources, and culture) and the requirements of its external environment. This perspective highlights the importance of organizational adaptation, emphasizing that each organization must develop specific configurations based on its constraints and opportunities.

From this viewpoint, the organization–environment fit cannot be reduced to a simple passive adjustment. Rather, it follows a dynamic and co-evolutionary logic in which the organization not only adapts to its environment but also influences it. This interaction relies on mechanisms of learning, interpretation, and resource reconfiguration that enable firms to maintain their competitiveness in uncertain contexts. Consequently, high-performing organizations are those capable of developing adaptive and anticipatory capabilities, allowing them to navigate effectively within complex and constantly changing environments.

Finally, in a context marked by digital transformation and the emergence of advanced technologies such as artificial intelligence, this interaction becomes even more strategic. Organizations must not only understand their environment but also leverage analytical and cognitive tools to strengthen their capacity for decision-making and action. This highlights the importance of an integrated approach in which organizational management relies on a deep understanding of environmental dynamics and a continuous capacity for adaptation.

1.1. Key Theoretical Perspectives at the Organizational Level

The current business landscape requires organizations to continuously adapt to changes in their environment. Companies constantly adjust their strategies, organizational structures, and objectives in order to respond to economic, technological, and social transformations. This reality has been theorized through the dynamic capabilities theory, according to which the sustainable performance of an organization depends on its ability to integrate, build, and reconfigure its internal and external competencies in response to an ever-changing environment (Teece et al., 1997; Teece & Pisano, 1994). Thus, dynamic capability does not simply refer to the possession of resources, but rather to an organizational ability to continuously renew these resources in order to maintain a competitive advantage (Eisenhardt & Martin, 2000).

According to Teece (2018), dynamic capabilities enable firms to identify market developments, seize emerging opportunities, and adjust their resources and organizational structures accordingly. From this perspective, the business model appears as the operational translation of these capabilities, since it organizes how the firm creates, delivers, and captures value. Winter (2003) emphasizes that a firm cannot simply choose a business model and expect it to remain effective over time; instead, it must continuously evolve. This evolution is made possible through dynamic capabilities, which allow organizations to identify environmental changes, respond to them, and modify their resources and internal processes accordingly.

The Technology–Organization–Environment (TOE) model, introduced by Tornatzky and Fleischer (1990), provides a relevant analytical framework for understanding the dynamics of organizational adaptation. According to the authors, this model highlights that the adoption and adjustment of organizations to change rely on three interrelated dimensions: (a) the technological context (availability and characteristics of technologies), (b) the organizational context (resources, structure, and culture), and (c) the environmental context (competitive pressure, regulations, and partners). It establishes a direct link between the contextual conditions of the firm and its capacity to integrate and deploy innovations, emphasizing that these transformations result from a dynamic interaction between these three dimensions. Therefore, the TOE model does not merely analyze technological adoption; it also provides a relevant framework for understanding the organizational determinants of innovation implementation within firms, situating technological change within a broader perspective of the innovation process (Baker, 2012).

Teece (2007) further emphasizes that technological innovation alone does not guarantee firm performance. Instead, performance depends on the ability to integrate technology within a structured managerial approach, particularly through processes that enable organizations to design appropriate business models, coordinate complementary assets, and effectively capture the value created. Consequently, competitive advantage results from the strategic orchestration of resources and their alignment with effective organizational processes.

From this perspective, organizational adaptation through the TOE model involves a multidimensional transformation that goes beyond simple technological integration. It requires the reconfiguration of internal processes, the development of human resource capabilities, and strategic alignment with external environmental requirements. In this context, high-performing firms are those capable of effectively orchestrating these three dimensions by mobilizing their dynamic capabilities to integrate technological innovations while considering organizational constraints and environmental pressures, thereby strengthening their agility and resilience in the face of contemporary changes.

The emergence of artificial intelligence fits fully within this logic of multidimensional adaptation highlighted by the Technology–Organization–Environment framework (Pramod et al., 2024). Indeed, these intelligent systems are no longer limited to the simple automation of tasks; they actively contribute to the transformation of organizational change processes (Aakula et al., 2024; Najana et al., 2024). In this context, the integration of AI cannot be understood solely from a technological perspective; it requires a reconfiguration of organizational structures, a redefinition of managerial roles, and an adaptation to the demands of the competitive and institutional environment (Alkhatib et al., 2025). AI therefore emerges as a strategic lever that connects technological, organizational, and environmental dimensions, while reinforcing the dynamic capabilities of firms to manage and optimize their business processes in an intelligent and evolving manner (Mpanza, 2025).

1.2. Key Theoretical Perspectives at the Individual Level

The emergence of artificial intelligence has profoundly transformed the way individuals think during decision-making processes (Chaturvedi & Dasgupta, 2024). This perspective aligns with the foundations of distributed cognition theory, developed by Edwin Hutchins (1995), according to which cognitive processes are distributed among individuals, tools, and information systems within a given environment. In this sense, intelligence no longer resides solely in individuals or technologies considered separately, but rather in the dynamic interaction between human actors and technological systems (Hollan et al., 2000).

In this context, technologies based on artificial intelligence differ from traditional information systems, which are generally designed to automate repetitive tasks according to predefined rules. AI systems instead rely on capabilities such as learning, contextual analysis, and the interpretation of large volumes of data (Davenport & Ronanki, 2018). These capabilities strengthen individuals' analytical abilities by enabling a better understanding of complex situations and facilitating the identification of patterns or relationships that are difficult to detect through traditional methods (Reicherts et al., 2025). Intelligent systems therefore act as cognitive partners, capable of providing information, recommendations, or alternative scenarios that support decision-making processes (Alshammari & Alshammari, 2026).

At the managerial level, the role of managers has progressively evolved from a primarily operational function to that of a strategic actor responsible for interpreting, evaluating, and integrating analyses generated by intelligent systems into decision-making processes (Canals, 2020; Raisch & Krakowski, 2021). Similarly, the integration of these technologies has transformed the way organizational processes are designed and managed (Abbasi et al., 2025).

Traditionally, business processes are structured as linear sequences of predefined activities executed in a standardized manner (Dumas et al., 2018). The integration of advanced analytical capabilities now enables the design of more adaptive processes capable of evolving according to contextual conditions and available data. Intelligent systems can detect anomalies, adjust workflows, or recommend corrective actions, thereby contributing to the emergence of a more flexible form of Business Process Management oriented toward organizational learning (Dumas et al., 2023).

Finally, the integration of artificial intelligence into organizational processes raises significant managerial and organizational challenges. It requires transformations in skills, organizational structures, and governance mechanisms in order to fully leverage the potential of augmented intelligence. Organizations must therefore develop a culture of human–machine collaboration based on trust, transparency, and a clear understanding of intelligent systems (Fountaine et al., 2019; Wilson & Daugherty, 2018). In this context, decisions are increasingly co-produced by human and technological actors, illustrating a dynamic of distributed cognition and augmented intelligence that redefines the foundations of management and organizational process management.

2. UNDERSTANDING AGENTIC AI: FROM HUMAN-IN-THE-LOOP TO AUTONOMOUS ACTION

Agentic AI refers to an advanced evolution of artificial intelligence characterized by the ability of systems to act autonomously, pursue explicit goals, and dynamically adapt their actions according to their environment (Widad, 2025). Unlike traditional AI systems, which are limited to executing predefined tasks, these intelligent agents incorporate mechanisms of perception, reasoning, planning, and continuous learning. These capabilities enable them to take initiatives and orchestrate complex strategies oriented toward the achievement of specific objectives (Bandi et al., 2025).

Figure 1. Levels of Artificial Intelligence Evolution by Degree of Sophistication

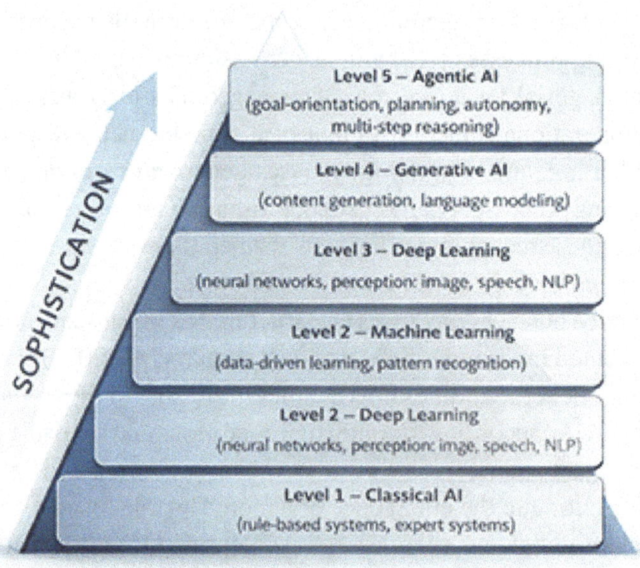

The pyramid diagram provides a synthetic representation of the evolution of artificial intelligence in terms of increasing levels of sophistication, ranging from classical systems to the most autonomous and adaptive forms. The base, corresponding to Level 1 – Classical AI, includes rule-based systems and expert systems characterized by limited deterministic capabilities. The progression to higher levels reflects the gradual integration of learning and perception methods. Machine Learning (Level 2) exploits data to detect patterns, while Deep Learning (Level 3) introduces neural networks capable of processing complex information such as images, speech, or natural language. Level 4 – Generative AI represents the ability to create content and generate advanced language models. Finally, at the top of the pyramid, Agentic AI (Level 5) embodies a new generation of autonomous AI capable of multi-step reasoning, planning, and pursuing goals adaptively in complex environments. The upward arrow symbolizes the increasing technical and cognitive sophistication, highlighting the progression toward more intelligent, autonomous, and decision-oriented systems that have the potential to transform organizational and strategic processes.

The concept of Agentic AI builds upon earlier work defining an intelligent agent as an entity capable of perceiving its environment and acting in ways that maximize the achievement of its objectives (Russell et al., 2021). It is further supported by the perspective that conceptualizes agent-based systems as autonomous entities endowed with distributed decision-making capabilities (Shoham & Leyton-Brown, 2008). In this context, recent research on AI emphasizes that modern agents are no longer simple executors but systems capable of planning, using tools, acting, reasoning, and learning continuously, marking a significant departure from traditional static architectures (Huang, 2025).

Figure 2. Graphical Representation of the 4C Framework Proposed by Abuadbba et al. (2026)

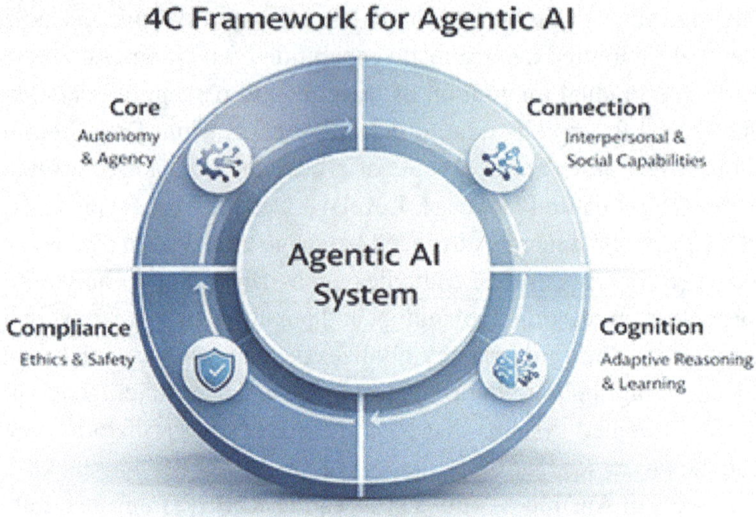

Abuadbba et al. (2026) propose an analytical framework based on four key components that characterize a model for analyzing Agentic AI systems:

Core (Technical Infrastructure and Operational Environment): This component corresponds to the technical foundation of the agentic system, including the execution infrastructure, models, memory systems, API interfaces, tools, and data flows. It represents the "digital body" of the agent, meaning the set of resources that enable it to operate within a given environment. This dimension is essential because any vulnerability at this level (e.g., data poisoning, poor access management, or tool misuse) can compromise the entire system, regardless of the quality of the agent's reasoning.

Connection (Interactions, Coordination, and Trust Among Agents): This dimension refers to the mechanisms of communication, coordination, and delegation between agents, as well as the trust relationships that structure their interactions. In multi-agent systems, decisions are not isolated but result from information exchanges and dynamic collaborations. This layer reflects the "social dimension" of Agentic AI, where risks such as manipulation, misinformation, or the propagation of errors may arise through interactions between agents.

Cognition (Internal Mechanisms of Reasoning and Decision-Making): This component relates to the internal processes of the agent, including belief formation, goal definition, reasoning, planning, and learning. It represents the "digital brain" of the system. This dimension is critical because cognitive failures (e.g., belief drift, biases, misinterpretation of data, or poorly aligned objectives) may lead to incorrect decisions even when the technical infrastructure and agent interactions function properly.

Compliance (Governance, Rules, and Regulatory Alignment): This component encompasses the governance mechanisms that regulate the behavior of agents, including ethical standards, legal requirements, organizational policies, and auditing and monitoring mechanisms. It ensures that the actions of agents remain aligned with human values and institutional constraints. This layer is particularly essential in autonomous systems, as it helps limit potential deviations related to autonomy by ensuring traceability, accountability, and regulatory compliance.

2.1. Technical Characteristics of Agentic AI

The transition from Human-in-the-Loop (HITL) systems to forms of increased autonomy follows an evolutionary trajectory in the development of artificial intelligence. HITL approaches rely on constant human supervision, ensuring the reliability, compliance, and control of algorithmic decisions, particularly in critical contexts. This logic is rooted in the work of Mitchell (1997), which established the foundations of supervised learning and human–machine interaction. However, these models present limitations in terms of scalability, responsiveness, and the ability to manage the growing complexity of digital environments. The emergence of intelligent agents therefore marks a transition toward human-on-the-loop systems, in which humans maintain a strategic supervisory role rather than direct operational intervention. As Huang (2025) emphasizes, this evolution enables a shift from rigid workflows to dynamic systems capable of anticipation, adaptation, and autonomous coordination.

Figure 3. Simplified Comparison Between the Functionalities of a Simple AI Agent and Agentic AI

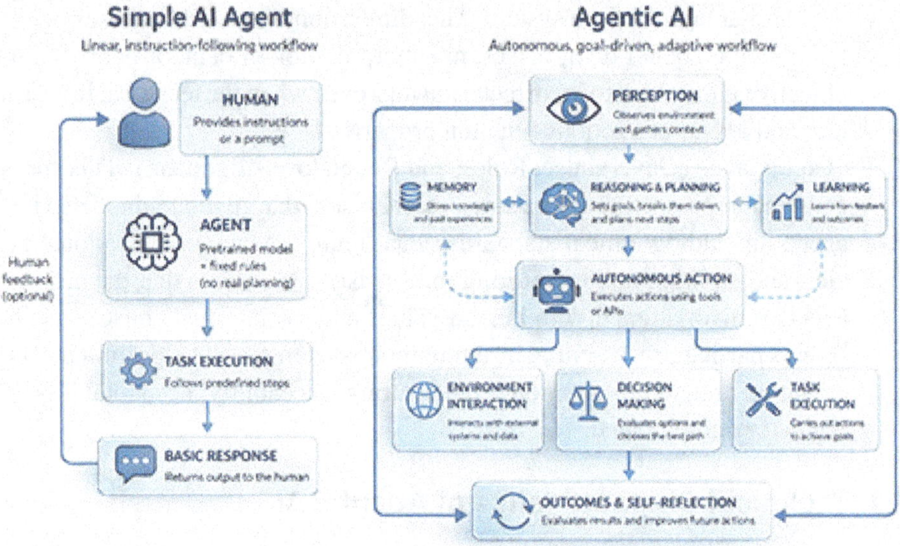

The evolution toward Autonomous Action systems represents a major paradigmatic shift, in which artificial intelligence becomes capable of perceiving, deciding, and acting without direct human intervention. This transformation relies on advances in deep learning and neural architectures, particularly the work of LeCun et al. (2015), which enabled the development of models capable of generalization and autonomous learning. From this perspective, Agentic AI represents the culmination of a technological continuum leading to self-organizing systems capable of self-improvement and complex decision-making in uncertain environments. As Huang (2025) emphasizes, these agents embody a new generation of systems capable of managing memory, planning actions, and continuously optimizing their performance. However, this increased autonomy also raises critical challenges related to governance, ethics, and algorithmic accountability, requiring the establishment of robust frameworks to ensure transparency, explainability, and alignment with human values.

2.2. Agentic AI in Organizations

Agentic AI has emerged as a major strategic lever for organizations (van Esch, 2026). By automating not only repetitive tasks but also complex organizational processes involving information collection, multi-step reasoning, and coordination between different systems, Agentic AI systems are profoundly transforming the way work is designed and executed within organizations. Indeed, these systems can take over activities that previously required continuous human coordination, such as information processing, decision routing, operational monitoring, and exception management. This capability paves the way for the automation of entire workflows and the reorganization of work structures toward models in which networks of intelligent agents collaborate to execute complex business processes (Bandara et al., 2026).

According to Sukharevsky, Krivkovich, et al. (2025), the rise of these technologies contributes to the emergence of what is referred to as the agentic organization, a new organizational paradigm based on collaboration between human actors and artificial intelligence agents capable of acting autonomously within business processes. In this model, Agentic AI systems no longer simply assist individuals but actively participate in the design, orchestration, and execution of complex organizational workflows. Organizations therefore evolve toward more flexible and results-oriented structures built on hybrid teams, where humans primarily perform functions related to supervision, strategic management, and exception handling, while intelligent agents handle the execution and optimization of processes.

Thus, Agentic AI can no longer be considered merely an automation tool but increasingly emerges as a genuine organizational actor, actively contributing to value creation, process optimization, and the strengthening of decision-making capabilities. From this perspective, it becomes relevant to analyze the impact of these systems on organizational learning and knowledge management mechanisms (Islam et al., 2026). To enrich this perspective, this chapter proposes examining the role of Agentic AI through the lens of organizational absorptive capacity, highlighting its contribution to the four dimensions identified by Zahra and George (2002): acquisition, assimilation, transformation, and exploitation of knowledge. In this context, Agentic AI can be viewed as a technological catalyst capable of amplifying these organizational cognitive processes by facilitating the identification and collection of new information, its analysis and interpretation, its recombination with existing knowledge, and its effective mobilization in strategic decisions and organizational innovation.

Table 1. Contribution of Agentic AI to the Dimensions of Organizational Absorptive Capacity

Dimension	Definition	Contribution of Agentic AI	Organizational Effects
Acquisition	Ability to identify and acquire external knowledge that is critical for the organization	Facilitates the detection and automated collection of relevant data (digital environments, databases, real-time data streams) and improves environmental scanning mechanisms	Increased monitoring capability and broader, faster access to relevant information sources
Assimilation	Ability to analyze, process, and understand acquired knowledge	Enables structured data processing (e.g., classification, synthesis, information extraction) and supports the interpretation of complex information	Improved understanding of data and reduction of informational complexity
Transformation	Ability to develop and refine routines that combine existing knowledge with new knowledge	Supports the reconfiguration and recombination of knowledge through cross-analyses and adaptive models, without replacing organizational routines	Strengthening of organizational learning and innovation capabilities
Exploitation	Ability to apply knowledge for commercial and operational purposes	Contributes to decision support and the automation of certain actions by mobilizing analyzed knowledge within business processes	Improvement of operational efficiency and decision-making processes

This table clearly shows that Agentic AI is not limited to a purely technological role but acts as a transversal lever of organizational absorptive capacity. It contributes to each of these dimensions by facilitating access to massive volumes of data, enabling advanced analytical capabilities, supporting the recombination of knowledge, and improving the quality of decision-making. Consequently, organizations with strong absorptive capacity are better positioned to leverage the potential of AI in managing business processes. Agentic AI intervenes at every stage of the organizational cognitive process, from the detection of information to its strategic exploitation. This reinforces the idea that firms endowed with strong absorptive capacity are those most capable of transforming the autonomous capabilities of Agentic AI into a competitive advantage, particularly in the management and optimization of business processes.

3. COGNITIVE MANAGEMENT: TOWARD AN INTELLIGENT AND ADAPTIVE VIEW OF MANAGEMENT

Cognitive management is part of a profound transformation in management sciences, in which the organization is now viewed as a cognitive system capable of perceiving, interpreting, and acting upon its environment (Nikulina & Khomenko,

2015). From this perspective, the firm can be considered a collective actor that carries cognitions, meaning that it is endowed with representations, interpretive schemes, and action logics that guide its decisions (Thoenig & Michaud, 2007). Consequently, cognitive management is no longer limited to allocating resources or coordinating tasks; it also involves shaping shared cognitive frameworks that enable collective action.

This perspective is reinforced by the concept of cognitive knowledge management, which views the organization as an integrated information-processing system aimed at improving decision-making, innovation, and problem-solving capabilities (Sarrasin & Ramangalahy, 2007). In this sense, management has gradually evolved from a behaviorist approach toward a cognitivist perspective, which seeks to mobilize and structure individual and collective cognitive capabilities in order to support decision-making processes and enhance organizational performance (Ahmadi et al., 2026). The table below presents a comparison between these two managerial approaches:

Table 2. Comparison Between Behavioral Management and Cognitive Management

Dimension	Behavioral Management	Cognitive Management
Scientific Field	Work psychology and human relations.	Cognitive sciences and organizational decision-making theory.
Object of Analysis	Behaviors, attitudes, and motivation of individuals within the organization.	Cognitive processes involved in perception, interpretation, and decision-making.
View of the Organization	A social system based on human interactions.	An information processing and interpretation system.
Role of the Manager	Influence behaviors and promote employee engagement.	Structure thinking processes and improve the quality of decisions.
Organizational Objective	Improve performance through employee engagement and well-being.	Improve the quality of strategic reasoning and decision-making.

The functioning of cognitive management refers to a managerial approach based on the mobilization of advanced cognitive functions (Liu & Yin, 2025). Organizational perception enables the detection of internal and external signals, which are often ambiguous or incomplete, while interpretation transforms this information into shared representations through processes of sensemaking (Weick, 1995). Decision-making and action subsequently translate these interpretations into strategic and operational choices. These mechanisms operate within a dynamic cycle similar to the perception–action cycle, in which perception, memory, attention, and intelligence continuously interact to adjust behaviors (Liu & Yin, 2025).

Within organizations, this dynamic translates into the ability to learn continuously, revise representations, and adapt actions in response to uncertainty. Furthermore, cognitive management mobilizes specific processes such as the codification and dissemination of knowledge, the articulation between tacit and explicit knowledge, and cognitive bricolage, which allows action frameworks to be adjusted according to contextual conditions (Michaud & Thoenig, 2007).

3.1. The Theoretical Foundations of Cognitive Management

Historically, cognitive management builds upon earlier research on knowledge management and learning organizations. The diagram below provides an analytical mapping of the main organizational approaches, structured around the conceptual framework developed by March (1991), which distinguishes two fundamental dimensions in the management of information and knowledge within organizations: exploration and exploitation. This representation makes it possible to position different theoretical perspectives according to their contribution to the creation, interpretation, and use of knowledge in organizational processes.

Figure 4. Conceptual Mapping of Organizational Theories within the Exploration–Exploitation Framework

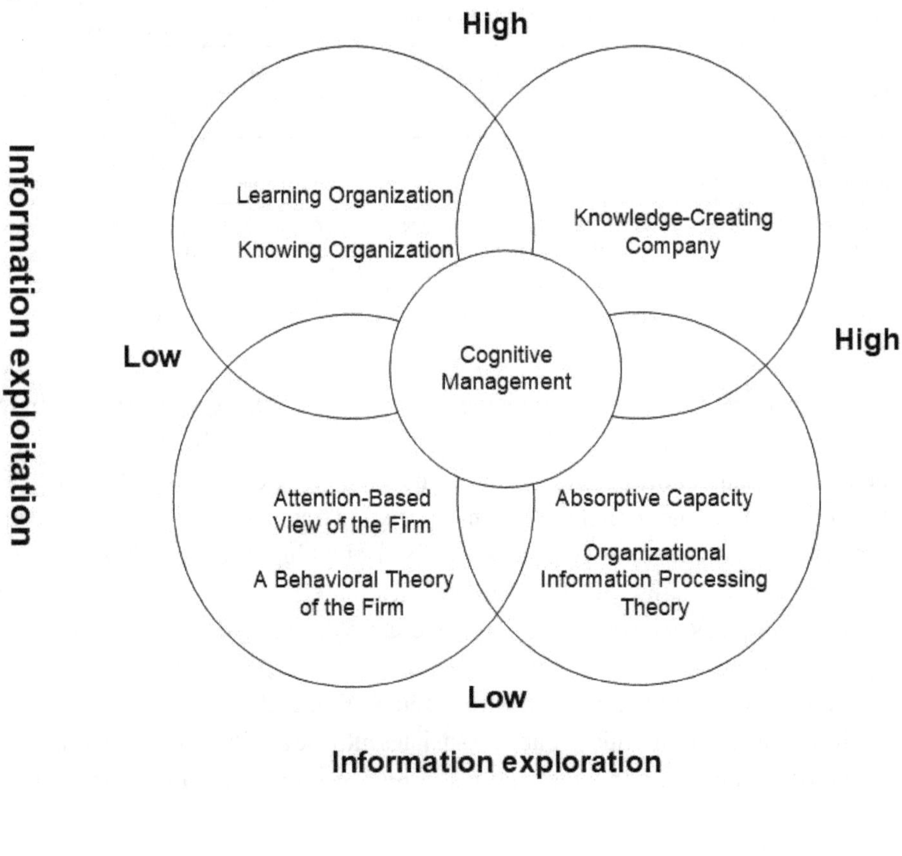

In approaches located in the low exploration and low exploitation area, such as A Behavioral Theory of the Firm or the Attention-Based View of the Firm, information is primarily processed through the cognitive and attentional limitations of organizational actors. These perspectives emphasize that organizations do not always possess the necessary capabilities to explore new knowledge or fully exploit available information, as decisions are constrained by bounded rationality, established routines, and selective attention. In contrast, some approaches place strong emphasis on the exploitation of information, such as the Learning Organization (Senge, 2006) and the Knowing Organization (Choo, 1996). Within these frameworks, organizations develop mechanisms that enable them to capitalize on existing information, improve collective learning processes, and transform accumulated knowledge into sustainable organizational capabilities. Exploitation thus becomes a central driver of performance and continuous improvement. Other perspectives are more strongly positioned within a logic of high informational exploration, particularly the Knowledge-Creating Company (Nonaka & Takeuchi, 1995), which emphasizes the dynamic creation of new knowledge through the interaction between tacit and explicit knowledge. This approach highlights that organizational competitiveness depends on the ability to continuously produce new information and renew the cognitive foundations of the organization. Finally, certain theories, such as absorptive capacity (Cohen & Levinthal, 1990; Zahra & George, 2002) and Organizational Information Processing Theory (Galbraith, 1973; Tushman & Nadler, 1978), occupy a position characterized by both high exploration and high exploitation of information. These perspectives argue that organizational performance depends on the ability to acquire, assimilate, transform, and exploit knowledge originating from the environment. They therefore highlight the importance of balancing exploration and exploitation of information in order to support adaptation, innovation, and decision-making in complex environments.

Overall, the diagram highlights the complementary contributions of these organizational theories to the understanding of information management within organizations. Each perspective emphasizes specific mechanisms through which organizations explore, interpret, and exploit knowledge. Together, these approaches illustrate how organizational parameters collectively contribute to the development of cognitive management. By integrating these theoretical foundations, organizations can better combine information exploration and exploitation, thereby strengthening their capacity for learning, adaptation, and strategic decision-making in complex environments.

3.2. Cognitive Management in Organizations

The integration of cognitive management within organizations implies a transformation of structures, processes, and managerial practices. Organizations thus become distributed cognitive systems, in which individuals share common representations and collectively construct meaning through their interactions. This dynamic requires structures that foster collaboration and learning, a culture oriented toward experimentation and critical reflection, and technological systems that support cognitive processes. From this perspective, cognitive management transforms the firm into an adaptive organization, capable of reconciling the exploitation of existing resources with the exploration of new opportunities in uncertain environments.

Helfat and Peteraf (2015) introduce the concept of managerial cognitive capability to emphasize that dynamic capabilities largely rely on cognitive microfoundations. From this viewpoint, the effectiveness of organizational adaptation processes depends significantly on the mental abilities of managers to perceive, interpret, and reconfigure information derived from their environment. In other words, organizational dynamics are not solely determined by structural or technological resources but are also rooted in managers' cognitive schemas, attentional capacities, and interpretive mechanisms, which shape how organizations identify opportunities, assess threats, and transform their resources to maintain a competitive advantage. Consequently, differences in organizational performance can partly be explained by variations in managers' cognitive capabilities, which directly influence their ability to lead strategic change. In this perspective, management can be understood as a cognitive process influencing the functioning of organizations (Hodgkinson, 2001; Reuter, 2024):

> First, it relies on the principle of bounded rationality, according to which individuals possess limited cognitive capacities to process information. Since managers cannot analyze all possible alternatives or integrate all available data, they rely on heuristics, routines, and cognitive schemas to simplify decision-making, particularly in contexts characterized by information overload.
>
> Second, it focuses on the cognitive processes underlying strategic decision-making. Faced with uncertainty and complexity, leaders often rely on simplified decision rules that may generate cognitive biases, thereby influencing their evaluation of situations and strategic choices.
>
> Third, it highlights the role of mental representations and cognitive maps in interpreting the competitive environment. Managers construct cognitive models that guide their attention and shape the formulation of organizational strategies.

Finally, cognitive management emphasizes the importance of organizational mechanisms that help manage the cognitive limitations of individuals. From this perspective, the organization can be viewed as a collective cognitive system, in which processes of perception, interpretation, and decision-making are distributed among multiple actors and organizational mechanisms.

4. THE CONTRIBUTION OF AGENTIC AI TO COGNITIVE MANAGEMENT

The emergence of Agentic AI represents a major shift in the evolution of cognitive management, introducing systems capable of acting autonomously, pursuing goals, and adapting to their environment. Unlike traditional artificial intelligence systems, these agents possess advanced cognitive capabilities that enable them to actively participate in organizational processes (Nisa et al., 2026). They now go beyond their role as analytical tools to become operational systems that contribute to strengthening and evolving the cognitive mechanisms mobilized in management processes (World Economic Forum & Capgemini, 2025).

In order to better understand the contribution of this technology, this chapter proposes an in-depth analysis of the role of Agentic AI within organizations. By adopting a systemic perspective, the objective is to examine the mechanisms and action systems mobilized by these intelligent technologies within managerial processes.

4.1. Agentic AI as a Reasoning-Based Process Management System (Thinker–Planner–Doer)

A Reasoning-Based Process Management System can be defined as a system for managing organizational processes that integrates reasoning mechanisms to analyze, interpret, and dynamically adapt the execution of processes. In an Agentic AI architecture designed as a Reasoning-Based Process Management System, process management relies on a structured reasoning logic that articulates analysis, planning, and execution.

The system therefore operates through three complementary components: the AI Thinker, the AI Planner, and the AI Doer. The continuous interaction between these three functions transforms the decision-making process into an adaptive cognitive cycle, in which reasoning guides action and action, in turn, feeds back into the reasoning process.

Figure 5. Functions of the Reasoning-Based Agentic AI Management Model

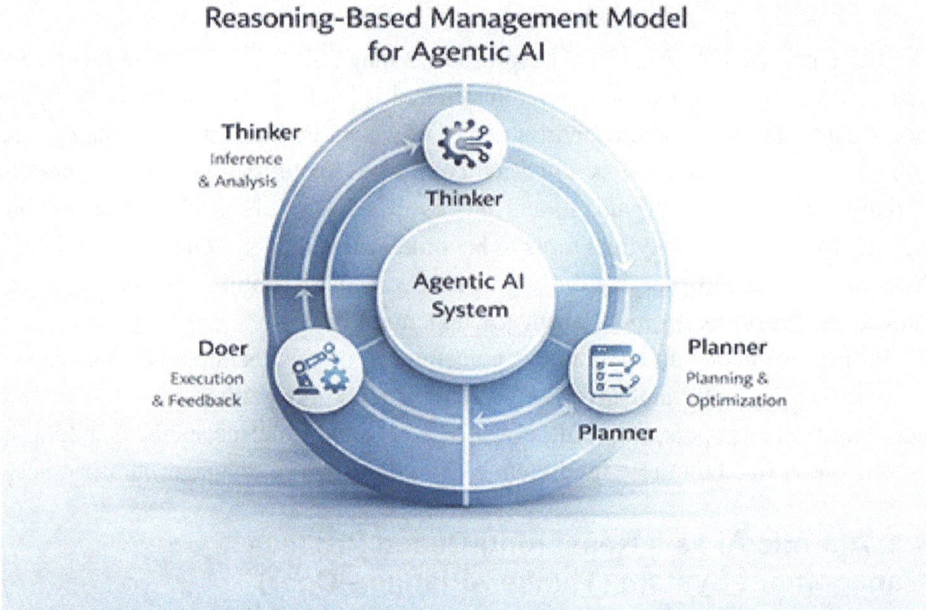

The Thinker represents the first layer of the cognitive process. This component is responsible for interpreting the informational context and formulating the problem. It mobilizes mechanisms of semantic analysis, inference, and contextual understanding in order to identify relevant objectives and associated constraints. In doing so, it transforms raw information flows into structured cognitive representations, enabling the system to determine what needs to be done and why. From this perspective, it functions as a cognitive orientation mechanism, comparable to processes of attention and sensemaking in organizations, by filtering relevant signals from the environment and generating hypotheses for action. This phase is critical, as the quality of subsequent reasoning depends on the system's ability to construct a coherent representation of the situation and to identify opportunities or problems that require attention.

The Planner represents the second stage of the process and is responsible for translating reasoning into operational strategies. Based on the objectives and hypotheses generated by the Thinker, this component organizes action into structured and hierarchical sequences by decomposing complex tasks into achievable sub-goals. It mobilizes mechanisms of algorithmic planning, optimization, and evaluation of alternatives in order to select the most appropriate courses of action. In this way, it plays a mediating role between cognition and action, transforming strategic intent into a coherent execution plan. From an organizational perspective, this function resembles processes of design and coordination that align resources and activities around common objectives. It also ensures the temporal and logical coherence of actions by anticipating task interdependencies and adapting planning to contextual constraints.

The Doer represents the operational dimension of the system, responsible for the effective execution of planned actions. This component implements the tasks defined by the Planner by interacting with information systems, databases, or other digital agents. However, the role of the Doer is not limited to the mechanical execution of instructions; it also generates informational feedback loops that continuously feed the reasoning and planning phases. The outcomes of actions, potential errors, or contextual changes are therefore reintegrated into the system to adjust future decisions. This dynamic transform execution into an adaptive process in which action becomes a source of learning and strategic reassessment. Within this model, the Doer thus contributes to the creation of a continuous cognitive cycle linking reasoning, planning, and action, enabling Agentic AI to function as a management system capable of adapting to complex and evolving environments.

4.2. Agentic AI in Business process management cycle

The integration of Agentic AI into Business Process Management (BPM) represents a major evolution in the way organizations design, execute, and improve their business processes. Historically, BPM systems have primarily relied on rule-based automation mechanisms aimed at orchestrating the execution of activities within structured workflows (Dumas et al., 2018). However, recent advances in artificial intelligence, particularly with the emergence of intelligent agents, open the way for systems capable of going beyond automation by introducing capabilities such as autonomy, reasoning, and dynamic adaptation in the management of organizational processes (Dumas et al., 2026).

Today, systems are evolving toward what the literature refers to as Agentic Business Process Management Systems, in which human or artificial agents become operational entities capable of perceiving the state of a process, analyzing data generated during its execution, and triggering actions aimed at achieving organizational objectives (Calvanese et al., 2026). Agentic AI introduces a data-driven and adaptive logic, enabling agents to exploit execution traces and process analysis techniques to detect improvement opportunities and adjust execution trajectories according to contextual conditions. In this perspective, agents generally operate through a continuous perceive–reason–act cognitive loop, allowing them to perceive their environment, reason about possible actions, and act to influence the state of the process while pursuing objectives aligned with those of the organization (Calvanese et al., 2026). The integration of these capabilities profoundly transforms the logic of BPM by introducing systems that are capable not only of supporting process execution but also of contributing to their analysis, optimization, and real-time adaptation (Kamal et al., 2025).

Figure 6. Presentation of Business Process Management Cycle

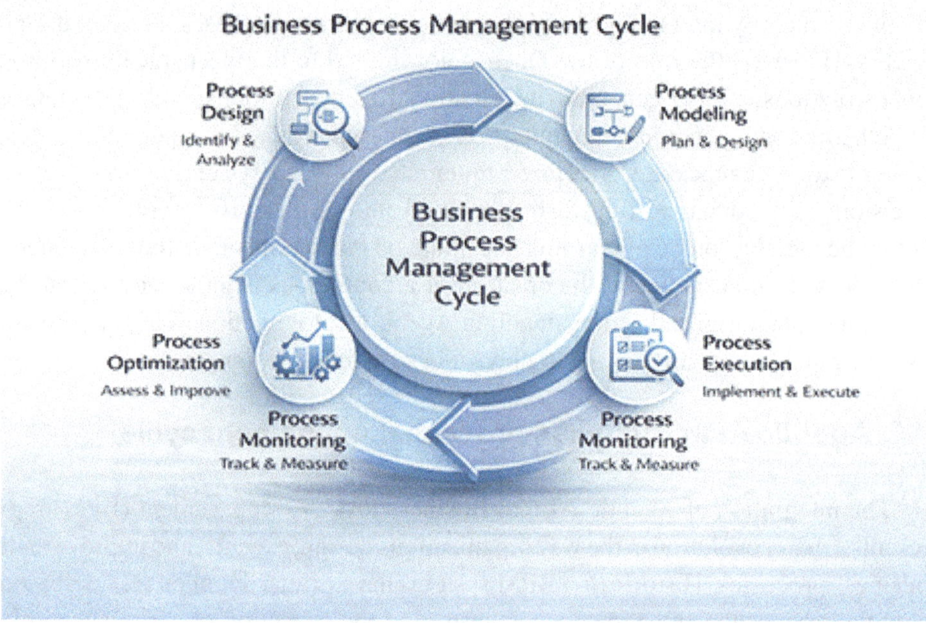

From this perspective, Agentic AI can be analyzed as a strategic lever operating at different stages of the Business Process Management cycle, ranging from process discovery and modeling to execution, monitoring, and continuous improvement.

Design Phase or Identification / Documentation

In the design phase, the AI Automatic Assistant primarily acts as a documentation support tool capable of collecting, structuring, and formalizing existing "as-is" processes. Its role remains largely reactive, focusing on organizing and presenting the information provided without fundamentally questioning the underlying process logic. In contrast, the AI Cognitive Partner adopts a proactive posture by participating in the co-construction of "to-be" processes. It identifies inefficiencies, proposes alternatives, and integrates optimization logic from the earliest stages. This evolution reflects a shift from simple operational assistance to a strategic contribution to process design.

Model Phase or Simulation

During the modeling phase, the AI Automatic Assistant enables the simulation of scenarios based on predefined parameters and rules. Its contribution remains mainly descriptive and strongly dependent on the initial assumptions introduced by the user. In contrast, the AI Cognitive Partner introduces predictive and adaptive analytical capabilities. It can explore multiple scenarios, anticipate risks, and dynamically adjust models based on available data. As a result, simulation evolves from a static exercise into an intelligent process capable of better capturing organizational complexity and uncertainty.

Execute Phase or Implementation

In the execution phase, the AI Automatic Assistant plays a central role in task automation through BPM or RPA systems, ensuring rapid, standardized execution consistent with predefined rules. However, its functioning remains rigid and poorly adaptable to unexpected situations. Conversely, the AI Cognitive Partner introduces a degree of decision-making autonomy by being able to adjust workflows in real time, manage exceptions, and adapt rules according to contextual conditions. This capability transforms execution into a more flexible, intelligent, and performance-oriented process.

Monitor Phase

In the monitoring phase, the AI Automatic Assistant is primarily limited to collecting, processing, and visualizing performance indicators, providing a descriptive overview of process performance. In contrast, the AI Cognitive Partner adopts a more advanced analytical approach by examining the gap between expected and observed performance. It identifies underlying causes, detects patterns, and generates actionable insights that support decision-making. Thus, it no longer simply presents information but actively contributes to its strategic interpretation.

Optimize Phase or Process Improvement

In the optimization phase, the AI Automatic Assistant supports continuous improvement initiatives by applying adjustments based on predefined rules or human analyses. However, its role remains limited to executing externally defined recommendations. By contrast, the AI Cognitive Partner operates within a continuous learning logic, leveraging data generated during monitoring to propose autonomous and dynamic improvements. In doing so, it feeds a continuous improvement loop capable of adapting processes in real time and anticipating future changes, thereby strengthening organizational adaptability.

5. FROM AI AUTOMATIC ASSISTANT TO AI COGNITIVE PARTNER

The logic of artificial intelligence, moving from the role of an automated assistant to that of a cognitive partner, reflects a profound transformation in managerial approach and in the ways decisions are produced within organizations (Alshammari & Alshammari, 2026; Park, 2026). According to Hughes et al. (2025), the evolution of artificial intelligence in organizations has occurred through several phases. In its initial phase, AI was primarily designed as an execution tool aimed at automating repetitive tasks and optimizing operational processes. This instrumental approach was grounded in a logic of efficiency, where AI functioned as a technical support system producing outputs based on predefined rules or models, while reasoning and decision-making remained entirely under the responsibility of managers.

However, recent technological advances have led to the emergence of artificial intelligence agents and agentic systems, characterized by their ability to learn, reason, and act autonomously in complex environments. In this context, AI now goes beyond its role as an automation tool and increasingly becomes an organizational actor capable of supporting decision-making and collaborating with human actors. Building on this evolution, this section analyzes two complementary perspectives: Agentic AI as a partner to managers in decision-making processes, and Agentic AI as a manager, where intelligent agents directly assume certain functions of coordination and organizational control.

5.1. Agentic AI – Manager interaction

The interaction between AI systems and human managers can take different forms depending on the level of engagement and the relational dynamics between the two actors. In order to account for this diversity, the figure below proposes a typology aimed at characterizing the different modes of human–Agentic AI interaction:

Figure 7. The 4C Human–Agentic AI Partnership Framework

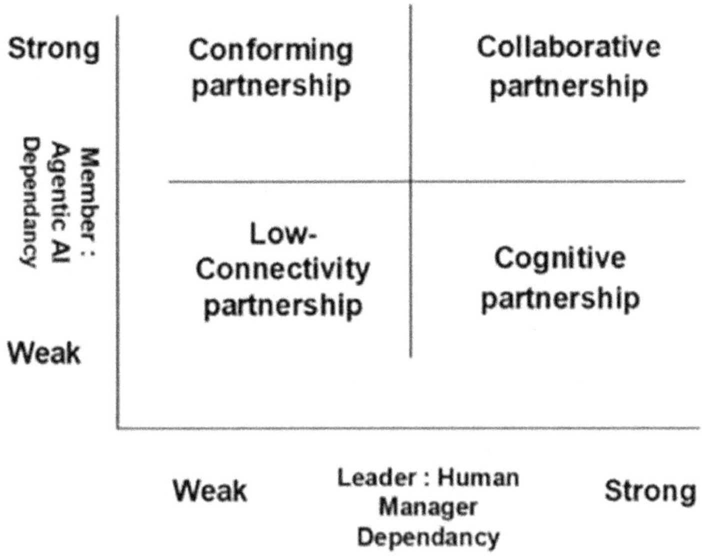

The construction of this typology primarily relies on the Leader–Member Exchange (LMX) theory introduced by Graen and Uhl-Bien (1995), which represents one of the major approaches to relational leadership within organizations. This theory emphasizes the quality of the exchange relationship developed between the leader and each of their members. It postulates that relationships within an organization are not homogeneous: some are characterized by high-quality exchanges based on trust, mutual support, and intensive information sharing, whereas others remain more formal and limited to role-based obligations.

In an organizational context increasingly shaped by the integration of artificial intelligence into decision-making and operational processes, this relational perspective provides a relevant analytical framework for understanding emerging interactions between human managers and AI systems. Indeed, similarly to human members in LMX theory, AI systems can be integrated into organizational structures at different levels of engagement, autonomy, and interaction with managers. This analogy therefore makes it possible to conceptualize the human–AI relationship as a form of organizational exchange whose intensity varies according to the degree of interaction, trust, and integration within decision-making processes.

In this context, the relationship between managers and artificial intelligence is evolving toward a form of cognitive collaboration. Patel (2025) shows that the emergence of Agentic AI systems introduces a new generation of technologies characterized by autonomy, adaptability, and goal-oriented behavior. Unlike traditional systems based on a command–response logic, these technologies are capable of taking initiatives, proposing alternatives, and actively participating in organizational reasoning processes. The interaction between the manager and AI thus becomes dialogical: managers mobilize their strategic judgment and contextual understanding, while AI contributes through the analysis of large volumes of data, the generation of scenarios, and the identification of complex patterns. This complementarity progressively transforms AI into an active partner in the construction of organizational decisions.

According to Alshammari and Alshammari (2026), AI systems no longer merely automate operational tasks or execute predefined instructions; they now directly participate in the cognitive processes mobilized by managers in analysis and decision-making. The authors explain that AI acts as a cognitive partner capable of enriching the informational environment within which managers interpret organizational situations. By providing intelligent suggestions, predictive analyses, and syntheses of complex data, these systems contribute to improving the quality of knowledge mobilized in managerial action. AI thus operates within a logic of cognitive augmentation, where its value lies less in task automation than in its ability to enhance reasoning and the quality of organizational decisions. The authors further emphasize that this transformation corresponds to an evolution toward a form of distributed organizational cognition. In this framework, cognitive processes are no longer located exclusively in the minds of managers but are distributed between human actors and technological systems. AI therefore actively participates in cognitive activity by structuring information, facilitating the exploration of new ideas, and providing adaptive feedback that refines managerial reasoning. This redistribution of cognitive functions transforms the nature of intellectual work and redefines the modalities of decision-making within digital organizations.

Gaddam and Raigoni (2026) further demonstrate that the integration of Agentic AI leads to a redefinition of roles and decision-making mechanisms within organizations. According to these authors, agentic AI systems no longer function solely as assistance tools but rather as actors capable of influencing decision-making processes and organizational dynamics. In this model, managers are no longer the sole holders of decision-making authority; they also become supervisors and interpreters of the analyses produced by AI. Meanwhile, AI contributes to the generation of strategic options, predictive analysis, and the optimization of decisions. This interaction gradually reshapes the boundaries between human cognition and computational capabilities within organizations.

5.2. Agentic AI as manager

The evolution of artificial intelligence has enabled Agentic AI systems to assume functions traditionally associated with management. This technology is no longer limited to the role of an executor or decision-support assistant but increasingly adopts a position of organizational orchestration (Chandrasekaran et al., 2025). In a manner analogous to the role of a manager, these systems supervise entire processes, distribute tasks among different agents or specialized modules, and monitor the progress of activities in order to achieve predefined objectives (Huang, 2025).

This orchestration logic relies on a multi-agent architecture in which several intelligent entities interact to accomplish common goals. Within such systems, certain agents specialize in specific tasks, while a higher-level agent ensures the coordination of the entire structure. This agent acts as a meta-agent responsible for planning activities, assigning responsibilities, and supervising task execution. This function corresponds to an emerging form of algorithmic management, in which the coordination of work is partially delegated to artificial intelligence (Nisa et al., 2026).

This conception of Agentic AI as a coordinating entity resonates with recent research on agent system architectures. Within this perspective, the orchestrator agent appears as a central component for managing the complexity of multi-agent systems and synchronizing interactions between different intelligent modules. By performing functions of planning, coordination, and supervision, this agent contributes to transforming artificial intelligence into a genuine organizational actor capable of structuring and directing the execution of complex processes (Chandrasekaran et al., 2025).

A study conducted by the American strategic consulting firm KPMG proposes a typology of artificial intelligence agents through the TACO Framework (Chandrasekaran et al., 2025):

Figure 8. Presentation of TACO Framework

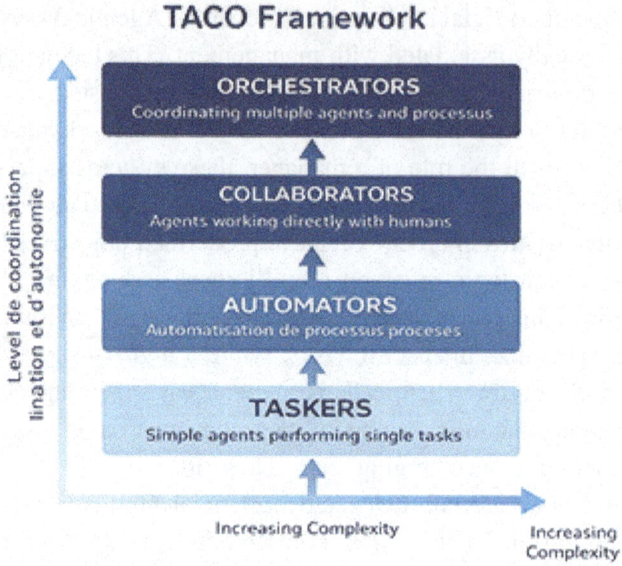

In this model, the authors distinguish several categories of agents according to their level of complexity and autonomy: Taskers, Automators, Collaborators, and Orchestrators. Orchestrators represent the most advanced level of this architecture. They are designed to coordinate multiple specialized agents, manage interconnected workflows, and supervise the execution of organizational processes on a large scale. In this configuration, the orchestrator agent acts as a true digital control tower, capable of allocating tasks, managing dependencies between activities, and optimizing the use of resources within the system.

Thus, the orchestrator agent can be interpreted as an advanced form of Agentic AI whose functions closely resemble the role of a manager within organizations. While lower-level agents execute specific tasks, the orchestrator assumes a strategic coordination role by ensuring the overall coherence of the system and the achievement of organizational objectives. This evolution suggests that artificial intelligence no longer merely supports managerial activities but can also contribute to structuring and directing organizational processes, paving the way for new forms of hybrid management that combine human supervision with algorithmic orchestration.

6. CHALLENGES AND GOVERNANCE ISSUES

The rise of Agentic AI in contemporary organizations cannot be reduced to a simple technological evolution. It is also accompanied by a set of multidimensional challenges—psychological, legal, economic, and organizational. In this context, the issue of governance no longer concerns only the technical regulation of intelligent systems but becomes a major strategic challenge aimed at ensuring the coherence, reliability, and legitimacy of decisions produced by these technologies.

6.1. Current Challenges of AI

At the technical level, the issue of algorithmic bias occupies a central place. Since AI systems are trained on historical data, they may incorporate social, economic, or cultural distortions (Barocas et al., 2023). In an agentic environment, these biases are not confined to a single module; they can circulate, reinforce one another, and influence the decisions of other agents through coordination mechanisms. Thus, the risk is no longer only individual but systemic. This situation requires governance mechanisms capable of identifying, measuring, and correcting these biases at different levels of the system (Mukherjee & Chang, 2025).

The opacity of models constitutes another critical challenge. Systems based on deep learning architectures are often described as "black boxes" due to the difficulty of interpreting their internal reasoning processes (Rudin, 2019; Rudin & Radin, 2019). This opacity is amplified in agentic systems, where decisions result from multiple interactions among autonomous agents. As Floridi et al. (2018) highlight, the absence of explainability can undermine user trust and limit the organizational acceptability of these technologies. Therefore, integrating explainability mechanisms becomes a fundamental governance requirement.

Data quality is also a crucial determinant of the performance of agentic systems. Batini and Scannapieco (2016) emphasize that the reliability of decisions directly depends on the quality of the data used. In a multi-agent environment, erroneous or incomplete data can quickly spread throughout the system, producing cascading effects. This phenomenon highlights the importance of implementing rigorous data governance policies, including data validation, updating, and traceability.

At the psychological level, overreliance on AI represents a frequently underestimated risk. Users may gradually delegate their judgment to automated systems, implicitly assuming their reliability (Klingbeil et al., 2024). In the case of Agentic AI, this dependence is reinforced by the autonomy of these systems and their ability to manage complex processes without human intervention. This phenomenon, known in the literature as automation bias, originally identified by Parasuraman and Riley (1997), can lead to the erosion of human skills and reduced vigilance toward potential errors (Romeo & Conti, 2025). Governance must therefore incorporate mechanisms of human supervision and encourage a critical stance toward automated decisions.

At the legal level, the issue of accountability becomes particularly complex in the context of Agentic AI. Unlike traditional systems, where responsibility can be attributed to a specific individual or algorithm, agentic systems rely on distributed decision-making. As Mittelstadt et al. (2016) note, this dispersion makes it difficult to identify responsibilities in the event of errors or damages. Consequently, legal and organizational frameworks must be reconsidered in order to clarify the roles and obligations of the different actors involved.

Security issues also concern Agentic AI systems. Due to their autonomy and ability to interact with various systems and digital tools, these technologies may be exposed to cyberattacks, manipulation attempts, or unauthorized access. Such risks may compromise the proper functioning of systems and influence the decisions generated by artificial intelligence (Lazer et al., 2026).

The alignment of objectives between AI systems and organizational goals also constitutes a critical challenge. As Russell (2019) emphasizes, an intelligent system may efficiently pursue a poorly defined objective, thereby producing outcomes contrary to human expectations. In a multi-agent system, this risk is amplified by the possibility of conflicts of objectives between agents or behavioral deviations. It is therefore essential to clearly define the objectives assigned to these systems and to implement control mechanisms ensuring their alignment with organizational values.

A specific challenge related to Agentic AI lies in the management of emergent behaviors. These systems may produce unexpected outcomes due to complex interactions among their components (Sapkota et al., 2026). While such behaviors may foster innovation, they may also generate unforeseen risks. This requires the implementation of continuous monitoring mechanisms capable of rapidly identifying deviations and adjusting control mechanisms accordingly.

6.2. Strategies and Development Actions for Agentic AI

In response to these challenges, the implementation of robust governance appears to be an essential condition for the successful integration of Agentic AI. Such governance should rely on principles of transparency, auditability, and traceability of decisions. As Floridi et al. (2018) argue, effective governance depends on the articulation between ethical frameworks, institutional regulations, and organizational practices. It must also remain adaptive in order to respond to the rapid pace of technological transformation.

Finally, building trust represents a transversal challenge. Trust does not depend solely on the technical performance of systems but also on their capacity to be understood, controlled, and perceived as fair. In this sense, the governance of Agentic

AI should not be viewed as a constraint but rather as a strategic lever capable of reconciling technological autonomy with human oversight. Only under these conditions can Agentic AI be sustainably integrated into organizational practices and contribute to a responsible and sustainable transformation of business processes.

CONCLUSION

The integration of artificial intelligence, and particularly Agentic AI, marks a profound transformation in the foundations of management and the sources of competitive advantage for firms. Beyond its initial role as an automation tool, artificial intelligence is increasingly emerging as a strategic lever for cognitive augmentation, enabling organizations to strengthen their analytical, anticipatory, and decision-making capabilities. This evolution reflects a shift in the competitive paradigm, in which organizational performance no longer relies solely on the possession of tangible or informational resources, but rather on the ability to mobilize and orchestrate hybrid systems combining human intelligence and artificial intelligence.

From this perspective, artificial intelligence contributes to value creation by improving the quality of knowledge mobilized within organizations, accelerating decision-making processes, and fostering organizational innovation. Firms capable of effectively integrating these technologies into their cognitive processes develop distinctive advantages that are difficult to imitate, particularly in terms of strategic agility, service personalization, and operational optimization. As suggested by resource-based and knowledge-based perspectives, the combination of human capabilities and algorithmic capabilities can constitute a source of sustainable competitive advantage, provided that it is coherently integrated into organizational structures and practices.

However, this advantage does not result solely from technological adoption; it also depends on the ability of organizations to establish appropriate governance structures and to effectively orchestrate interactions between human actors and intelligent systems. Organizations that move beyond a substitution logic and adopt a complementarity logic, in which artificial intelligence functions as a cognitive partner rather than merely an automation tool, are those most likely to fully benefit from this transformation. Conversely, poorly managed integration, characterized by algorithmic opacity, the presence of biases, or excessive dependence on automated systems, may undermine decision quality and compromise long-term organizational performance.

Furthermore, the rise of Agentic AI raises significant issues related to governance, ethics, and accountability. The increasing autonomy of these systems and their capacity to intervene in decision-making processes challenge existing mechanisms of control, supervision, and responsibility for decisions produced by artificial intelligence. In this context, the governance of intelligent systems cannot be limited to technical regulation but must be embedded in a broader framework that integrates organizational, legal, and ethical dimensions. Organizations must therefore implement mechanisms capable of ensuring algorithmic transparency, decision accountability, and stakeholder trust.

Ultimately, artificial intelligence does not constitute a competitive advantage in itself; rather, it becomes a key differentiating factor when it is integrated within a coherent organizational, strategic, and ethical framework. The challenge for firms is therefore not only to adopt these technologies but also to develop the organizational capabilities required to leverage them intelligently, responsibly, and in alignment with strategic objectives. From this perspective, the future of management lies in the ability to build forms of augmented intelligence, in which the synergy between humans and machines becomes the primary driver of performance, innovation, and sustainable competitiveness.

REFERENCES

Aakula, A., Saini, V., & Ahmad, T. (2024). The Impact of AI on Organizational Change in Digital Transformation. *Internet of Things and Edge Computing Journal*, *4*(1), 75–115.

Abbasi, M., Nishat, R. I., Bond, C., Graham-Knight, J. B., Lasserre, P., Lucet, Y., & Najjaran, H. (2025). A review of AI and machine learning contribution in business process management (process enhancement and process improvement approaches). *Business Process Management Journal*, *31*(4), 1414–1452. DOI: 10.1108/BPMJ-07-2024-0555

Abuadbba, A., Sultan, N., Nepal, S., & Jha, S. (2026). *Human Society-Inspired Approaches to Agentic AI Security : The 4C Framework* (arXiv:2602.01942). arXiv. https://doi.org//arXiv.2602.01942DOI: 10.48550

Ahmadi, M. M., Hajizadeh, P., & Valmohammadi, C. (2026). Cognitive Management : A Systematic Explication of Applying Cognitive Sciences in Management. *Emerging Technologies and Governance*, *1*(1), 66–92. DOI: 10.47176/ETG.2026.1003

Al-Mahdawi, E.Dr Emad Al-Mahdawi. (2025). From Generative to Agentic : How AI Shapes the Future of Work. *European Journal of Sports Science Technology*, *15*(59), 30–39. DOI: 10.58305/ejsst.v15i59.736

Alkhatib, A. W., Zaid, M., & Issa, A. (2025). Antecedents and outcomes of artificial intelligence adoption on the sustainable performance : The TOE framework perspective. *Information Discovery and Delivery*, •••, 1–16. DOI: 10.1108/IDD-04-2025-0076

Alshammari, K. H., & Alshammari, A. F. (2026). AI as a Cognitive Partner : Investigating Knowledge Augmentation and Its Role in Digital Transformation Outcomes. *Systems*, *14*(1), 65. DOI: 10.3390/systems14010065

Baker, J. (2012). The Technology–Organization–Environment Framework. In Dwivedi, Y. K., Wade, M. R., & Schneberger, S. L. (Eds.), *Information Systems Theory* (Vol. 28, pp. 231–245). Springer New York., DOI: 10.1007/978-1-4419-6108-2_12

Bandara, E., Gore, R., Shetty, S., Rajapakse, S., Kularathna, I., Karunarathna, P., Mukkamala, R., Foytik, P., Bouk, S. H., Rahman, A., Liang, X., Hass, A., Hewa, T., Keong, N. W., Zoysa, K. D., Withanage, A., & Loganathan, N. (2026). *A Practical Guide to Agentic AI Transition in Organizations* (arXiv:2602.10122). arXiv. https://doi.org//arXiv.2602.10122DOI: 10.48550

Bandi, A., Kongari, B., Naguru, R., Pasnoor, S., & Vilipala, S. V. (2025). The Rise of Agentic AI : A Review of Definitions, Frameworks, Architectures, Applications, Evaluation Metrics, and Challenges. *Future Internet*, *17*(9), 404. DOI: 10.3390/fi17090404

Barocas, S., Hardt, M., & Narayanan, A. (2023). *Fairness and machine learning : Limitations and opportunities*. MIT press.

Batini, C., & Scannapieco, M. (2016). *Data and Information Quality : Dimensions, Principles and Techniques*. Springer International Publishing., DOI: 10.1007/978-3-319-24106-7

Benhmama, A., & Sabiri, B. (2025). Can AI improve individual employee performance? *2025 International Conference on Innovation in Artificial Intelligence and Internet of Things (AIIT)*, 1-6. https://ieeexplore.ieee.org/abstract/document/11082834/

Brynjolfsson, E., & Mcafee, A. (2017). The business of artificial intelligence. *Harvard Business Review*, *7*(1), 1–2.

Calvanese, D., Casciani, A., Giacomo, G. D., Dumas, M., Fournier, F., Kampik, T., Malfa, E. L., Limonad, L., Marrella, A., Metzger, A., Montali, M., Amyot, D., Fettke, P., Polyvyanyy, A., Rinderle-Ma, S., Sardiña, S., Tax, N., & Weber, B. (2026). *Agentic Business Process Management : A Research Manifesto* (arXiv:2603.18916). arXiv. https://doi.org//arXiv.2603.18916DOI: 10.48550

Canals, J. (2020). The evolving role of general managers in the age of AI. *The Future of Management in an AI World: Redefining Purpose and Strategy in the Fourth Industrial Revolution*, 37-64.

Chandrasekaran, S., Edin, P., & Lohr, T. (2025). *The agentic AI advantage : Unlocking the next level of AI value*. KPMG., https://kpmg.com/us/en/articles/2025/the-agentic-ai-advantage.html

Chaturvedi, A., & Dasgupta, M. (2024). Managerial Perception of AI in Strategic Decision-Making. *Journal of Computer Information Systems*, •••, 1–13. DOI: 10.1080/08874417.2024.2418403

Choo, C. W. (1996). The knowing organization : How organizations use information to construct meaning, create knowledge and make decisions. *International Journal of Information Management*, *16*(5), 329–340. DOI: 10.1016/0268-4012(96)00020-5

Cohen, W. M., & Levinthal, D. A. (1990). Absorptive capacity : A new perspective on learning and innovation. *Administrative Science Quarterly*, *35*(1), 128–152. DOI: 10.2307/2393553

Davenport, T. H., & Ronanki, R. (2018). Artificial intelligence for the real world. *Harvard Business Review*, *96*(1), 1.

Dumas, M., Fournier, F., Limonad, L., Marrella, A., Montali, M., Rehse, J.-R., Accorsi, R., Calvanese, D., De Giacomo, G., Fahland, D., Gal, A., La Rosa, M., Völzer, H., & Weber, I. (2023). AI-augmented Business Process Management Systems : A Research Manifesto. *ACM Transactions on Management Information Systems*, *14*(1), 1–19. DOI: 10.1145/3576047

Dumas, M., La Rosa, M., Mendling, J., & Reijers, H. A. (2018). Introduction to Business Process Management. In Dumas, M., La Rosa, M., Mendling, J., & Reijers, H. A. (Eds.), *Fundamentals of Business Process Management* (pp. 1–33). Springer., DOI: 10.1007/978-3-662-56509-4_1

Dumas, M., Milani, F., & Chapela-Campa, D. (2026). *Agentic Business Process Management Systems* (arXiv:2601.18833). arXiv. /arXiv.2601.18833DOI: 10.1007/978-3-032-13426-4_1

Eisenhardt, K. M., & Martin, J. A. (2000). Dynamic capabilities : What are they? *Strategic Management Journal*, *21*(10-11), 1105–1121. DOI: 10.1002/1097-0266(200010/11)21:10/11<1105::AID-SMJ133>3.0.CO;2-E

Emery, F. E., & Trist, E. L. (1965). The causal texture of organizational environments. *Human Relations; Studies Towards the Integration of the Social Sciences*, *18*(1), 21–32. DOI: 10.1177/001872676501800103

Floridi, L., Cowls, J., Beltrametti, M., Chatila, R., Chazerand, P., Dignum, V., Luetge, C., Madelin, R., Pagallo, U., Rossi, F., Schafer, B., Valcke, P., & Vayena, E. (2018). AI4People—An Ethical Framework for a Good AI Society : Opportunities, Risks, Principles, and Recommendations. *Minds and Machines*, *28*(4), 689–707. DOI: 10.1007/s11023-018-9482-5

Fountaine, T., McCarthy, B., & Saleh, T. (2019). Building the AI-powered organization. *Harvard Business Review*, *97*(4), 62–73.

Gaddam, S. G. S., & Raigoni, G. S. (2026). Agentic Artificial Intelligence as a Strategic HR Partner : Redefining Decision-Making Authority and Strategic Roles. *International Journal of Innovative Research in Engineering*, *7*(1), 81–86. https://www.doi.org/10.59256/ijire.20260701010. DOI: 10.59256/ijire.20260701010

Galbraith, J. (1973). Designing complex organizations. *Reading, Mass.* https://search.proquest.com/openview/2124f3c691b3cca453388622c2a20df4/1?pq-origsite=gscholar&cbl=1817083

Gavetti, G., & Levinthal, D. (2000). Looking Forward and Looking Backward : Cognitive and Experiential Search. *Administrative Science Quarterly*, *45*(1), 113–137. DOI: 10.2307/2666981

Graen, G. B., & Uhl-Bien, M. (1995). Relationship-based approach to leadership : Development of leader-member exchange (LMX) theory of leadership over 25 years: Applying a multi-level multi-domain perspective. *The Leadership Quarterly*, *6*(2), 219–247. DOI: 10.1016/1048-9843(95)90036-5

Helfat, C. E., & Peteraf, M. A. (2015). Managerial cognitive capabilities and the microfoundations of dynamic capabilities. *Strategic Management Journal*, *36*(6), 831–850. DOI: 10.1002/smj.2247

Hodgkinson, G. P. (2001). Cognitive Processes in Strategic Management : Some Emerging. *Handbook of Industrial, Work & Organizational Psychology: Volume 2: Organizational Psychology*, 416.

Hollan, J., Hutchins, E., & Kirsh, D. (2000). Distributed cognition : Toward a new foundation for human-computer interaction research. *ACM Transactions on Computer-Human Interaction*, *7*(2), 174–196. DOI: 10.1145/353485.353487

Hosseini, S., & Seilani, H. (2025). The role of agentic AI in shaping a smart future : A systematic review. *Array (New York, N.Y.)*, *26*, 100399. DOI: 10.1016/j.array.2025.100399

Hrebiniak, L. G., & Joyce, W. F. (1985). Organizational Adaptation : Strategic Choice and Environmental Determinism. *Administrative Science Quarterly*, *30*(3), 336–349. DOI: 10.2307/2392666

Huang, K. (Ed.). (2025). *Agentic AI : Theories and Practices*. Springer Nature Switzerland., DOI: 10.1007/978-3-031-90026-6

Hughes, L., Dwivedi, Y. K., Li, K., Appanderanda, M., Al-Bashrawi, M. A., & Chae, I. (2025). AI Agents and Agentic Systems : Redefining Global it Management. *Journal of Global Information Technology Management*, *28*(3), 175–185. DOI: 10.1080/1097198X.2025.2524286

Hutchins, E. (1995). *Cognition in the Wild*. MIT press. DOI: 10.7551/mitpress/1881.001.0001

Islam, M. A., Somu, S., & Aldaihani, F. M. F. (2026). The Rise of Agentic AI : Synthesis of Current Knowledge and Future Research Agenda. *Global Business and Organizational Excellence*, *45*(3), 402–416. DOI: 10.1002/joe.70019

Kamal, A., Tanvir Ansari, M., & Chapaneri, K. (2025). *Agentic AI –The new frontier in GenAI: An executive playbook.* PricewaterhouseCoopers., https://www.pwc.ch/en/publications/2025/ch-accelerating-process-automation-playbook.pdf

Klingbeil, A., Grützner, C., & Schreck, P. (2024). Trust and reliance on AI — An experimental study on the extent and costs of overreliance on AI. *Computers in Human Behavior, 160*, 108352. DOI: 10.1016/j.chb.2024.108352

Korzynski, P., Edwards, A., Gupta, M. C., Mazurek, G., & Wirtz, J. (2025). Humanoid robotics and agentic AI : Reframing management theories and future research directions. *European Management Journal, 43*(4), 548–560. DOI: 10.1016/j.emj.2025.06.002

Lazer, S. J., Aryal, K., Gupta, M., & Bertino, E. (2026). *A Survey of Agentic AI and Cybersecurity : Challenges, Opportunities and Use-case Prototypes* (arXiv:2601.05293). arXiv. https://doi.org//arXiv.2601.05293DOI: 10.48550

LeCun, Y., Bengio, Y., & Hinton, G. (2015). Deep learning. *nature, 521*(7553), 436-444.

Liu, Z., & Yin, X. (2025). A Review of Cognitive Control : Advancement, Definition, Framework, and Prospect. *Actuators, 14*(1), 32. DOI: 10.3390/act14010032

Luthans, F., & Stewart, T. I. (1977). A General Contingency Theory of Management. *Academy of Management Review, 2*(2), 181. DOI: 10.2307/257902

March, J. G. (1991). Exploration and Exploitation in Organizational Learning. *Organization Science, 2*(1), 71–87. DOI: 10.1287/orsc.2.1.71

Miles, R. E., Snow, C. C., & Pfeffer, J. (1974). Organization-Environment [Object Identifier] [DOI]. *Concepts and Issues., 13*(3), 244–264. DOI: 10.1111/j.1468-232X.1974.tb00581.xDigital

Mintzberg, H. (2009). *Managing*. Berrett-Koehler Publishers.

Mitchell, T. M. (1997). Does machine learning really work? *AI Magazine, 18*(3), 11–11.

Mittelstadt, B. D., Allo, P., Taddeo, M., Wachter, S., & Floridi, L. (2016). The ethics of algorithms : Mapping the debate. *Big Data & Society, 3*(2), 2053951716679679. DOI: 10.1177/2053951716679679

Mpanza, S. S. (2025). Revisiting the Technological-Organizational-Environmental (TOE) Framework and Diffusion of Innovation (DOI) : A Theoretical Review for Artificial Intelligence (AI) Adoption. *International Journal of Applied Research in Business and Management*, *6*(5). Advance online publication. https://www.wr-publishing.org/index.php/ijarbm/article/view/441. DOI: 10.51137/wrp.ijarbm.441

Mukherjee, A., & Chang, H. H. (2025). *Agentic AI : Autonomy, Accountability, and the Algorithmic Society* (arXiv:2502.00289). arXiv. /arXiv.2502.00289DOI: 10.2139/ssrn.5123621

Murire, O. T. (2024). Artificial Intelligence and Its Role in Shaping Organizational Work Practices and Culture. *Administrative Sciences*, *14*(12), 316. DOI: 10.3390/admsci14120316

Najana, M., Bhattacharya, S., Kewalramani, C., & Pandiya, D. K. (2024). AI and organizational transformation : Navigating the future. *International Journal of Global Innovations and Solutions (IJGIS)*. https://ijgis.pubpub.org/pub/u5r681dy

Nayebi, F. (2025). *Foundations of Agentic AI for Retail : Concepts, Technologies, and Architectures for Autonomous Retail Systems (1re* (Divergence, G., Ed.).

Nikulina, I. E., & Khomenko, I. V. (2015). Cognitive Management : Theory and Practice in the Organization. *Procedia - Social and Behavioral Sciences, Proceedings of The International Conference on Research Paradigms Transformation in Social Sciences 2014 (RPTSS-2014), 166*, 441-445. DOI: 10.1016/j.sbspro.2014.12.550

Nisa, U., Shirazi, M., Saip, M. A., & Pozi, M. S. M. (2026). Agentic AI : The age of reasoning—A review. *Journal of Automation and Intelligence*, *5*(1), 69–89. DOI: 10.1016/j.jai.2025.08.003

Nonaka, I., & Takeuchi, H. (1995). *The Knowledge-Creating Company : How Japanese Companies Create the Dynamics of Innovation*. Oxford University Press., DOI: 10.1093/oso/9780195092691.001.0001

Nosova, S., Norkina, A., Makar, S., Gerasimenko, T., & Medvedeva, O. (2022). Artificial intelligence as a driver of business process transformation. *Procedia Computer Science, 2022 Annual International Conference on Brain-Inspired Cognitive Architectures for Artificial Intelligence: The 13th Annual Meeting of the BICA Society, 213*, 276-284. DOI: 10.1016/j.procs.2022.11.067

Ousghir, S., Benhmama, A., & Sabiri, B. (2024). HR Team Performance in the Realm of AI-HRM : A Case Study of Moroccan Small and Medium Companies. *International Conference on Digital Technologies and Applications*, 202-211. https://link.springer.com/chapter/10.1007/978-3-031-68650-4_20

Parasuraman, R., & Riley, V. (1997). Humans and Automation : Use, Misuse, Disuse, Abuse. *Human Factors, 39*(2), 230–253. DOI: 10.1518/001872097778543886

Park, M. J. (2026). AI as a cognitive collaborator : Assimilation and accommodation in human–machine teaming for innovation. *Journal of Innovation & Knowledge, 12*, 100892. DOI: 10.1016/j.jik.2025.100892

Parnell, J. A., Wright, P., & Tu, H. S. (1996). Beyond the Strategy-Performance Linkage : The Impact of the Strategy-Organization-Environment Fit on Business Performance. *American Business Review, 14*(2), 41.

Pastakia, K., Scotis, S., Scoble-Williams, N., Elliott, S., & Lettau, M. (2025). *Work, reworked : What it takes to win in the age of Agentic AI.* Deloitte., https://www.deloitte.com/cbc/en/services/consulting/perspectives/human-agentic-workforce.html

Patel, J. (2025). Exploring the role of agentic AI in enhancing Human-AI collaboration. *International Journal of Engineering Technology Research & Management, 9*(8), 51–61. DOI: 10.5281/zenodo.16739658

Pramod, D., Patil, K. P., Bharathi, S. V., Vijaykumar, B. V., Birabar, S., & Sahoo, B. (2024). Artificial intelligence and machine learning adoption in automotive organizations using technology-organization-environment framework. *Journal of Information and Optimization Sciences, 45*(7), 1963–1976. DOI: 10.47974/JIOS-1748

Raisch, S., & Krakowski, S. (2021). Artificial Intelligence and Management : The Automation–Augmentation Paradox. *Academy of Management Review, 46*(1), 192–210. DOI: 10.5465/amr.2018.0072

Ransbotham, S., Kiron, D., Khodabandeh, S., Iyer, S., & Das, A. (2025). *The Emerging Agentic Enterprise : How Leaders Must Navigate a New Age of AI.* MIT Sloan Management Review & Boston Consulting Group., DOI: 10.63383/jAXE2583

Reicherts, L., Zhang, Z. T., Von Oswald, E., Liu, Y., Rogers, Y., & Hassib, M. (2025). AI, Help Me Think—but for Myself : Assisting People in Complex Decision-Making by Providing Different Kinds of Cognitive Support. *Proceedings of the 2025 CHI Conference on Human Factors in Computing Systems*, 1-19. DOI: 10.1145/3706598.3713295

Reuter, E. (2024). *Cognitive Capacity in the Management and Organization Research : A Review and Agenda for Future Research* (SSRN Scholarly Paper 5040625). Social Science Research Network. DOI: 10.4018/979-8-3693-5777-4.ch001

Romeo, G., & Conti, D. (2025). Exploring automation bias in human–AI collaboration : A review and implications for explainable AI. *AI & Society*. Advance online publication. DOI: 10.1007/s00146-025-02422-7

Rudin, C. (2019). Stop explaining black box machine learning models for high stakes decisions and use interpretable models instead. *Nature Machine Intelligence*, *1*(5), 206–215. DOI: 10.1038/s42256-019-0048-x

Rudin, C., & Radin, J. (2019). Why are we using black box models in AI when we don't need to? A lesson from an explainable AI competition. *Harvard Data Science Review*, *1*(2), 1–9. DOI: 10.1162/99608f92.5a8a3a3d

Russell, S. (2019). *Human compatible : AI and the problem of control*. Penguin Uk.

Russell, S., Norvig, P., Popineau, F., Miclet, L., & Cadet, C. (2021). *Intelligence artificielle : Une approche moderne (4e édition)*. Pearson France. https://hal.science/hal-04245057

Sapkota, R., Roumeliotis, K. I., & Karkee, M. (2026). AI Agents vs. Agentic AI : A Conceptual taxonomy, applications and challenges. *Information Fusion*, *126*, 103599. DOI: 10.1016/j.inffus.2025.103599

Sarrasin, N., & Ramangalahy, C. (2007). La gestion cognitive des connaissances dans les organisations. *Documentation et Bibliothèques*, *53*(1), 43–51. DOI: 10.7202/1029216ar

Senge, P. M. (2006). *The fifth discipline : The art and practice of the learning organization*. Broadway Business.

Shoham, Y., & Leyton-Brown, K. (2008). *Multiagent systems : Algorithmic, game-theoretic, and logical foundations*. Cambridge University Press. DOI: 10.1017/CBO9780511811654

Sukharevsky, A., Kerr, D., Hjartar, K., Hämäläinen, L., Bout, S., Di Leo, V., & Dagorret, G. (2025). Seizing the agentic AI advantage. *McKinsey & Company*. https://www.fredericsanz.com/s/Seizing-the-Agentic-AI-Advantage.pdf

Sukharevsky, A., Krivkovich, A., Gast, A., Storozhev, A., Maor, D., Mahadevan, D., Hämäläinen, L., & Durth, S. (2025). *The agentic organization : Contours of the next paradigm for the AI era*. McKinsey & Company.

Teece, D., & Pisano, G. (1994). The dynamic capabilities of firms : An introduction. *Industrial and Corporate Change*, *3*(3), 537–556. DOI: 10.1093/icc/3.3.537-a

Teece, D., Pisano, G., & Shuen, A. (1997). Dynamic capabilities and strategic management. *Strategic Management Journal*, *18*(7), 509–533. DOI: 10.1002/(SICI)1097-0266(199708)18:7<509::AID-SMJ882>3.0.CO;2-Z

Teece, D. J. (2007). Explicating dynamic capabilities : The nature and microfoundations of (sustainable) enterprise performance. *Strategic Management Journal, 28*(13), 1319–1350. DOI: 10.1002/smj.640

Teece, D. J. (2018). Business models and dynamic capabilities. *Long Range Planning, 51*(1), 40–49. DOI: 10.1016/j.lrp.2017.06.007

Terreberry, S. (1968). The Evolution of Organizational Environments. *Administrative Science Quarterly, 12*(4), 590–613. DOI: 10.2307/2391535

Thoenig, J.-C., & Michaud, C. (2007). *Le management cognitif* [Article]. HAL open access archive. https://shs.hal.science/halshs-00142994/document

Tornatzky, L. G., & Fleischer, M. (1990). *The processes of technological innovation.* Heath & Company.

Tushman, M. L., & Nadler, D. A. (1978). Information Processing as an Integrating Concept in Organizational Design. *Academy of Management Review, 3*(3), 613. DOI: 10.2307/257550

van Esch, P. (2026). From agentic AI to AI-orchestrated organizations : Understanding the next surge in artificial intelligence. *Business Horizons.* Advance online publication. DOI: 10.1016/j.bushor.2026.03.003

Verma, L. (2025). The Transformative Impact of Artificial Intelligence on Business Process Management. *European Journal of Computer Science and Information Technology, 13*(48), 14–25. Advance online publication. DOI: 10.37745/ejcsit.2013/vol13n481425

Weick, K. E. (1995). *Sensemaking in organizations* (Vol. 3). Sage publications Thousand Oaks, CA. https://www.torrossa.com/gs/resourceProxy?an=5564159&publisher=FZO137#page=545

Widad, J. (2025). The Evolution of Autonomy : From Reactive AI to Agentic AI. In Reddy, C. K. K., Joseph, S., Joshi, H., Doss, S., & Ouaissa, M. (Eds.), *The Power of Agentic AI: Redefining Human Life and Decision-Making* (pp. 1–9). Springer Nature Switzerland., DOI: 10.1007/978-3-031-89424-4_1

Wilson, H. J., & Daugherty, P. R. (2018). Collaborative intelligence : Humans and AI are joining forces. *Harvard Business Review, 96*(4), 114–123.

Winter, S. G. (2003). Understanding dynamic capabilities. *Strategic Management Journal, 24*(10), 991–995. DOI: 10.1002/smj.318

Zahra, S. A., & George, G. (2002). Absorptive Capacity : A Review, Reconceptualization, and Extension. *Academy of Management Review*, *27*(2), 185. DOI: 10.2307/4134351

Chapter 3
Large Language Models in Business Process Management:
From Process Documentation to Intelligent Decision Support

Othman Boulitama
http://orcid.org/0000-0003-3541-1582
Hassan II University of Casablanca, Morocco

Amina Ouatiq
Hassan I University, Morocco

Driss Rahli
http://orcid.org/0000-0002-3306-4460
Hassan II University of Casablanca, Morocco

Karim Sabri
Hassan II University of Casablanca, Morocco

ABSTRACT

This chapter examines how large language models are reshaping business process management from process documentation to intelligent decision support. It shows that the relevance of LLMs in BPM lies not only in text generation but also in their ability to connect dispersed procedural knowledge, formal process representations, analytical interpretation, and managerial judgment. The chapter adopts a continuum-based perspective covering process documentation, process analysis, execution support, and decision support. It argues that LLMs can help formalize

DOI: 10.4018/979-8-2600-0050-2.ch003

fragmented process knowledge, interpret textual evidence on bottlenecks and deviations, support users during process execution, and improve decision preparation through contextualized explanations and recommendations. At the same time, their contribution remains dependent on contextual grounding, data quality, governance mechanisms, and human oversight. The chapter concludes that the future of BPM will likely rely on hybrid sociotechnical arrangements in which LLMs augment rather than replace human actors.

1. INTRODUCTION

Organizations have long documented, analyzed, improved, and governed the routines through which they create value through business process management (BPM). What has changed in the current technological moment is not the relevance of process thinking, but the nature of the intelligence that can now be mobilized around process work (Teixeira et al., 2024). Recent literature suggests that the emergence of artificial intelligence (AI), particularly generative AI (GenAI), is redirecting BPM from a limited emphasis on procedural efficiency to more expansive inquiries regarding process knowledge, enhancement, and adaptive organizational capability (Rosemann et al., 2024). Research on large language models (LLMs) has demonstrated that these systems have transcended their previous limitations in generic text generation. They can interpret procedural descriptions, extract structure from unstructured narratives, and assist with tasks that were previously handled through fragmented, task-specific natural language processing pipelines (Grohs et al., 2024; Szczepanek et al., 2025). Taken together, these developments make it timely to reconsider BPM through the lens of LLM-enabled process intelligence.

This change is important for both practical and theoretical reasons. In practice, many organizations still depend on process knowledge that is dispersed across policies, audit reports, emails, meeting minutes, user stories, compliance manuals, standard operating procedures, and the tacit expertise of employees. Much of this knowledge remains only partially formalized, which weakens process visibility and slows redesign efforts, making it difficult for organizations to efficiently adapt to changing business environments (Ledro et al., 2025). LLMs matter in this context because they can operate across heterogeneous textual artifacts and transform diffuse procedural knowledge into more usable forms. Emerging BPM research already shows that LLMs can support the creation and editing of formal process models from natural language, thereby lowering the barrier between business users who describe work in ordinary language and technical systems that require structured representations such as BPM and notation (Kourani et al., 2025; Licardo et al., 2026).

This is not a trivial automation gain. It signals a more profound reconfiguration of how organizations capture, maintain, and mobilize process knowledge.

However, the chapter starts from a deliberately cautious position. The increasing prominence of LLMs in corporate environments has generated considerable enthusiasm; however, academic evaluation necessitates more than mere demonstrations of fluency or promising prototypes. The central question is not whether LLMs can produce plausible process descriptions, but whether they can contribute reliably to process understanding and decision support under real organizational conditions (Lin, 2026). Existing studies suggest both promise and constraint. Grohs et al. (2024), for instance, indicate that LLMs can perform several text-related BPM tasks at levels comparable to or better than prior specialized approaches. At the same time, more recent evaluation work indicates substantial performance variation across models and emphasizes that output quality, error handling, and model optimization remain critical concerns in business process modeling contexts (Kourani et al., 2025). This mixed evidence implies that LLM adoption in BPM should be approached neither as a simple extension of earlier automation nor as a substitute for methodological rigor, domain expertise, and governance.

The case for examining LLMs in BPM becomes even stronger when one moves beyond documentation toward explanation and decision support. Business processes are not managed only through diagrams and procedures. They are also interpreted, justified, escalated, adapted, and redesigned through language. In this respect, LLMs open new possibilities for helping managers and analysts synthesize process evidence, compare alternatives, explain deviations, and formulate context-sensitive recommendations. Research in decision support now treats LLMs as a more general-purpose support technology than earlier bespoke tools, precisely because they can assist with information gathering, alternative generation, and deliberative reasoning across many business contexts (Handler et al., 2024). Yet this same versatility introduces new behavioral, technical, and economic questions. Similar concerns appear in BPM-oriented explainability research, where LLM-based explanations may improve perceived fidelity when properly guardrailed but may also reduce interpretability and raise concerns about trust, hallucination, and reasoning limits (Fahland et al., 2025). The movement from process documentation to intelligent decision support is, therefore, not linear or risk-free. It is a sociotechnical transition that requires careful analytical framing.

Against this background, the purpose of this chapter is to develop a rigorous and integrative understanding of how LLMs can reshape BPM across a continuum extending from process documentation and knowledge formalization to process analysis, execution support, and intelligent decision support. The chapter argues that the real contribution of LLMs to BPM lies not merely in automating textual work but in connecting previously fragmented layers of procedural knowledge, analytical interpretation, and managerial judgment. At the same time, it maintains that such contribution is contingent on governance quality, data integrity, contextual grounding, and meaningful human oversight, especially given the wider governance risks associated with generative AI, including hallucination, opacity, information leakage, and control challenges (Taeihagh, 2025). The discussion that follows is therefore designed to be both forward-looking and critical: forward-looking in identifying where LLMs may extend BPM capabilities, and critical in clarifying the technical, organizational, and ethical conditions under which such extensions become academically defensible and managerially credible.

2. CONCEPTUAL FOUNDATIONS

2.1. Business Process Management: Scope, Evolution, and Core Logics

BPM is a structured managerial discipline that deals with finding, designing, documenting, analyzing, improving, carrying out, and overseeing organizational processes. Its central premise is that organizational performance is shaped not only by isolated tasks or functions but also by end-to-end processes through which resources, decisions, and responsibilities are coordinated to create value. Contemporary BPM literature also makes clear that the field has moved well beyond its earlier procedural and workflow-centered formulations. While efficiency, standardization, compliance, and control remain foundational concerns, BPM is increasingly framed as a broader organizational capability that connects strategy, technology, people, and data across the process lifecycle.

This evolution is especially visible in recent work on BPM in the age of AI. Rosemann et al. (2024) argue that the field is undergoing important shifts in its dominant assumptions, including a movement from process exploitation toward more adaptive and intelligent process configurations. In that perspective, BPM is no longer limited to documenting stable routines or optimizing repetitive workflows. It is increasingly concerned with how organizations perceive process conditions, mobilize distributed knowledge, and redesign process architectures under conditions of uncertainty and digital change. This broader view is particularly important for the present chapter because it creates the conceptual room needed to analyze LLMs not as peripheral automation tools but as emerging components of process intelligence (Zhang et al., 2025).

A further implication follows from this shift. Traditional BPM has often relied on explicit representations such as process models, procedures, rules, and performance indicators. Yet in many organizations, a considerable share of process-relevant knowledge remains dispersed across informal narratives, emails, reports, support tickets, meeting notes, audit documents, and domain-specific conversations. This means that BPM has always had to operate at the boundary between structured and unstructured knowledge, even if its classical tools were better suited to the former than the latter. That tension becomes decisive when considering LLMs, because their value proposition lies precisely in their ability to work with unstructured language while contributing to process-related tasks that were previously difficult to scale.

2.2. LLMs: Principles, Capabilities, and Limits

LLMs are probabilistic foundation models trained on massive corpora to predict and generate language sequences. Their practical importance does not stem from a single narrow function but from their ability to perform multiple language-intensive tasks within one general architecture, including summarization, information extraction, classification, reformulation, explanation, question answering, and text generation. In organizational contexts, this generality matters because many knowledge processes do not arrive in clean analytical formats. They emerge through documents, discussions, and fragmented textual traces. LLMs, therefore, offer a qualitatively different technological basis from earlier task-specific natural language processing systems, which often required separate models and pipelines for each business use case (Naveed et al., 2025; Peykani et al., 2025; Yang et al., 2025).

From a BPM perspective, the most relevant capabilities of LLMs include their ability to interpret procedural language, infer structure from descriptions of work, translate semi-formal narratives into modeling elements, and provide context-sensitive assistance around process-related questions. Existing BPM research already shows that LLMs can support tasks such as mining process models from textual descriptions and assessing the suitability of tasks for robotic process automation (Yang et al., 2025). More recent benchmarking work also indicates that LLMs can generate business process models at meaningful levels of quality, although performance varies substantially across model families and prompting configurations. These findings are significant because they suggest that LLMs can contribute not only to documentation but also to analytical and design activities that sit closer to the core of BPM practice (Wang & Chen, 2026).

At the same time, academic rigor requires a clear recognition of the limitations of these models. Their outputs may be fluent and persuasive while still containing factual errors, omitted constraints, invented details, or structurally weak reasoning. In BPM contexts, these weaknesses are significant. They can distort process models, misrepresent business rules, or generate recommendations that appear plausible but are not operationally or legally sound. Recent evaluation research in business process modeling confirms that output quality is uneven and that error handling remains a decisive factor in model usefulness. Related work on users' perceptions of LLM based process explanations similarly shows that perceived helpfulness does not remove concerns regarding fidelity, interpretability, and trust. For that reason, LLMs should be conceptualized as augmentative and probabilistic systems whose value depends on contextual grounding, validation mechanisms, and human oversight (Peykani et al., 2025).

2.3. Why LLMs Matter for BPM

The relevance of LLMs for BPM lies in a structural mismatch that has long existed in organizations. On the one hand, BPM depends on explicit representations of processes, roles, rules, events, and performance logic. On the other hand, a substantial proportion of operational knowledge remains embedded in natural language rather than in formal models. This includes procedure manuals, exception handling notes, quality reports, audit observations, customer complaints, incident logs, internal communications, and expert interpretations (Lazaropoulos, 2026). Traditional BPM methods have often struggled to absorb this material at scale. LLMs matter because they offer a mechanism for bridging that divide. They can convert dispersed textual knowledge into more structured, process-relevant forms and thereby extend the informational base on which process understanding and redesign depend (Kourani et al., 2025).

This matters not only for process documentation, but also for process analysis and decision support. Process problems are usually not visible in transactional data alone. Textual evidence produced around the process reveals many bottlenecks, ambiguities, control failures, and coordination breakdowns (Khatoon et al., 2025). Careful use of LLMs can help synthesize that evidence, identify recurring themes, explain deviations, and support more informed managerial interpretation. In that sense, their contribution to BPM is epistemic, relating to knowledge and understanding, as much as operational. They expand what can be known, interpreted, and acted upon within the process domain. This is one reason why recent BPM literature increasingly treats AI as a source of deeper change in the field rather than as a marginal add-on to existing workflow technologies (Cazorla et al., 2026).

However, the chapter adopts a deliberately non-deterministic position. LLMs do not automatically create better BPM. Their contribution depends on whether organizations can integrate them into process governance without weakening accountability, traceability, and methodological discipline. The real issue is therefore not whether LLMs can produce process-related output but under what conditions that output becomes sufficiently reliable, explainable, and useful for organizational action. This question justifies the analytical direction of the chapter: to examine LLMs across a continuum that begins with process documentation and knowledge formalization and extends toward intelligent decision support, while maintaining a critical focus on quality, control, and sociotechnical design.

3. LITERATURE POSITIONING AND ANALYTICAL FRAMING

The literature on AI in BPM has expanded rapidly, but it remains uneven in structure and fragmented in emphasis. One stream examines AI-augmented systems for managing business processes as part of a broader architectural and managerial vision in which process-aware systems become more adaptive, proactive, context-sensitive, and understandable (Raina et al., 2026). In this view, AI is not treated as a narrow automation add-on but as a means of reconfiguring how process knowledge is captured, interpreted, and operationalized across the lifecycle of business processes. This line of work is important because it shifts the conversation from isolated tools toward systemic process intelligence, which enables organizations to leverage data-driven insights for continuous improvement and innovation in their business processes (Ritala et al., 2024). At the same time, recent BPM research suggests that the field is undergoing deeper conceptual shifts under the influence of AI, with increasing emphasis on intelligence, interaction, and adaptability rather than on procedural control alone (Adamakis & Rachiotis, 2025; Capolupo et al., 2025).

A second stream focuses more specifically on language-based and LLM-enabled BPM tasks. Early studies in this area indicate that LLMs can perform several process-related tasks directly based on textual descriptions, including the mining of imperative and declarative process models and the assessment of task suitability for robotic process automation (Feng et al., 2026). More recent benchmarking research on business process modeling confirms that these models have substantial potential yet also shows pronounced differences in performance across model families and prompting strategies. This emerging body of work is especially relevant because it demonstrates that LLMs are not confined to generic text generation. They are increasingly being evaluated against core BPM activities that require structure, domain interpretation, and representational accuracy (Mikalef et al., 2023).

A third stream focuses on explanation, user interaction, and decision support in AI-enabled BPM settings. Research on situation-aware explainability for business processes suggests that LLMs can improve the perceived fidelity of process explanations when they are embedded within guardrailed frameworks and supplied with structured contextual inputs (Lazaropoulos, 2026). However, the same research also shows that gains in perceived usefulness may come with losses in interpretability, which is a crucial warning for BPM scholars and practitioners, as it suggests that while users may perceive AI systems more helpful, they might struggle to understand the rationale behind the decisions made by these systems (Bello et al., 2025). Related contributions on conversational AI in AI-augmented BPM systems similarly indicate a future where process participants do not merely execute predefined workflows but engage with intelligent systems that elucidate states, present recommendations, and facilitate context-dependent decisions. These studies collectively indicate that the role of LLMs in BPM is expanding from representation and documentation toward interpretation and action support (Fahland et al., 2025).

Despite this progress, the literature still tends to treat documentation, automation, explanation, and decision support as partially separate conversations. Some studies examine natural language processing for model translation. Others address understandability, conversational support, or process-aware decision assistance (Collins et al., 2021). Still others discuss AI in BPM at a strategic or architectural level. What remains less developed is an integrative analytical lens that connects these contributions into a coherent continuum (Hughes et al., 2025; Kanbach et al., 2024; Peykani et al., 2025). This gap matters because the practical value of LLMs in BPM does not emerge from one isolated task alone. It emerges when organizations move from capturing dispersed procedural knowledge to structuring it, interpreting it in relation to process performance, and, finally, using it in support of human judgment and decision-making. Recent work identifying LLM use cases across the BPM lifecycle reinforces this point by showing that relevant applications span multiple lifecycle phases rather than one narrow stage.

Accordingly, this chapter adopts a continuum-based analytical framing. Unlike prior BPM studies that often examine documentation, automation, explanation, or decision support separately, this chapter integrates these dimensions within a unified continuum-based framework of LLM-enabled BPM capabilities. It positions LLMs in BPM across four connected layers: process documentation and knowledge formalization, process analysis and improvement, process execution support, and intelligent decision support. The rationale for this framing is both theoretical and practical. Theoretically, it reflects the fact that BPM depends on the progressive transformation of organizational knowledge from diffuse descriptions into actionable process understanding. Practically, it recognizes that LLMs create value precisely because they can operate across those layers, linking unstructured language, formal process representations, contextual explanations, and managerial recommendations. This framing also makes it possible to assess the technology critically, because each layer raises distinct questions about quality, trust, explainability, governance, and human oversight.

The continuum framework proposed in this chapter was conceptually developed through an integrative synthesis of recent literature on AI-augmented BPM, LLM-enabled process modeling, process-aware conversational systems, explainable AI, and intelligent decision support. Rather than treating these streams as isolated domains, the framework organizes them according to the progressive role of LLMs across the

BPM lifecycle, extending from process knowledge formalization to process-informed managerial support. This conceptual structuring was designed to provide a coherent analytical lens capable of connecting documentation, analysis, execution assistance, and decision support within a unified BPM perspective.

Within this analytical perspective, the chapter departs from both technological determinism and narrow task-based evaluation. It does not assume that stronger language generation automatically leads to better BPM outcomes. Nor does it reduce the relevance of LLMs (LLMs) to whether a model can generate a syntactically correct process diagram. Instead, it treats LLM enabled BPM as a sociotechnical configuration in which organizational value depends on contextual grounding, integration with process knowledge repositories, validation mechanisms, user trust, and governance design. This positioning is consistent with the broader ABPMS agenda, which emphasizes that intelligent process systems must remain trustworthy, understandable, and aligned with organizational control requirements if they are to support meaningful and accountable process action.

4. A CONTINUUM PERSPECTIVE: FROM PROCESS DOCUMENTATION TO INTELLIGENT DECISION SUPPORT

To understand how LLMs contribute to BPM, it is useful to treat their role as a continuum rather than a single isolated application. In BPM practice, value rarely comes from one disconnected capability. It emerges when organizations can move from dispersed textual knowledge to structured process representations, from those representations to analytical interpretation, and from interpretation to context-sensitive operational and managerial support (Grohs et al., 2024). Recent BPM research increasingly supports this broader perspective. Studies on process modeling, process mining, explanation, and process-aware conversational systems all suggest that LLMs are beginning to intervene across multiple layers of the BPM lifecycle rather than in one narrow technical niche (Feng et al., 2026).

This continuum perspective is analytically important because each layer involves a different relationship between language, structure, and judgment. At the documentation layer, the main issue is knowledge capture and formalization. At the analysis layer, the issue becomes interpretation, diagnosis, and improvement (Theodorakopoulos et al., 2025). At the execution layer, the focus shifts toward contextual assistance and human augmentation. At the decision support layer, LLMs begin to shape how alternatives are framed, how evidence is synthesized, and how recommendations are articulated (Li et al., 2026). The sections below develop this progression in a way that remains attentive to both potential and limitation. To synthesize this continuum perspective, Figure 1 illustrates the progressive evolution of LLM-enabled capabilities across the BPM lifecycle.

Figure 1. **Continuum of LLM-Enabled Capabilities Across the BPM Lifecycle**

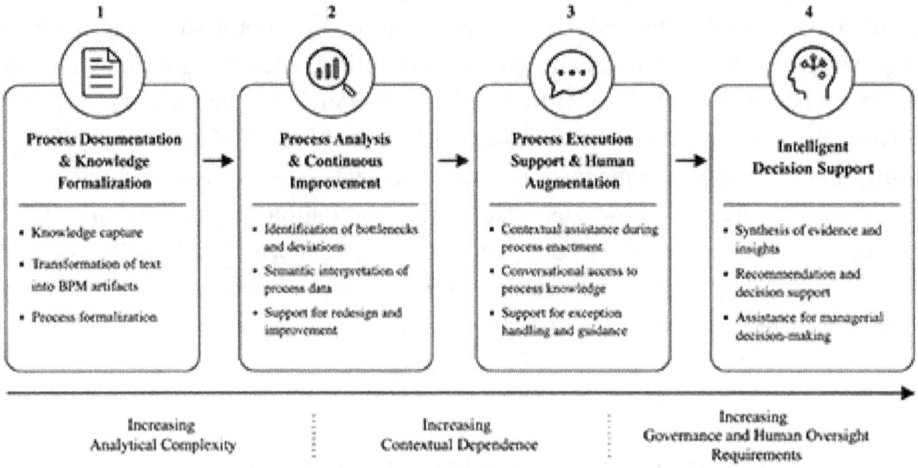

4.1. LLMs for Process Documentation and Knowledge Formalization

The first and most immediate contribution of LLMs to BPM lies in process documentation. Organizations often possess extensive procedural knowledge, but much of it is distributed across policy manuals, emails, audit observations, meeting notes, onboarding materials, user instructions, and informal expert explanations. Before traditional BPM methods can operationally utilize this knowledge, they necessitate its translation into more structured representations. LLMs are relevant because they can help transform diffuse natural language descriptions into clearer, more standardized, and more explicit process artifacts. This includes summarizing procedures, extracting roles and activities, clarifying task sequences, and supporting the conversion of textual descriptions into formal process models (Grohs et al., 2024; Theodorakopoulos et al., 2025).

Recent work provides concrete evidence for this claim. Grohs et al. (2024) showed that LLMs can already perform BPM relevant language tasks, such as mining imperative and declarative process models from textual input. More recent studies have moved beyond proof of concept toward model generation and editing support. (Kourani et al., 2025) evaluated LLMs on tasks related to business process modeling and found meaningful, though uneven, performance across models. Licardo et al. (2026) further demonstrated an LLM based BPMN assistant capable of creating and editing BPMN diagrams through natural language interaction, which is particularly significant for reducing the gap between business users and formal modeling environments.

The deeper significance of this documentation function should not be underestimated. It is not merely about faster drafting. It concerns the formalization of process knowledge that would otherwise remain fragmented, ambiguous, or locally embedded. In that sense, LLMs may strengthen one of the long-standing weak points of BPM, namely, the difficulty of absorbing unstructured organizational knowledge into process design and governance (Feng et al., 2026). Still, this promise depends on rigorous validation, because formalization errors at this early stage can propagate into downstream analysis and decision support, potentially leading to flawed process designs and misguided organizational decisions. A fluent process description is not necessarily an accurate one, as it may overlook critical details or nuances that are essential for effective process design and governance.

To provide a concise overview of this continuum perspective, Table 1 summarizes the four interconnected layers of LLM-enabled BPM, together with their primary functions, contributions, benefits, and associated risks.

Table 1. Continuum of LLM-Enabled BPM Capabilities

Continuum Layer	BPM Focus	LLM Contribution	Main Benefit	Main Risk
Process Documentation	Capture and formalize process knowledge	Transform textual information into structured process artifacts	Better process visibility	Inaccurate formalization
Process Analysis	Identify issues and improvement opportunities	Analyze textual evidence and detect recurring patterns	Improved analytical insight	Misinterpretation
Process Execution Support	Assist users during process execution	Provide contextual and conversational guidance	Faster operational support	Incorrect guidance
Intelligent Decision Support	Support managerial interpretation and decisions	Synthesize evidence and generate recommendations	Enhanced decision preparation Hallucination and overreliance	Hallucination and overreliance

4.2. LLMs for Process Analysis and Continuous Improvement

The second layer of the continuum concerns process analysis. Organizations must identify bottlenecks, ambiguities, unnecessary steps, deviations, and improvement opportunities after documenting or formalizing process knowledge. This is an area where conventional BPM has often relied heavily on structured event data, formal process models, or expert workshops. However, textual material such as incident reports, audit comments, user complaints, exception justifications, support tickets, and retrospective reviews also encode many process problems. LLMs become valuable here because they can interpret and synthesize those textual traces in ways that complement more traditional process analytics (Teixeira et al., 2024).

Recent research points in this direction. The PM-LLM-Benchmark work indicates that LLMs can support process mining-related tasks, although results vary by model and implementation strategy (Ray & Pradhan, 2025). Other emerging work has explored the use of LLMs for automated identification of unnecessary process steps, suggesting that semantic analysis can support redesign by uncovering inefficiencies that are not always obvious from structural representations alone. This is particularly relevant for knowledge-intensive or exception-rich processes, where deviations are often explained in language rather than fully captured in event logs (Berti et al., 2024).

From a BPM standpoint, this work means that LLMs can extend the evidence base for continuous improvement. They can help consolidate findings from heterogeneous textual sources, identify recurring problem themes, and generate preliminary improvement suggestions for analyst review. However, this contribution should be understood as augmentative rather than autonomous. Improvement analysis requires causal judgment, domain knowledge, and sensitivity to organizational constraints. LLM outputs may surface useful patterns, but they do not remove the need for human interpretation, especially when process redesign affects compliance, risk exposure, or cross-functional coordination (Dong et al., 2024).

4.3. LLMs for Process Execution Support and Human Augmentation

The third layer concerns the use of LLMs during process execution itself. The primary objective here is to support actors during the execution of their work, rather than modeling or analyzing the process after the fact. In many organizations, employees regularly need clarification on procedures, exception handling, policy interpretation, case status, and next best actions. Traditional workflow systems can route tasks and enforce rules, but they often provide limited contextual assistance when the situation deviates from standard paths. LLM-based interfaces can address this gap by offering conversational access to process relevant knowledge, summarizing case context, answering procedural questions, and guiding users through non-routine situations (Siino, 2025).

This possibility is especially visible in work on business process-aware LLMs. Bernardi et al. (2024) propose a framework in which LLMs interact with human workers through retrieval-augmented and process-specific knowledge structures, enabling actionable conversations linked to process context (Bernardi et al., 2024). The significance of this development is that it moves LLM use beyond static content generation toward situated assistance. Rather than producing generic text, the system becomes a process-aware conversational layer that can help employees navigate tasks, understand constraints, and respond more effectively to case-specific variation (Tirulo et al., 2026).

In analytical terms, this changes the role of intelligence in BPM. Instead of residing only in predefined models, static rule sets, or post hoc dashboards, support becomes available during enactment, where uncertainty, interruption, and ambiguity are most visible. Yet this does not imply that human discretion disappears. On the contrary, LLMs are more plausibly understood as augmentative companions that reduce search effort, improve contextual access to knowledge, and help users articulate and interpret process situations. Their value is highest when they strengthen rather than replace human judgment.

4.4. LLMs for Intelligent Decision Support in BPM

The fourth and most advanced layer of the continuum is intelligent decision support. At this point, LLMs do more than just write down or explain processes. They begin to assist with the interpretation of process evidence, the framing of alternatives, and the articulation of recommendations relevant to process-related decisions. This may include summarizing compliance implications, comparing redesign options, clarifying potential downstream effects of a process change, synthesizing stakeholder inputs, or helping managers reason through operational exceptions. The move toward decision support is therefore not a trivial extension of documentation. It represents a shift from process representation to process-informed judgment (Hao et al., 2025).

The literature suggests both promise and caution. Users perceive LLM-generated explanations as helpful and accessible, particularly when combined with structured knowledge ingredients, according to research on process explanation. Likewise, process-aware conversational architectures indicate that LLMs can support more actionable human interaction around live process contexts (Al-Busaidi et al., 2024). However, the step from explanation to recommendation raises stronger governance concerns. Decision support in BPM often affects cost, service levels, compliance, accountability, and risk. In such settings, persuasive language is not enough. Recommendations must be grounded, traceable, and open to scrutiny. The current literature therefore points toward a hybrid model in which LLMs support reasoning and synthesis, while final authority remains with human actors and organizational control structures (Hao et al., 2025; Tirulo et al., 2026).

For this reason, the real strategic value of LLMs in BPM lies not in autonomous decision-making but in enhanced decision preparation. They can consolidate fragmented evidence, surface relevant considerations, and make complex process-related knowledge more accessible to decision makers. When properly grounded in enterprise knowledge and embedded in governance mechanisms, the technology can improve the quality and speed of managerial sensemaking. When deployed without such safeguards, it can just as easily amplify error, overconfidence, and opaque reasoning, leading to poor decision-making and potentially detrimental outcomes for the organization. The distinction is decisive for both research and practice.

5. APPLICATION DOMAINS AND AN INTEGRATIVE FRAMEWORK FOR LLM-ENABLED BPM

The organizational relevance of LLMs in BPM is best assessed through the concrete process domains in which they operate and the conditions under which their outputs become reliable and useful. In BPM, value does not arise from language generation alone. It arises when LLMs are embedded in process contexts where dispersed textual knowledge must be formalized, interpreted, and mobilized for action. Recent research therefore suggests that LLMs should be understood as enabling technologies that can extend BPM across documentation, analysis, execution support, and explanation, provided that they are properly grounded in organizational knowledge and governed through appropriate control mechanisms (Beerepoot et al., 2023; Berti et al., 2025).

A first application domain is process documentation and modeling. Many organizations still rely on fragmented procedural knowledge distributed across manuals, emails, internal policies, audit reports, and expert narratives. In such contexts, LLMs can support the transformation of natural language descriptions into more structured process representations (Sapkota et al., 2026). Recent studies show that they can assist in generating BPMN models from text, editing BPMN artifacts through natural language interaction, and externalizing process knowledge that is only weakly formalized. Their contribution in this domain is therefore not limited to drafting support. More fundamentally, it concerns the codification of organizational process knowledge in forms that are more accessible for analysis, communication, and governance (Ehrlich-Sommer et al., 2025).

A second application domain concerns process analysis and coherence assessment. Once process knowledge is documented, organizations must evaluate whether process descriptions are internally consistent, sufficiently complete, and aligned across levels of abstraction. Here, LLMs can support semantic checking, inconsistency detection, and interpretation of process changes (Angulo Martínez et al., 2026). This use is particularly relevant because many process weaknesses are not purely structural. They also arise from ambiguity, fragmentation, and semantic drift within the underlying documentation. In this sense, LLMs can contribute not only to process representation but also to the semantic quality of the process knowledge base on which BPM depends (Dormehl et al., 2026).

A third application domain is process execution support. Traditional workflow systems are effective in routing tasks and enforcing predefined logic, but they provide limited support when users must handle exceptions, clarify rules, or interpret case-specific situations during enactment. Process-aware conversational architectures seek to address this gap by coupling LLMs with process knowledge and retrieval mechanisms so that users can access contextualized support while performing work. In such settings, the LLM functions as an interaction layer between formal process logic and situated human judgment. Its value is particularly high in knowledge-intensive and service-oriented processes, where execution quality depends on both procedural compliance and contextual interpretation.

A fourth application domain is process explanation and decision preparation. BPM also involves interpreting deviations, clarifying causal mechanisms, and preparing managerial decisions regarding redesign, compliance, or operational intervention. Research indicates that LLM-supported explanations can improve the accessibility and perceived usefulness of process interpretations when they are embedded within structured explanatory settings (Viteri-Sánchez & Novillo-Villegas, 2025). However, this domain is also the most sensitive from a governance perspective, since outputs may shape organizational judgments with operational or regulatory consequences. For that reason, LLMs are more appropriately viewed here as instruments of assisted judgment than as autonomous decision-makers (Miedema et al., 2026).

These domains can be integrated into a concise framework built around four interdependent dimensions. The first is capability configuration, referring to model specialization, prompting strategy, and the use of retrieval or domain adaptation. The second is process-task fit, since the requirements of documentation, modeling, execution support, and decision preparation differ substantially. The third is knowledge grounding, because BPM outputs must remain aligned with organizational rules, repositories, and vocabularies. The fourth is governance and human oversight, which becomes increasingly important as LLM use moves from documentation toward explanation and decision support. Together, these dimensions explain why LLM value in BPM is conditional rather than automatic. It depends on the alignment between model capability, process task, knowledge environment, and control architecture.

Taken together, this perspective suggests that the most promising BPM applications of LLMs are found in environments that are text-rich, interpretation-intensive, and still dependent on human accountability. Their strategic significance lies less in full automation than in their ability to bridge unstructured organizational language and structured process action. Under those conditions, LLMs extend the analytical and operational reach of BPM by making process knowledge more formalizable, interpretable, and usable across multiple layers of process management.

6. CHALLENGES, FUTURE RESEARCH DIRECTIONS, AND PRACTICAL IMPLICATIONS

The integration of LLMs into BPM raises a central tension: the same capabilities that make these models attractive for process documentation, explanation, and support also introduce new risks for reliability, accountability, and governance. In BPM settings, this tension is especially consequential because model outputs may influence process representations, operational guidance, and managerial judgment rather than merely produce generic text. Recent work in AI-augmented BPM therefore converges on a cautious view: value creation depends less on model fluency than on the quality of contextual grounding, control mechanisms, and human supervision embedded in the process environment.

A first challenge concerns knowledge reliability. The quality of the procedural material, including policies, manuals, repositories, and case-specific information, shapes LLM outputs in BPM. When these sources are incomplete, outdated, or weakly integrated, the model may still generate coherent responses that are substantively misleading. This problem is amplified in BPM because apparently minor semantic inaccuracies can distort process descriptions, weaken explanations, or bias redesign discussions. Research on process-aware conversational systems and LLM-supported process explanation suggests that reliable use therefore requires structured knowledge ingredients and contextual retrieval rather than decontextualized prompting alone.

A second challenge is explainability and accountability. As LLM use shifts from documentation toward explanation and decision preparation, the acceptable margin of error narrows considerably. In process settings, recommendations or explanations that appear persuasive but are weakly grounded may affect compliance, service quality, and operational risk. This is why recent literature increasingly treats explainability, auditability, and human oversight as design requirements rather than optional enhancements. In BPM, the relevant question is not whether the model can produce an answer, but whether that answer can be justified, scrutinized, and governed within an accountable organizational structure.

A third challenge concerns responsible implementation. LLM deployment in BPM cannot be reduced to a technical adoption decision. It is a sociotechnical design problem involving validation procedures, escalation rules, access boundaries, and the definition of appropriate human-in-the-loop controls. The BPM literature on responsible AI makes this point clearly by emphasizing control points and audit structures as necessary conditions for trustworthy process support. In this respect, governance is not external to LLM-enabled BPM. It is one of its constitutive conditions.

These challenges also define the main directions for future research. First, the field needs stronger empirical evidence on LLM performance across BPM tasks under realistic organizational conditions. Current studies demonstrate feasibility in process-aware interaction and process explanation, but broader evidence is still required on robustness, error patterns, transferability, and task-specific limitations. Future work should therefore move beyond proof-of-concept demonstrations toward comparative and longitudinal evaluation in real process settings.

Second, future research should address governance design more directly. This includes the study of validation architectures, explainability mechanisms, escalation logic, and accountability arrangements suitable for BPM use cases. More generally, BPM literature would benefit from translating the emerging literature on LLM-enhanced explainability into process-specific models that clarify what kinds of explanations are actually useful for analysts, managers, and frontline users operating under organizational constraints.

Third, the field should examine task and domain contingency more systematically. LLMs are unlikely to create equivalent value across process documentation, execution support, coherence checking, and decision preparation. Their effectiveness depends on the nature of the process task, the degree of contextual ambiguity, and the level of risk associated with error. A more mature research agenda should therefore identify where LLMs are most beneficial, where they remain unreliable, and how their role changes across different process environments.

From a practical standpoint, the current evidence supports a restrained implementation logic. Organizations should initiate use cases that involve text-rich process knowledge, repetitive interpretive effort, and the preservation of human validation without excessive friction. Documentation support, bounded explanation, and controlled execution assistance are therefore more suitable starting points than autonomous decision support. In parallel, managers should invest not only in models but also in the process knowledge infrastructure needed to ground them, including curated repositories, retrieval mechanisms, and governance rules. The main implication is straightforward: in BPM, LLMs create value when they are embedded within disciplined knowledge and control architectures; absent those foundations, they are more likely to amplify ambiguity than improve process intelligence.

7. CONCLUSION

This chapter has argued that the contribution of LLMs to BPM should not be reduced to text generation or interface innovation. Their deeper significance lies in their capacity to extend BPM into knowledge spaces that have traditionally remained difficult to formalize, analyze, and mobilize. In that respect, LLMs matter because they strengthen the connection between unstructured organizational language and structured process action, thereby expanding the informational and interpretive foundations of process management. Recent BPM literature supports this broader view by showing that AI is already reshaping the field toward more intelligent, interactive, and adaptive process configurations.

At the same time, the chapter has maintained a deliberately cautious position. The fact that LLMs can support process documentation, process-aware interaction, and explanation does not mean that their outputs are automatically reliable, governable, or suitable for high-stakes process decisions. Existing research instead suggests that their organizational value is conditional on contextual grounding, structured knowledge inputs, and meaningful human oversight. In BPM, this condition is decisive because process-related outputs influence not only communication, but also compliance, accountability, coordination, and operational judgment.

The central implication is therefore clear. LLM-enabled BPM should be understood as a sociotechnical design project rather than as a simple technological upgrade. Its success depends on the alignment of model capability, process-task fit, knowledge architecture, and governance design. Where this alignment is achieved, LLMs can enhance process visibility, improve access to procedural knowledge, and support more informed interpretation across the process lifecycle. Where it is absent, they risk amplifying ambiguity under the appearance of fluency.

From this perspective, the future of BPM is unlikely to be defined by autonomous process intelligence alone. More plausibly, it will be shaped by hybrid arrangements in which language models augment human actors, enrich process knowledge, and support judgment without displacing responsibility. This is where the real promise of LLMs in BPM resides: not in replacing process management, but in making it more knowledge-aware, more responsive, and potentially more intelligent under conditions of disciplined organizational control.

REFERENCES

Adamakis, M., & Rachiotis, T. (2025). Artificial Intelligence in Higher Education: A State-of-the-Art Overview of Pedagogical Integrity, Artificial Intelligence Literacy, and Policy Integration. *Encyclopedia, 5*(4), 180. DOI: 10.3390/encyclopedia5040180

Al-Busaidi, A. S., Raman, R., Hughes, L., Albashrawi, M. A., Malik, T., Dwivedi, Y. K., Al- Alawi, T., AlRizeiqi, M., Davies, G., Fenwick, M., Gupta, P., Gurpur, S., Hooda, A., Jurcys, P., Lim, D., Lucchi, N., Misra, T., Raman, R., Shirish, A., & Walton, P. (2024). Redefining boundaries in innovation and knowledge domains: Investigating the impact of generative artificial intelligence on copyright and intellectual property rights. *Journal of Innovation & Knowledge, 9*(4), 100630. DOI: 10.1016/j.jik.2024.100630

Angulo Martínez, I., Garcia, A., González-Potes, A., Martínez-Castro, D., Paredes, C. M., Ochoa-Brust, A., Mena, L. J., Martínez-Peláez, R., Félix, V. G., & Félix-Cuadras, R. A. (2026). Hybrid AI and LLM-Enabled Agent-Based Real-Time Decision Support Architecture for Industrial Batch Processes: A Clean-in-Place Case Study. *AI 2026, Vol. 7, Page 51, 7*(2), 51. DOI: 10.3390/ai7020051

Beerepoot, I., Di Ciccio, C., Reijers, H. A., Rinderle-Ma, S., Bandara, W., Burattin, A., Calvanese, D., Chen, T., Cohen, I., Depaire, B., Di Federico, G., Dumas, M., van Dun, C., Fehrer, T., Fischer, D. A., Gal, A., Indulska, M., Isahagian, V., Klinkmüller, C., & Zerbato, F. (2023). The biggest business process management problems to solve before we die. *Computers in Industry, 146*, 103837. DOI: 10.1016/j.compind.2022.103837

Bello, M., Bello, R., García, M. M., Nowé, A., Sevillano-García, I., & Herrera, F. (2025). A Three-level Framework for LLM-enhanced Explainable AI: From Technical Explanations to Natural Language. *Information Systems Frontiers*, •••, 1–22. DOI: 10.1007/s10796-025-10668-1

Bernardi, M. L., Casciani, A., Cimitile, M., & Marrella, A. (2024). Conversing with business process-aware Large Language Models: the BPLLM framework. *Journal of Intelligent Information Systems*. DOI: 10.21203/rs.3.rs-4125790/v1

Berti, A., Kourani, H., & van der Aalst, W. M. P. (2024). PM-LLM-Benchmark: Evaluating Large Language Models on Process Mining Tasks. *Lecture Notes in Business Information Processing, 533*, 610–623. DOI: 10.1007/978-3-031-82225-4_45

Berti, A., Wang, X., Kourani, H., & Van der Aalst, W. M. P. (2025). Specializing large language models for process modeling via reinforcement learning with verifiable and universal rewards. *Process Science 2025 2:1, 2*(1), 26-. DOI: 10.1007/s44311-025-00034-4

Capolupo, N., Romeo, E., Giampaola, V., & Adinolfi, P. (2025). To AI or not to AI? Unpacking the Hamletian dilemma in public organizations' work processes through a socio-technical perspective. *Business Process Management Journal*, •••, 1–29. DOI: 10.1108/BPMJ-07-2025-1188

Cazorla, A., Flogie, A., Woodruff, E., & Hewitt, J. (2026). Epistemic Agency in the Age of Large Language Models: Design Principles for Knowledge-Building AI. *AI 2026, Vol. 7, Page 99, 7*(3), 99. DOI: 10.3390/ai7030099

Collins, C., Dennehy, D., Conboy, K., & Mikalef, P. (2021). Artificial intelligence in information systems research: A systematic literature review and research agenda. *International Journal of Information Management*, *60*, 102383. DOI: 10.1016/j.ijinfomgt.2021.102383

Dong, M. M., Stratopoulos, T. C., & Wang, V. X. (2024). A scoping review of ChatGPT research in accounting and finance. *International Journal of Accounting Information Systems*, *55*, 100715. DOI: 10.1016/j.accinf.2024.100715

Dormehl, J. A., Andrews, R., Kratsch, W., Röglinger, M., Wynn, M. T., & Zetzsche, F. (2026). Domain experts in the loop: Leveraging generative artificial intelligence for interactive data validation in process mining. *Information Systems*, *140*, 102715. DOI: 10.1016/j.is.2026.102715

Ehrlich-Sommer, F., Eberhard, B., & Holzinger, A. (2025). ForestGPT and Beyond: A Trustworthy Domain-Specific Large Language Model Paving the Way to Forestry 5.0. *Electronics 2025, Vol. 14, Page 3583, 14*(18), 3583. DOI: 10.3390/electronics14183583

Fahland, D., Fournier, F., Limonad, L., Skarbovsky, I., & Swevels, A. J. E. (2025). How well can a large language model explain business processes as perceived by users? *Data & Knowledge Engineering*, *157*, 102416. DOI: 10.1016/j.datak.2025.102416

Feng, P., Peng, X., Zhang, Z., Lu, Z., Drakopoulos, P., Malousoudis, P., Nousias, N., Tsakalidis, G., & Vergidis, K. (2026). Do LLMs Speak BPMN? An Evaluation of Their Process Modeling Capabilities Based on Quality Measures. *Computation 2026, Vol. 14, Page 10, 14*(1), 10. DOI: 10.3390/computation14010010

Grohs, M., Abb, L., Elsayed, N., & Rehse, J. R. (2024). Large Language Models Can Accomplish Business Process Management Tasks. *Lecture Notes in Business Information Processing, 492 LNBIP*, 453–465. DOI: 10.1007/978-3-031-50974-2_34

Handler, A., Larsen, K. R., & Hackathorn, R. (2024). Large language models present new questions for decision support. *International Journal of Information Management, 79*, 102811. DOI: 10.1016/j.ijinfomgt.2024.102811

Hao, X., Demir, E., & Eyers, D. (2025). Beyond human-in-the-loop: Sensemaking between artificial intelligence and human intelligence collaboration. *Sustainable Futures : An Applied Journal of Technology, Environment and Society, 10*, 101152. DOI: 10.1016/j.sftr.2025.101152

Hughes, L., Mavi, R. K., Aghajani, M., Fitzpatrick, K., Gunaratnege, S. M., Shekarabi, S. A. H., Hughes, R., Khanfar, A., Khatavakhotan, A., Mavi, N. K., Li, K., Mahmoud, M., Malik, T., Mutasa, S., Nafar, F., Yates, R., Alahmad, R., Jeon, I., & Dwivedi, Y. K. (2025). Impact of artificial intelligence on project management (PM): Multi-expert perspectives on advancing knowledge and driving innovation toward PM2030. *Journal of Innovation & Knowledge, 10*(5), 100772. DOI: 10.1016/j.jik.2025.100772

Kanbach, D. K., Heiduk, L., Blueher, G., Schreiter, M., & Lahmann, A. (2024). The GenAI is out of the bottle: Generative artificial intelligence from a business model innovation perspective. *Review of Managerial Science, 18*(4), 1189–1220. DOI: 10.1007/s11846-023-00696-z

Khatoon, U. T., Babgi, M., Hadi, N. T., Mir, R. N., & Velidandi, A. (2025). Technology-Driven Change in Human Resource Management: Reshaping Talent Management and Organizational Design. *Administrative Sciences 2025, Vol. 15, Page 452, 15*(11), 452. DOI: 10.3390/admsci15110452

Kourani, H., Berti, A., Schuster, D., & van der Aalst, W. M. P. (2025). Evaluating large language models on business process modeling: Framework, benchmark, and self-improvement analysis. *Software & Systems Modeling*, •••, 1–36. DOI: 10.1007/s10270-025-01318-w

Lazaropoulos, A. G. (2026). Simulation-Based Analysis of Performance Role Transformation in SLA-Aware BPMN IT Processes: The Enabling Role of AI in Sustainable Process Governance. *Sustainability 2026, Vol. 18, Page 3369, 18*(7), 3369. DOI: 10.3390/su18073369

Ledro, C., Nosella, A., Vinelli, A., Dalla Pozza, I., & Souverain, T. (2025). Artificial intelligence in customer relationship management: A systematic framework for a successful integration. *Journal of Business Research, 199*, 115531. DOI: 10.1016/j.jbusres.2025.115531

Li, H., Feng Tian, ·, Tian, F., Li, H., & Tian, F. (2026). Advancing Decision-Making through AI-Human Collaboration: A Systematic Review and Conceptual Framework. *Group Decision and Negotiation 2026 35:2, 35*(2), 26-. DOI: 10.1007/s10726-026-09980-1

Licardo, J. T., Tanković, N., & Etinger, D. (2026). BPMN Assistant: An LLM-Based Approach to Business Process Modeling. *Applied Sciences 2026, Vol. 16, Page 2213, 16*(5), 2213. DOI: 10.3390/app16052213

Lin, Z. (2026). Large Language Models as Psychological Simulators: A Methodological Guide. *Advances in Methods and Practices in Psychological Science, 9*(1), 25152459251410153. Advance online publication. DOI: 10.1177/25152459251410153

Miedema, E., Waschull, S., & Emmanouilidis, C. (2026). Towards trustworthy artificial intelligence for decision-making: A lifecycle perspective on knowledge- and data-driven artificial intelligence systems. *Computers in Industry, 174*, 104409. DOI: 10.1016/j.compind.2025.104409

Mikalef, P., Islam, N., Parida, V., Singh, H., & Altwaijry, N. (2023). Artificial intelligence (AI) competencies for organizational performance: A B2B marketing capabilities perspective. *Journal of Business Research, 164*, 113998. DOI: 10.1016/j.jbusres.2023.113998

Naveed, H., Khan, A. U., Qiu, S., Saqib, M., Anwar, S., Usman, M., Akhtar, N., Barnes, N., & Mian, A. (2025). A Comprehensive Overview of Large Language Models. *ACM Transactions on Intelligent Systems and Technology, 16*(5), 1–72. Advance online publication. DOI: 10.1145/3744746

Peykani, P., Ramezanlou, F., Tanasescu, C., & Ghanidel, S. (2025). Large Language Models: A Structured Taxonomy and Review of Challenges, Limitations, Solutions, and Future Directions. *Applied Sciences 2025, Vol. 15, Page 8103, 15*(14), 8103. DOI: 10.3390/app15148103

Raina, K., Sharma, G. D., Taheri, B., Dev, D., & Chavriya, S. (2026). Artificial intelligence-driven management: Bridging innovation, knowledge creation, and sustainable business practices. *Journal of Innovation & Knowledge, 11*, 100860. DOI: 10.1016/j.jik.2025.100860

Ray, P. P., & Pradhan, M. P. (2025). An evaluation framework for measuring prompt wise metrics for large language models in resource-constrained edge. *BenchCouncil Transactions on Benchmarks. Standards and Evaluations*, *5*(4), 100249. DOI: 10.1016/j.tbench.2025.100249

Ritala, P., Aaltonen, P., Ruokonen, M., & Nemeh, A. (2024). Developing industrial AI capabilities: An organisational learning perspective. *Technovation*, *138*, 103120. DOI: 10.1016/j.technovation.2024.103120

Rosemann, M., Brocke, J. vom, Van Looy, A., & Santoro, F. (2024). Business process management in the age of AI – three essential drifts. *Information Systems and E-Business Management 2024 22:3*, *22*(3), 415–429. DOI: 10.1007/s10257-024-00689-9

Sapkota, R., Roumeliotis, K. I., & Karkee, M. (2026). AI Agents vs. Agentic AI: A Conceptual taxonomy, applications and challenges. *Information Fusion*, *126*, 103599. DOI: 10.1016/j.inffus.2025.103599

Siino, M. (2025). Exploring the use of LLMs in the Italian legal domain: A survey on recent applications. *Computer Law & Security Review*, *58*, 106164. DOI: 10.1016/j.clsr.2025.106164

Szczepanek, M., Rudnik, K., Deptuła, A., & Estrada, Q. (2025). Development of approaches to business processes improvements in a digital transformation. *IFAC-PapersOnLine*, *59*(24), 257–262. DOI: 10.1016/j.ifacol.2025.11.874

Taeihagh, A. (2025). Governance of Generative AI. *Policy and Society*, *44*(1), 1–22. DOI: 10.1093/polsoc/puaf001

Teixeira, A. R., Ferreira, J. V., & Ramos, A. L. (2024). Optimization of Business Processes Through BPM Methodology: A Case Study on Data Analysis and Performance Improvement. *Information 2024, Vol. 15, Page 724*, *15*(11), 724. DOI: 10.3390/info15110724

Theodorakopoulos, L., Theodoropoulou, A., & Halkiopoulos, C. (2025). Cognitive Bias Mitigation in Executive Decision-Making: A Data-Driven Approach Integrating Big Data Analytics, AI, and Explainable Systems. *Electronics 2025, Vol. 14, Page 3930*, *14*(19), 3930. DOI: 10.3390/electronics14193930

Tirulo, A., Yadav, M., Lolamo, M., Chauhan, S., Siano, P., & Shafie-khah, M. (2026). Beyond automation: Unveiling the potential of agentic intelligence. *Renewable & Sustainable Energy Reviews*, *226*, 116218. DOI: 10.1016/j.rser.2025.116218

Viteri-Sánchez, C., & Novillo-Villegas, S. (2025). A Framework for a Sustainable Adoption of Business Process Management. *Sustainability 2025, Vol. 17, Page 9827, 17*(21), 9827. DOI: 10.3390/su17219827

Wang, C., & Chen, Y. (2026). Evaluating large language models for evidence-based clinical question answering. *Patterns (New York, N.Y.), 101519*(5), 101519. Advance online publication. DOI: 10.1016/j.patter.2026.101519 PMID: 42130942

Yang, W., Some, L., Bain, M., & Kang, B. (2025). A comprehensive survey on integrating large language models with knowledge-based methods. *Knowledge-Based Systems, 318*, 113503. DOI: 10.1016/j.knosys.2025.113503

Zhang, Z., Kang, Y., Lu, Y., & Li, P. (2025). The Role of Artificial Intelligence in Business Model Innovation of Digital Platform Enterprises. *Systems 2025, Vol. 13, Page 507, 13*(7), 507. DOI: 10.3390/systems13070507

Chapter 4
Intelligent Business Decisions With Large Language Models

Renugadevi Ramalingam
http://orcid.org/0000-0001-7462-9205
Saveetha Engineering College, Chennai, India

V. S. Saranya
http://orcid.org/0009-0001-6527-7022
SRM Institute of Science and Technology, Chennai, India

J. Shobana
http://orcid.org/0000-0001-9754-2604
SRM Institute of Science and Technology, Chennai, India

ABSTRACT

The incorporation of Large Language Models (LLMs) into Decision Support Systems (DSS) signifies a major transformation in the domain of Business Intelligence (BI). Traditionally, DSS relied primarily on structured datasets and rule-driven algorithms to generate insights and assist decision-makers. With the emergence of LLMs, organizations now have the capability to examine unstructured information, identify complex patterns in textual data, and produce responses that resemble human communication. This chapter examines the paradigm created by integrating the advanced functionalities of LLMs with DSS, thereby strengthening their ability to deliver meaningful insights, support effective decision-making, and enhance organizational performance. It highlights real-time natural language interaction, improved contextual comprehension, and the capacity to process information from

DOI: 10.4018/979-8-2600-0050-2.ch004

varied data sources. In addition, the chapter discusses implementation challenges, including issues related to data privacy, model interpretability, and seamless integration with existing organizational workflows.

1. INTRODUCTION

In the modern business landscape, characterized by rapid change and extensive data generation, the ability to make prompt and well-informed decisions has become a defining factor for organizational success. Although traditional decision-making frameworks have played an essential role in guiding business operations, they often struggle to cope with the growing volume, diversity, and complexity of information produced in today's digital economy. In this context, artificial intelligence—particularly Large Language Models (LLMs)—is emerging as a powerful tool that is reshaping the way organizations analyze information and make strategic decisions.

LLMs, including models such as OpenAI's GPT architecture and other similar systems, are sophisticated artificial intelligence technologies trained on extensive collections of textual data derived from books, online resources, articles, and other digital content. These models possess the ability to comprehend context, generate coherent human-like responses, and perform certain reasoning tasks. Unlike earlier AI techniques that focused mainly on structured datasets, LLMs can effectively interpret and derive insights from unstructured data, which constitutes a significant portion of the information generated by modern businesses.

Across various organizational functions—including strategic planning, customer relationship management, market intelligence, and operational management—LLMs are increasingly being adopted to support decision-making processes. Rather than replacing human expertise, these systems enhance decision-making by providing deeper analytical insights, accelerating information processing, and facilitating more efficient communication. For instance, an LLM can analyze large volumes of customer feedback to identify sentiment patterns, evaluate real-time news data to anticipate market changes, or convert raw business data into structured reports suitable for executive review. Such capabilities significantly reduce the time and cognitive effort traditionally required for data analysis and reporting.

A particularly valuable contribution of LLMs lies in their capacity to provide decision-makers with context-aware information whenever it is required. Business leaders and managers are no longer restricted by the analytical limitations of human teams alone. LLMs can model potential scenarios, identify possible risks, and suggest alternative strategies while continuously improving their performance through feedback and updated data. As a result, these systems serve as an essential component in the development of advanced decision support systems that are capable of not only reacting to existing conditions but also anticipating future trends and opportunities.

Despite the promising advantages of this transformative technology, organizations must address several important challenges when integrating LLMs into business environments. Concerns related to data privacy, algorithmic transparency, potential bias in model outputs, and excessive dependence on automated systems must be carefully managed. Successful implementation requires not only technological capability but also ethical governance frameworks, organizational preparedness, and a clear awareness of the limitations associated with AI-driven systems.

In recent years, Artificial Intelligence (AI), especially large language models, has experienced rapid advancement and widespread adoption. These models, trained on extensive datasets, have demonstrated remarkable abilities in understanding and producing human-like language. When combined with knowledge graphs—structured frameworks that represent relationships among entities—AI systems gain access to both linguistic understanding and semantic knowledge. This combination enables more intelligent reasoning and has contributed to significant innovations across multiple sectors, including healthcare, defense, engineering, education, and environmental management.

Within defense applications, the integration of AI models with knowledge graphs has improved strategic planning and operational decision-making. For example, Liang et al. (2019) proposed a semantic matching approach based on military knowledge graphs that enhances the evaluation of combat plans by aligning them with operational objectives. By interpreting complex military information through semantic relationships, AI systems contribute to improved planning accuracy and operational effectiveness.

Similarly, the tourism and cultural industries are experiencing significant changes through the application of AI technologies. Hu (2023) examined how large AI models can be utilized to create personalized digital experiences for tourists, thereby improving service quality and user engagement. The combination of AI models with knowledge graphs enables a deeper understanding of traveler preferences, which enhances destination recommendations and supports the preservation of cultural heritage resources.

In the healthcare sector, the convergence of AI technologies and knowledge graphs is driving progress in personalized medicine. By analyzing patient data in conjunction with structured medical knowledge bases, AI systems can generate customized treatment suggestions. Liu et al. (2021) demonstrated that such systems enhance clinical decision-making by leveraging medical knowledge graphs to improve diagnostic accuracy and treatment planning, ultimately contributing to better patient outcomes. Furthermore, the integration of AI with blockchain technology is transforming water quality management by ensuring transparency, strengthening data security, and enabling predictive analytics for improved resource management in urban environments (Renugadevi, 2025).

Education is another domain benefiting from AI-driven innovations. Wang et al. (2023) illustrated how large language models integrated with knowledge graphs can improve library reference services. The ability of these systems to process and retrieve information based on contextual understanding increases the efficiency of information delivery in academic and research environments, helping users locate the most relevant resources quickly.

Environmental management also demonstrates the potential of AI and knowledge graphs in supporting sustainable practices. Li and Li (2023) developed a digital twin platform for watershed management that combines AI with knowledge graph technologies to simulate and analyze environmental systems. This approach supports monitoring of water quality, forecasting ecological changes, and enabling more informed environmental decision-making.

In industrial and construction sectors, AI systems integrated with knowledge graphs are contributing to improved safety standards and operational performance. Zhang et al. (2023) explored the use of AI-based question-answering systems to manage safety procedures and maintenance tasks in high-risk industrial settings. By merging knowledge graphs with large language models, these systems can deliver accurate and timely responses, minimizing operational disruptions and enhancing workplace safety. In addition, AI-driven communication robots are increasingly being utilized for safety inspections in specialized industries. Xu et al. (2023) investigated how such robots, powered by knowledge bases and language models, can assist in monitoring complex machinery and hazardous environments, thereby improving efficiency while safeguarding workers in high-risk operational settings.

2. LITERATURE REVIEW: LARGE LANGUAGE MODELS IN BUSINESS DECISION-MAKING

Large Language Models (LLMs), including advanced systems such as GPT-4, have increasingly influenced business decision-making in recent years. Their capability to analyze extensive volumes of unstructured information, produce human-like language, and perform complex reasoning tasks has accelerated their adoption across multiple industries. This section reviews recent developments and practical applications of LLMs in business decision-making, focusing on significant research contributions and frameworks published between 2022 and 2025. A notable study conducted by researchers at the University of Chicago demonstrated that GPT-4 can evaluate corporate financial statements with a level of accuracy comparable to professional financial analysts. By applying chain-of-thought reasoning techniques, the model predicted future earnings growth with an accuracy rate of approximately 60.35%, outperforming several traditional analytical models and human evaluators.

The comparison over the Traditional Decision Support System and LLM Based Decision Support System is Depicted in the following figure 1.

Figure 1. Traditional DSS Vs LLM Based DSS

Comparison: Traditional DSS vs LLM-Based Decision Support System

Dimension	Traditional Decision Support System (DSS)	LLM-Based Decision Support System
Knowledge Source	Structured data in databases, data warehouses, predefined models and rules	Structured + unstructured data, enterprise documents, emails, chats, web, and external knowledge via LLMs and RAG
Data Handling	Works well with structured and numeric data; limited handing of unstructured data	Understands and reasons over unstructured, semi-structured and structured data (text, tables, images, etc.)
Analysis Approach	Rule-based logic, statistical models, OLAP, what-if analysis	Natural language understanding, reasoning, pattern discovery, contextual analysis, and generative insights
User Interaction	Dashboards, forms, reports; requires user knowledge of system/query languages	Conversational interface (natural language); no coding or complex queries required
Flexibility	Limited to pre-defined models, schemas and workflows	Highly flexible; adapts to new questions, contexts and domains with minimal reconfiguration
Insight Generation	Descriptive analytics and predefined reports	Descriptive + exploratory + generative insights, summaries, explanations and recommendations
Decision Support	Supports analysis and evaluation of alternatives based on predefined criteria	Supports end-to-end decision support: information retrieval, analysis, option generation, reasoning, and recommendation
Integration	Integrates with internal systems via ETL, APIs and data pipelines	Integrates via RAG, APIs, tools/functions, and connectors to multiple enterprise and external sources
Knowledge Maintenance	Manual updates to rules, models and databases by IT or domain experts	Continuously learns from new data (fine-tuning/RAG refresh), knowledge stays up-to-date with sources
Explainability	High (models and rules are explicit and traceable)	Improving but context-dependent; explanations can be generated, with traceability via source citations (RAG)
Scalability	Limited by system design, hardware and data model	Highly scalable using cloud infrastructure and LLM services
Cost Structure	High upfront development and maintenance costs; licensing based on users/modules	Lower upfront; usage-based (token/API) costs; cost-effective for rapid deployment and iteration
Security & Governance	Role-based access and database-level security	Enterprise-grade security with data governance, PII masking, access controls, and prompt guardrails
Use Cases	Standard reports, KPIs, dashboards, what-if analysis, forecasting	Knowledge Q&A, research, complex analysis, scenario planning, recommendations, automation, and more

The broader integration of Artificial Intelligence (AI) technologies is also driving major innovations across sectors such as cybersecurity, healthcare, water management, and finance. In cybersecurity, AI-based Security Information and Event Management (SIEM) systems improve threat detection and incident response through real-time data processing and automated analysis, allowing organizations to identify and mitigate potential security threats more efficiently, particularly within critical infrastructures (Ramalingam et al., 2025). In the healthcare domain, AI technologies are enhancing nuclear medicine by improving diagnostic accuracy and enabling personalized treatment planning, ultimately contributing to better patient outcomes (Renugadevi, 2025). Similarly, the finance sector is witnessing substantial transformation through AI-driven automation of routine processes, improved financial forecasting, and enhanced decision-support capabilities. These developments are reshaping accounting and financial operations by increasing operational efficiency and minimizing human errors (Vaishnavi et al., 2025). Overall, the adoption of AI across these fields promotes innovation by improving precision, expanding automation, and supporting more individualized solutions. Continued research suggests that the convergence of AI with emerging technologies such as blockchain and LLMs will further transform areas such as cybersecurity, healthcare, resource management, and business decision-making.

Another significant contribution in this domain is the DeLLMa framework proposed by Liu et al. (2024), which integrates principles of decision theory with large language models to support decision-making under uncertain conditions. This framework improves decision accuracy by as much as 40% compared with conventional approaches and offers a structured and auditable mechanism for handling complex business decisions. In addition, Forrester's 2023 industry report emphasized the growing integration of LLMs into Business Intelligence (BI) platforms. These integrations enable natural language querying and automated report generation, allowing even non-technical users to interact with data systems more effectively. Such capabilities broaden access to organizational data insights and accelerate the overall decision-making process.

Li et al. (2023) introduced OptiGuide, an LLM-driven framework designed to interpret optimization results within supply chain management contexts. By translating complex optimization outcomes into easily understandable text explanations, the system bridges the gap between technical optimization models and human understanding, thereby promoting greater transparency and supporting better-informed decisions. Similarly, Beheshti et al. (2023) developed ProcessGPT, a generative AI solution intended to model, evaluate, and optimize business processes. By automating repetitive tasks and recommending improvements in operational workflows, ProcessGPT helps organizations increase efficiency and respond more effectively to changing business environments.

Practical applications of LLMs can also be observed in the financial technology sector. Ant Group's models, Zhixiaozhu 1.0 and Zhixiaobao 2.0, demonstrate how LLM-based systems can enhance consumer-facing financial services. These models assist users in performing market analysis, comparing financial products, and making asset allocation decisions. Reports indicate that the systems achieve approximately 95% accuracy in identifying financial intentions, thereby significantly improving the consumer decision-making experience. In addition, Schoenegger et al. (2024) examined the integration of LLMs with traditional analytical methods such as Monte Carlo simulations. Their findings showed that incorporating LLMs into simulation models reduced prediction errors by around 40%, resulting in more reliable and strategic business decisions.

Furthermore, LLMs are increasingly contributing to the enhancement of threat intelligence and cyber threat detection systems. These models support more accurate identification of evolving cyber threats and improve automated decision-making capabilities in cybersecurity environments (Renugadevi, Muthumeenakshi, & Karthika, 2025). However, the growing reliance on LLMs in decision-making processes also introduces several important concerns, including issues related to data privacy, transparency of AI models, and ethical governance. Existing research emphasizes the necessity of establishing clear frameworks for responsible AI implementation, including mechanisms to address algorithmic bias and maintain appropriate human oversight to minimize potential risks.

Overall, the integration of LLMs into business decision-making processes has produced significant improvements across multiple domains, including financial analysis, supply chain optimization, and strategic planning. Although these technologies offer considerable advantages, it remains essential to address associated challenges such as ethical considerations, accountability, and model interpretability. Future research should therefore focus on refining LLM-based applications and developing comprehensive governance frameworks to ensure their responsible and effective deployment within business environments.

The integration of Artificial Intelligence models, particularly large language models, with knowledge graphs has also produced notable advancements across a range of disciplines. Liang et al. (2019) investigated the use of military knowledge graphs to support combat planning. Their study proposed a semantic matching technique that enhances decision-making in military operations by aligning combat plans with operational requirements. By utilizing the structured relational information available within military knowledge graphs, this approach improves the evaluation and coordination of strategic actions, illustrating the potential of AI technologies in defense contexts.

Hu (2023) explored the application of large AI models within the digital culture and tourism industry. The research demonstrated how AI-driven technologies can personalize tourism experiences, support cultural heritage preservation, and enable interactive tourism applications. These systems contribute to improved customer engagement and operational efficiency within the digital tourism sector.

In the field of information services, Wang et al. (2023) examined the use of large language models in library reference systems. Their research showed that AI technologies can automate and enhance information retrieval processes, thereby improving both the efficiency and accuracy of library services. The study also discussed challenges and possible solutions associated with implementing AI-driven systems for knowledge dissemination and user interaction.

Environmental management represents another domain benefiting from the combination of LLMs and knowledge graphs. Li and Li (2023) proposed a digital twin knowledge platform for watershed management that integrates large language models with knowledge graph technologies. The platform utilizes real-time environmental data to simulate water management systems, enabling more effective decision-making for water conservation and environmental monitoring. This work highlights the potential of AI-driven systems in supporting sustainable environmental management and infrastructure planning.

Zhang et al. (2023) examined the integration of large language models with knowledge graphs in automated question-answering systems. Their research demonstrated that combining structured knowledge from graphs with the language understanding capabilities of AI significantly enhances the accuracy and relevance of automated responses. The study highlighted applications ranging from customer support services to complex research assistance, emphasizing the potential of integrated AI systems.

In the construction and engineering sectors, Qin et al. (2023) analyzed the application of large language models in construction project management. Their research investigated how AI technologies can automate design and planning tasks, improve project management processes, and optimize the allocation of resources. The results suggest that integrating AI models with project knowledge graphs can significantly enhance efficiency and coordination within construction projects.

Xu et al. (2023) further explored the development of AI-powered communication robots designed for safety monitoring in specialized industrial equipment environments. Their study demonstrated that integrating knowledge graphs with AI models enables responsive communication systems that assist in equipment maintenance and safety inspections. The findings highlight the potential of such technologies to improve operational safety and efficiency in industries involving complex machinery.

Collectively, these studies illustrate the growing importance of large language models and knowledge graphs in transforming numerous industries, including defense, tourism, information services, environmental management, construction engineering, and industrial safety. By enabling more intelligent, adaptive, and data-driven systems, these technologies enhance decision-making, optimize operational processes, and provide personalized user experiences. As research in AI and knowledge graph technologies continues to progress, their applications are expected to expand further, driving innovation and efficiency across a wide range of sectors.

3. METHODS OF APPLYING LARGE LANGUAGE MODELS IN BUSINESS DECISION-MAKING

The integration of Large Language Models (LLMs) into business environments is transforming organizational decision-making by enabling faster data analysis, improved insight generation, and stronger strategic support. These models are applied through several established methods that allow organizations to utilize AI capabilities across multiple decision-making domains. Such approaches include natural language processing, advanced analytics, and knowledge management techniques, all of which help organizations increase operational efficiency, reduce uncertainty, and make more informed data-driven decisions. The architecture model of LLM Based Decision making modules and Enterprise Knowledge System is presented in Figure 2.

Figure 2. Architecture of LLM based Decision making modules and Enterprise Knowledge system

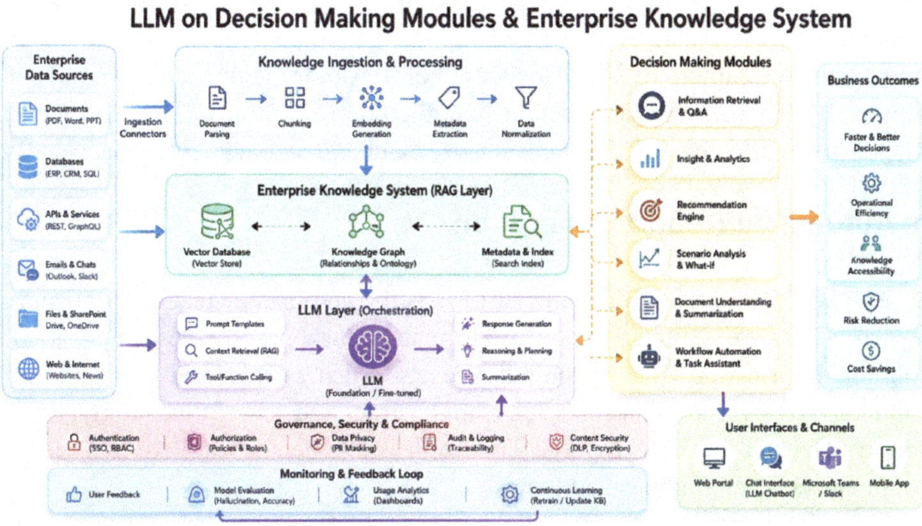

a. Text Summarization and Sentiment Analysis

One of the most common applications of LLMs in business decision-making is the summarization of large volumes of unstructured information. Businesses generate extensive textual data from sources such as customer reviews, service requests, social media interactions, internal reports, and feedback forms. LLMs can efficiently condense this information into concise summaries that highlight key insights, enabling organizations to improve product development, refine marketing strategies, and enhance customer support services.

In addition to summarization, LLMs are highly effective in performing sentiment analysis. These models can identify emotional tone, opinions, and attitudes expressed within textual content. By analyzing customer feedback at scale, organizations can monitor public perception, evaluate brand reputation, and assess customer satisfaction in real time. Such insights help businesses respond quickly to consumer needs and maintain positive customer relationships.

Modern LLMs such as GPT-4, LLaMA 2, and FinBERT demonstrate strong capabilities in processing large volumes of textual information with high contextual understanding. For example, LLaMA 2 has been applied in financial contexts to analyze corporate disclosures and generate sentiment assessments that correlate strongly with stock market performance. Studies indicate that such models often outperform traditional sentiment analysis tools in predictive accuracy.

Furthermore, LLMs can summarize extensive financial documents including earnings reports, market analyses, and news articles, enabling decision-makers to quickly understand essential insights without reviewing lengthy documents. Organizations such as Sprinklr employ LLM-based technologies to monitor social media discussions and capture real-time insights about customer sentiment and brand perception. By enabling rapid and accurate interpretation of textual information, LLMs support more effective forecasting, strategic planning, and market responsiveness.

b. Automated Report Generation

Another important application of LLMs is the automated creation of business reports, executive summaries, and performance dashboards. These models can combine information from structured datasets—such as sales statistics or financial metrics—with insights extracted from unstructured sources including analyst commentary and business reports. The result is a coherent narrative that clearly communicates key findings and trends.

Financial institutions, for example, increasingly use LLMs to generate compliance documentation, investment summaries, and earnings analyses. Automating such tasks reduces manual effort, shortens report preparation time, and ensures consistency and accuracy across organizational reports.

Several real-world implementations demonstrate the value of this approach. JPMorgan Chase has introduced an internal AI system known as **LLM Suite**, which supports nearly 200,000 employees in activities such as document summarization, content creation, and idea generation within its asset and wealth management divisions. This initiative has significantly improved productivity and streamlined internal workflows.

Similarly, Tencent developed the **SiriusBI** platform, which integrates LLM capabilities to enhance business intelligence operations. The system automates SQL query generation and supports conversational interactions with data systems, enabling users to retrieve insights through natural language queries. These innovations allow organizations to obtain real-time analytical insights and improve decision-making across sectors such as finance, marketing, and digital advertising.

Advances in financial analytics have also led to specialized LLM systems designed to analyze earnings reports and financial disclosures. These models utilize techniques such as retrieval-augmented generation and instruction-based fine-tuning to produce accurate financial summaries, often outperforming traditional analytical tools in both efficiency and precision.

c. Conversational Decision Support Systems

Large Language Models also serve as the foundation for conversational decision support systems (CDSS), including intelligent chatbots and virtual assistants that support business operations. These AI-driven tools enable employees and managers to interact with organizational data systems using natural language, allowing them to obtain information, analyze scenarios, and receive recommendations quickly.

Unlike conventional decision support systems that require predefined queries or static reports, conversational systems allow users to explore data dynamically. Employees can ask questions, request insights, and receive contextual explanations through interactive dialogue interfaces.

LLMs such as GPT-4 are particularly effective in these systems because they can interpret natural language input, maintain conversational context, and generate detailed explanations or recommendations. These models can be trained or fine-tuned for specific industries, ensuring that responses align with domain-specific knowledge and organizational requirements.

In a conversational decision support system, the objective is to assist users in making well-informed choices. For example, business leaders may use such systems to evaluate strategic options based on financial performance indicators, while healthcare professionals might rely on them to analyze patient information and determine appropriate treatment options. The systems can provide features such as risk assessment, predictive analysis, and scenario evaluation using both historical and real-time data.

Integration with APIs and enterprise databases further enhances the capabilities of conversational systems. These systems can access real-time information such as stock market data, operational metrics, or customer records, enabling them to provide up-to-date insights. By maintaining contextual memory of previous interactions, conversational systems can also personalize their recommendations according to user preferences and organizational contexts, thereby improving their practical usefulness.

d. Scenario Simulation and Predictive Modeling

LLMs can also be applied to scenario simulation and predictive analysis, supporting organizations in evaluating possible future outcomes. Businesses often face uncertainty due to market volatility, changing consumer behavior, and technological disruption. Scenario simulation enables organizations to explore potential outcomes by testing different strategic decisions under varying conditions.

Although LLMs are not traditional predictive models like statistical regressions or machine learning classifiers, they can complement these techniques by interpreting analytical outputs and generating context-aware explanations. For example, a retail organization may analyze the potential effects of different pricing strategies across multiple regions. Traditional models may calculate projected outcomes, while LLMs can translate these results into understandable insights and recommendations.

Scenario simulation involves creating hypothetical situations—such as market fluctuations, supply chain disruptions, or new product launches—to evaluate possible impacts on business performance. LLMs enhance this process by generating potential scenarios, processing natural language queries, and summarizing predicted outcomes.

Predictive modeling relies on historical data and statistical algorithms to forecast future trends, such as product demand, inventory requirements, customer behavior, or financial performance. LLMs assist in this process by helping interpret complex data patterns and presenting results in a conversational format that decision-makers can easily understand.

e. Data-to-Text Applications

Data-to-Text (D2T) systems represent another significant application of LLMs in business decision-making. These systems automatically convert structured data—such as spreadsheets, databases, and dashboards—into natural language explanations and narratives.

By transforming numerical information into readable text, D2T applications make analytical insights more accessible to executives and managers who may not have technical expertise in data analysis. This approach is widely used in business intelligence dashboards, investor communications, and internal performance reporting.

Natural Language Generation (NLG) technology forms the foundation of Data-to-Text systems. NLG algorithms analyze structured datasets and generate narratives that describe trends, highlight anomalies, and summarize key insights. For example, a sales dataset might be transformed into a statement such as: *"The organization experienced a 15% increase in quarterly sales, primarily driven by a 20% growth in Product A, which contributed 30% of total revenue."*

These systems typically use structured input data such as sales figures, customer engagement metrics, inventory records, and operational performance indicators. The goal is to convert these data points into meaningful stories that help decision-makers quickly understand important trends and opportunities.

Data-to-Text systems also support personalized decision-making by tailoring insights to the specific roles of stakeholders. For instance, a Chief Financial Officer might receive summaries emphasizing financial indicators, while a marketing manager might receive insights related to customer engagement and campaign performance.

Additional capabilities of Data-to-Text systems include:

1. Contextual Understanding:

LLMs analyze business context to generate meaningful interpretations of data. For example, if sales decline in a particular region, the system may identify potential causes such as competitor promotions or seasonal demand variations.

2. Real-Time Data Interpretation:

When integrated with live data streams such as sales dashboards or financial markets, LLMs can generate immediate summaries and highlight emerging trends.

3. Dynamic Report Generation:

LLMs can automatically generate periodic reports, such as weekly sales summaries or performance evaluations, by combining information from multiple enterprise data sources.

f. Knowledge Base Enhancement

Large Language Models are also widely used to enhance enterprise knowledge management systems. By integrating with corporate knowledge bases, LLMs improve the ability of organizations to store, retrieve, and interpret information effectively.

These models can act as intelligent interfaces for enterprise knowledge repositories, enabling employees to retrieve relevant documents, policies, and best practices through natural language queries. As a result, organizations can leverage their collective knowledge more effectively when making operational or strategic decisions.

Knowledge base enhancement typically involves several techniques, including fine-tuning, transfer learning, and continuous data updates. Fine-tuning allows organizations to train LLMs using domain-specific datasets such as financial records, healthcare reports, or engineering documentation. This enables the model to generate more accurate and contextually relevant responses.

In many cases, knowledge bases contain both structured data—such as databases and reports—and unstructured information such as research documents and internal communications. LLMs can integrate these sources into hybrid systems that provide deeper analytical insights.

For example, an LLM connected to a company's CRM or ERP system can retrieve customer interaction histories and generate recommendations for sales strategies or customer relationship management.

Continuous knowledge updates further enhance system performance. By incorporating new information such as market trends, customer feedback, and competitor activities, LLMs remain aligned with current business environments and can adapt their recommendations accordingly.

Context-aware reasoning is particularly valuable for complex business decisions. When evaluating a new market entry strategy, for instance, an LLM can analyze historical company strategies, competitor performance, regional demographics, and economic conditions to generate a tailored recommendation.

Additionally, training models with domain-specific terminology improves their understanding of industry language. This is particularly important in specialized fields such as healthcare, finance, legal services, and engineering, where accurate interpretation of technical terms is essential for decision-making.

Enhancing Decision-Making through Knowledge Bases

Improving knowledge bases with LLMs offers several advantages for business decision-making:

Improved Decision Contextualization:

Enhanced models can generate insights tailored to specific business situations by drawing on comprehensive internal and external data sources.

Rapid Access to Relevant Information:

Decision-makers can quickly retrieve relevant knowledge from organizational databases, enabling faster responses to operational challenges.

Data-Driven Strategy Development:

LLMs can synthesize large volumes of internal and market data to generate strategic recommendations for business growth.

Enhanced Predictive Capabilities:

By continuously learning from new data, LLMs can improve their ability to forecast market trends, demand patterns, and operational performance.

Automated Reporting and Decision Support:

LLM-powered systems can automatically generate analytical reports and summaries, ensuring that managers always have access to timely and relevant insights.

4. DESIGNING A DECISION-MAKING ASSISTANT KNOWLEDGE BASE SYSTEM BASED ON LARGE LANGUAGE MODELS

Relational Data Model

The Decision-Making Assistant Knowledge Base System utilizes advanced information retrieval technologies to support efficient decision-making. Database technology has matured significantly and performs effectively when managing structured knowledge data. Based on the principles of set theory, the relational data model establishes connections between data entities, their attributes, and the relationships that link them. Within this model, both entities and the connections among them are represented through relationships in a unified format. A relational database built on this framework is used to store validated knowledge information. Such databases provide several advantages, including simplicity, structured organization, clarity in data representation, and stable management of information. Additionally, they offer reliable data access while consuming minimal system resources, making them practical for many enterprise applications.

Despite these benefits, relational databases have certain limitations. Their retrieval mechanisms are mainly based on content matching, which makes them less effective when dealing with long-form text or unstructured documents. As a result, these databases are not well suited for storing or retrieving knowledge from unstructured textual sources. Therefore, additional technologies are required to effectively extract and manage knowledge from such data.

Large Language Models (LLMs)

While relational databases efficiently store structured knowledge, maintaining and updating this information can require significant human effort and operational costs. Moreover, the updates in such systems may lag behind real-time developments. In contrast, valuable and up-to-date knowledge often exists in open and unstructured documents, which are difficult to process using traditional database approaches due to their complexity and lack of structure.

Large Language Models provide an effective solution to address this challenge. These models can first interpret a user's query and identify the relevant knowledge domain. They then retrieve and synthesize scattered pieces of information from multiple documents, presenting them as coherent and natural language responses that are easy to understand.

LLMs possess strong semantic comprehension and natural language generation capabilities. To process unstructured text efficiently, a text-embedding model converts segmented text into multidimensional vectors containing semantic information. These vectors represent relationships between textual elements in a vector space that reflects natural language associations. Using vector similarity matching, a prompt is constructed and provided to the LLM, which then generates refined and human-readable responses. This method significantly improves the system's ability to retrieve knowledge from diverse and unstructured data sources.

Robotic Process Automation

Robotic Process Automation (RPA) is a technology designed to automate repetitive software tasks by imitating human interactions with digital systems. In the Decision-Making Assistant Knowledge Base System, RPA is applied differently from conventional automation tools. Rather than simulating mouse clicks and keyboard actions directly, the system employs a browser plug-in combined with web crawler techniques to automate interactions with web pages.

This automation is achieved by injecting JavaScript code into targeted webpages. The RPA component monitors real-time updates on these pages, identifies relevant keywords or phrases, and activates the knowledge retrieval function to gather information useful for decision-makers. The RPA browser plug-in also acts as a communication bridge between third-party business systems and the Decision-Making Assistant Knowledge Base System. Importantly, this integration occurs without altering the original business logic or underlying code of the external systems, enabling users to access domain-specific knowledge efficiently.

Access Logic of Structured Knowledge

In this system, structured knowledge is stored in an encyclopedia-style database. The storage framework follows a **"keyword + interpretation + association"** structure. Each knowledge element is divided into specific keywords or knowledge points, which are stored as individual records. Every record includes an explanation of the knowledge point as well as links to related knowledge entities, each stored in separate database fields.

The primary mechanism used for querying structured knowledge is fuzzy keyword matching. When a user submits a query, the system identifies the database entry with the highest similarity to the requested keyword and selects it as the most relevant knowledge point. This storage structure is intuitive and straightforward, making it easy for users to understand and manage.

However, the verification and validation of structured knowledge entries require manual effort. Both during the initial setup of the knowledge base and during long-term maintenance, human intervention is necessary to ensure data accuracy and reliability.

Processing Logic of Unstructured Documents

For unstructured knowledge sources, the system utilizes the capabilities of Large Language Models to preprocess and analyze textual data. Initially, documents are segmented and converted into semantic vectors to enable efficient knowledge retrieval.

When a decision-maker submits a query, the system first interprets the request and transforms the keyword or sentence into a comparable semantic vector. The system then identifies the content segment with the highest similarity score. This relevant content is used as a prompt for the LLM, which generates a clear and informative response in natural language. Additionally, the system maintains a link to the original source document, enabling users to trace the retrieved information for further verification.

Functional Logic of Knowledge Query

The knowledge question-and-answer function provides a human-computer interaction method that closely resembles natural human communication. Instead of using simple keywords, decision-makers can submit complete questions or sentences describing their information needs.

The system analyzes the query, identifies the key elements of the request, and searches for relevant information based on the sentence structure. The LLM then uses the retrieved content as a prompt to generate a natural language response that directly addresses the user's question.

During multi-round interactions, the system continuously improves its understanding of the user's intent. With each subsequent interaction, it refines the context of the conversation, allowing it to provide more precise and relevant responses.

System Structure

The LLM-based Decision-Making Assistant Knowledge Base System is designed to operate within a **private cloud environment**. The infrastructure mainly consists of a **Web Access Server, High-Performance Computing Server, Browser Plug-in, Network Switch Devices**, and other supporting components.

The **Web Server** provides the operating environment and database access services required by the system. It hosts application services that require relatively low computational resources. The **High-Performance Computing Server** supplies the GPU-based computing power necessary for executing large language models.

The **Web Browser** acts as the primary interface through which users interact with the system. Meanwhile, the **Browser Plug-in**, built with RPA technology, automatically extends the system's capabilities and facilitates integration with third-party business applications. Network routers and switches establish a secure internal network that enables communication between all system components.

System Functional Model

The system architecture consists of four primary functional modules:

1. **System Management Module** – Responsible for user management, access permissions, system configuration, and other administrative tasks that ensure stable system operation.
2. **Knowledge Search Module** – Provides knowledge retrieval services, including database field matching, document-based knowledge extraction through LLMs, and conversational question-answering capabilities.
3. **Knowledge Management Module** – Supports updating and maintaining domain-specific knowledge resources, including database management, verification of knowledge entries, and management of document repositories.
4. **RPA Assistant Module** – Controls the Web-RPA browser plug-in and supports the configuration and execution of automated services.

The LLM-based Decision-Making Assistant Knowledge Base System retrieves knowledge from both structured databases and unstructured document repositories using traditional database search methods combined with intelligent document retrieval techniques. The system then organizes the retrieved information and generates clear natural language explanations, providing decision-makers with reliable and timely knowledge support.

By utilizing the high-performance computing capabilities of local servers, the system can fully leverage the strengths of large language models and provide extended functionality. The system is expected to be applied widely across various business scenarios, including the following.

A. Knowledge Correlation to Support Decision-Making

Knowledge correlation focuses on identifying and analyzing relationships among different entities. By connecting knowledge entities within a large relational network, the system forms a knowledge base that supports search, traceability, and clustering of related information.

With the integration of LLMs and advancements in knowledge graph technologies, the system is expected to further enhance its knowledge correlation capabilities. This integration will enable the system to link the logical relationships among entities, events, development trends, and contextual information. Consequently, decision-makers will be able to access integrated knowledge resources that provide reliable and efficient decision support.

B. Foreign Language Translation and Information Extraction

Another potential application of the Decision-Making Assistant Knowledge Base System is in translating and analyzing foreign language documents. Using LLM capabilities, the system can perform translation and proofreading of technical or academic literature.

By leveraging domain-specific knowledge stored within the knowledge base and incorporating user preferences, the system can significantly improve the quality of machine translation. In addition, the system can extract relevant information from foreign language documents through semantic matching techniques based on text embeddings.

By combining these extraction capabilities with the reasoning functions of large language models, the system can analyze underlying causes, evaluate potential impacts, and even suggest possible solutions or countermeasures. This functionality can greatly assist decision-makers who rely on global information sources for strategic planning and analysis.

Below is a **rephrased version with approximately the same length and academic tone**, suitable for a **book chapter**. The structure and meaning are preserved while changing wording and sentence construction.

5. USE CASES OF COMPANIES USING LARGE LANGUAGE MODELS (LLMS) IN BUSINESS DECISION-MAKING

Large Language Models (LLMs) such as OpenAI's GPT-4, Google's Gemini, and other transformer-based architectures are significantly transforming organizational decision-making processes. These advanced AI systems can comprehend, generate, and interpret human-like text, enabling their application across multiple decision-making scenarios in different industries. This section presents several practical examples of how organizations employ LLMs to strengthen business decision-making, spanning sectors such as finance, e-commerce, customer service, legal compliance, and technology.

1. Morgan Stanley: Improving Financial Advisory Services

Morgan Stanley, a global leader in financial services, utilizes a GPT-based platform to assist its financial advisors. Through the integration of LLMs into internal knowledge repositories, advisors can access extensive collections of research reports, market analyses, and proprietary insights using natural language queries. This capability enables advisors to make more informed investment recommendations for clients, reduces the time required to locate relevant information, and improves responsiveness in financial decision-making.

2. Klarna: Customer Service Automation and Compliance Monitoring

Klarna, a financial technology company, has incorporated LLMs into its customer support systems to manage queries from users across different languages and regions. The model analyzes user intent, sentiment, and previous interaction history to determine the most appropriate response or escalation path. Additionally, Klarna applies LLMs to automate the review of communications for regulatory compliance, thereby enhancing decision accuracy and ensuring adherence to industry regulations.

3. PwC: Legal Decision Support through Harvey AI

PwC has collaborated with the legal AI startup Harvey to improve legal and tax advisory services. Built on GPT technology, Harvey assists legal professionals by analyzing complex legal documents, generating draft legal texts, and identifying potential regulatory risks. This integration significantly reduces manual research efforts while improving the consistency and efficiency of legal decision-making processes.

4. Coca-Cola: Supporting Marketing Strategy and Creative Development

Coca-Cola has partnered with OpenAI and Bain & Company to explore the use of LLMs in marketing strategy and creative design. The technology generates advertising copy, evaluates possible customer reactions to campaigns, and suggests promotional ideas. These capabilities assist marketing teams in making strategic decisions by testing creative concepts at scale and adapting campaigns based on consumer sentiment and feedback.

5. Shopify: AI-Driven Business Assistance for Entrepreneurs

E-commerce platform Shopify employs LLM-powered assistants to support merchants in various business tasks. The AI helps generate product descriptions, compose marketing emails, and analyze store performance. It can also provide strategic recommendations, such as identifying products with declining sales or suggesting pricing adjustments based on market trends. This functionality enables small business owners to make data-driven decisions without requiring specialized analytics teams.

6. Amazon: Product Development through Customer Feedback Analysis

Amazon uses LLMs to process and analyze extensive volumes of customer reviews and feedback. By summarizing large datasets of textual information, the models identify frequently reported issues, customer preferences, and desired product features. These insights assist product development teams in improving existing offerings, prioritizing updates, and making informed inventory decisions aligned with consumer demand.

7. Netflix: Content Strategy and Audience Engagement (Experimental)

Netflix is experimenting with LLM technologies to evaluate scripts, reviews, and viewer feedback. These models help detect patterns and themes that resonate with audiences and estimate the potential success of new content. The insights generated can support decisions related to content investment, release timing, and audience engagement strategies. Although still under development, LLMs offer a scalable approach to analyzing viewer preferences and entertainment trends.

8. Duolingo: Personalized Learning Experiences

Language learning platform Duolingo has integrated GPT-4 into its premium subscription services to provide personalized educational experiences. The model evaluates learner progress, explains complex topics conversationally, and simulates realistic dialogue scenarios. These features allow the platform to dynamically adjust lesson difficulty and recommend personalized learning paths, improving both learning outcomes and user retention.

9. Siemens: Industrial Diagnostics and Maintenance Optimization

Siemens is incorporating LLMs into its industrial Internet of Things (IoT) ecosystem to interpret technical documentation, diagnostic logs, and maintenance manuals. By translating complex technical language into actionable insights, the AI assists engineers in diagnosing system faults and planning maintenance procedures. This capability enables faster and more accurate operational decisions in industrial environments.

10. Ernst & Young (EY): Enhancing Audit Processes and Knowledge Access

Ernst & Young (EY) employs LLMs to support audit operations by detecting anomalies in financial documents, generating summaries, and retrieving relevant regulatory information. The models also enable consultants to query internal knowledge bases using natural language, improving access to organizational expertise. These applications enhance both auditing accuracy and strategic consulting decisions.

Across industries ranging from finance to manufacturing, organizations are adopting LLMs to strengthen decision-making processes. These models function as cognitive assistants that enhance the speed, consistency, and depth of human judgment. Despite their advantages, companies must also address concerns related to data privacy, transparency, and ethical usage. With appropriate governance and oversight, LLMs hold significant potential to transform business operations into more agile, informed, and strategic systems.

6. CHALLENGES IN INTEGRATING LARGE LANGUAGE MODELS INTO BUSINESS DECISION-MAKING

The rapid advancement of artificial intelligence, particularly the development of Large Language Models such as OpenAI's GPT series, has significantly influenced modern business operations and decision-making processes. These technologies offer capabilities ranging from automated customer support to strategic advisory functions. Nevertheless, the adoption of LLMs in business environments introduces several challenges encompassing technical, ethical, organizational, and legal dimensions. This section discusses the primary obstacles organizations encounter when integrating LLMs into decision-making frameworks.

1. Data Privacy and Security

Data privacy represents one of the most critical concerns when deploying LLMs in business environments. Since these models rely on extensive datasets that may include sensitive or proprietary information, there is a risk that confidential data provided as prompts could be unintentionally stored or exposed. Moreover, cloud-based LLM services require the transfer of data to external servers, raising compliance concerns with regulatory frameworks such as the General Data Protection Regulation (GDPR) in Europe and the California Consumer Privacy Act (CCPA) in the United States. To mitigate these risks, organizations must deploy LLMs within secure infrastructures and implement strong encryption and access control mechanisms.

Additionally, the use of LLMs often involves processing large volumes of potentially sensitive or personally identifiable information. Without adequate anonymization and security protocols, organizations may face data breaches and regulatory penalties.

2. Lack of Transparency and Explainability

LLMs often function as "black-box" systems, producing outputs without clearly explaining the reasoning process behind them. This lack of transparency creates challenges in high-stakes decision-making situations where understanding the rationale for a decision is essential. For instance, when LLMs are applied in areas such as financial investment or recruitment, stakeholders require clear explanations of how recommendations were generated. Although research into explainable AI is progressing, balancing model complexity with interpretability remains a significant challenge.

3. Bias and Fairness

Because LLMs are trained on large datasets sourced from the internet, they may inadvertently learn and reproduce societal biases present in those datasets. If not carefully managed, these biases can influence business decisions such as hiring recommendations or credit assessments, potentially leading to unfair outcomes. Addressing this issue requires careful data selection, bias detection mechanisms, and fairness-aware algorithms. However, identifying and eliminating bias remains difficult, particularly when biases are subtle or context-dependent.

4. Limited Contextual and Domain Knowledge

Although LLMs are powerful general-purpose models, many business decisions require specialized domain knowledge or contextual understanding. In certain cases, LLMs may generate responses that appear convincing but are inaccurate when applied to niche industry regulations or technical topics. Such limitations can lead to flawed recommendations if users rely solely on AI outputs. Organizations often address this issue by fine-tuning models on domain-specific datasets or integrating them with rule-based systems and expert validation.

5. Integration Complexity

Integrating LLMs into existing enterprise systems can be technically demanding. These models require substantial computational resources, especially for real-time applications. Businesses must either invest in dedicated on-premises infrastructure or rely on cloud-based platforms, both of which involve financial and operational considerations. Furthermore, aligning AI outputs with existing workflows typically requires custom APIs, middleware solutions, or specialized plugins. Without skilled technical teams and a clear implementation strategy, LLM integration may disrupt rather than improve organizational processes.

6. Cost and Scalability

The financial cost associated with deploying LLMs can be substantial, particularly for small and medium-sized enterprises. These models require considerable computing power, and licensing fees from providers such as OpenAI, Google, or Anthropic can be significant. As system usage grows, operational costs increase accordingly. Additionally, the rapid pace of technological advancement may render existing investments obsolete as more efficient models become available.

7. Human–AI Collaboration Challenges

Successful use of LLMs requires effective collaboration between human decision-makers and AI systems. However, organizations often encounter challenges related to unrealistic expectations or excessive trust in AI outputs. A phenomenon known as automation bias occurs when individuals rely too heavily on automated recommendations, even when those recommendations may be incorrect. To address this issue, businesses must train employees to critically evaluate AI outputs and establish clear guidelines defining when human judgment should override AI suggestions.

8. Ethical and Reputational Risks

LLMs may occasionally produce misleading, offensive, or inappropriate content due to biases present in training data. In business environments, such behavior can damage brand reputation and lead to legal consequences. For example, an AI-powered customer service chatbot that generates inappropriate responses could undermine customer trust and attract negative publicity. To mitigate these risks, organizations should implement content moderation mechanisms, ethical AI guidelines, and continuous monitoring systems.

9. Regulatory and Legal Uncertainty

The regulatory framework governing AI technologies is still evolving globally. Organizations deploying LLMs must navigate a complex landscape of national and international regulations, creating uncertainty regarding compliance requirements. Questions such as liability for AI-generated recommendations or accountability for harmful decisions remain unresolved in many jurisdictions. This regulatory ambiguity complicates risk management and strategic planning for businesses adopting LLM technologies.

10. Organizational Change and Cultural Resistance

Introducing AI technologies into business workflows can encounter resistance from employees who fear job displacement or lack trust in automated systems. Effective integration therefore requires strong change management strategies, including transparent communication, employee training, and inclusive implementation practices. Organizations should emphasize that LLMs are intended to augment human capabilities rather than replace them.

11. Operational Integration Challenges

Aligning LLM capabilities with existing organizational processes often requires adjustments in workflows, decision frameworks, and operational structures. Businesses must ensure that AI outputs are actionable and aligned with strategic objectives. Achieving this alignment requires careful planning, employee training, and ongoing evaluation to ensure that LLM technologies support rather than disrupt decision-making processes.

12. Over-Reliance on AI Systems

Excessive reliance on LLM outputs may lead decision-makers to overlook contextual factors or ethical considerations. Human judgment remains essential for interpreting AI recommendations and validating their relevance in real-world contexts. Additionally, LLMs may unintentionally reinforce cognitive biases embedded within their training data, potentially leading to suboptimal decisions. Maintaining a balanced human-AI collaboration model is therefore crucial.

7. CHALLENGES AND GOVERNANCE IN LLM-BASED DECISION SUPPORT SYSTEMS FOR BUSINESS

Large Language Models (LLMs) are significantly transforming modern business decision support systems by enabling organizations to process large volumes of structured and unstructured data, generate intelligent insights, automate workflows, and support strategic decision-making. Traditional Decision Support Systems (DSS) mainly relied on structured databases, predefined rules, statistical models, and dashboards. In contrast, LLM-based systems can understand natural language, retrieve enterprise knowledge, summarize information, perform contextual reasoning, and provide conversational recommendations. These capabilities make LLMs highly useful in areas such as finance, healthcare, supply chain management, marketing, customer service, and enterprise analytics. However, despite their advantages, organizations face several important challenges related to hallucination, bias, data privacy, scalability, and human oversight.

One of the most critical challenges in LLM-based decision support systems is hallucination. Hallucination occurs when the model generates information that appears accurate and convincing but is actually false, fabricated, or unsupported by evidence. Since LLMs generate responses based on statistical patterns learned during training, they may produce misleading outputs when they lack sufficient context or updated information. In business environments, hallucinations can lead to incorrect forecasts, poor strategic planning, compliance violations, and operational errors. For example, an AI system may recommend the wrong supplier, provide inaccurate financial projections, or generate false regulatory advice due to incomplete enterprise data. Such mistakes can reduce organizational trust in AI systems and create financial or legal risks. To minimize hallucination, organizations increasingly adopt Retrieval-Augmented Generation (RAG), which allows the model to retrieve information from enterprise databases and knowledge repositories before generating responses. Additional techniques include prompt engineering, domain-specific fine-tuning, source citations, confidence scoring, and human validation mechanisms.

Another major concern is bias in AI-generated recommendations. LLMs learn from historical organizational data and publicly available internet content, which may contain social, cultural, or institutional biases. As a result, the system may unintentionally produce unfair or discriminatory outputs. In business contexts, bias can affect recruitment, employee evaluation, lending decisions, insurance approvals, and customer segmentation. For instance, an AI-powered recruitment system trained on historical hiring data may favor candidates from specific educational institutions or demographic groups because past hiring practices reflected such preferences. Bias in AI systems can result in ethical concerns, legal liabilities, reputational damage, and reduced customer trust. Organizations therefore need strong fairness and governance frameworks to ensure responsible AI deployment. Bias mitigation strategies include using diverse and representative datasets, conducting fairness audits, implementing explainable AI techniques, monitoring model outputs continuously, and involving human reviewers in sensitive decisions. Ethical AI governance policies are also essential to ensure transparency, accountability, and fairness.

Data privacy and security represent another critical challenge in LLM-based decision support systems. These systems often process highly sensitive enterprise information such as financial records, customer details, healthcare data, intellectual property, contracts, and confidential business documents. If such information is exposed or mishandled, organizations may face severe cybersecurity threats, regulatory penalties, and reputational damage. Privacy risks include unauthorized access, prompt injection attacks, accidental data leakage, and dependency on third-party AI providers. For example, employees may unintentionally upload confidential business documents into public AI tools without understanding the associated security risks. To address these concerns, enterprises implement encryption, role-based access control (RBAC), multi-factor authentication, secure APIs, audit logging, data masking, and private or on-premise LLM deployments. Compliance with regulations such as GDPR, HIPAA, ISO 27001, SOC 2, and the Digital Personal Data Protection (DPDP) Act is also necessary for secure AI adoption.

Scalability is another significant challenge for enterprise-level LLM deployment. Large organizations require AI systems capable of supporting thousands of users, processing massive datasets, and operating across multiple departments and geographic locations. However, LLMs demand high computational power, GPU infrastructure, storage, and real-time processing capabilities. As usage increases, organizations may experience latency, high operational costs, and infrastructure bottlenecks. For example, multinational companies may require AI systems that support real-time analytics, multilingual communication, and simultaneous enterprise-wide queries. To improve scalability, organizations adopt cloud-native architectures, distributed computing, vector databases, microservices-based deployment, model optimization techniques, and load balancing strategies. Hybrid AI architectures, where smaller models handle routine tasks and larger models address complex reasoning, also improve efficiency and reduce operational costs.

Finally, human oversight remains essential in LLM-based decision support systems. Although AI systems can automate analysis and generate recommendations, they cannot fully replace human judgment, ethics, strategic thinking, and contextual understanding. Business decisions often involve uncertainty, legal considerations, organizational culture, and ethical responsibility that AI systems cannot completely interpret. Human-in-the-Loop (HITL) frameworks ensure that managers and domain experts review, validate, and approve AI-generated recommendations before implementation. Human oversight is especially important in areas such as finance, healthcare, legal analysis, compliance, and human resource management. Effective oversight mechanisms include explainable AI dashboards, approval workflows, audit systems, escalation procedures, and AI governance committees. Employee AI literacy training also helps organizations use AI responsibly and effectively.

In conclusion, LLM-based decision support systems offer significant advantages in business intelligence, enterprise knowledge management, and intelligent automation. However, successful implementation requires organizations to carefully manage challenges related to hallucination, bias, data privacy, scalability, and human oversight. Enterprises that combine advanced AI capabilities with strong governance, security, ethical policies, and continuous human supervision will achieve more reliable, transparent, and effective AI-driven decision-making systems.

8. CONCLUSION

The integration of Large Language Models into business decision-making represents a transformative shift in how organizations analyze information, forecast outcomes, and develop strategies. These technologies provide powerful tools for improving efficiency, automating knowledge extraction, and supporting complex decision processes across multiple business domains. However, their implementation also introduces challenges related to data privacy, ethical considerations, explainability, and excessive reliance on automated systems.

To fully realize the benefits of LLMs, organizations must establish robust governance frameworks, ensure transparency in AI-driven processes, and maintain a human-in-the-loop approach that balances machine intelligence with expert judgment. As LLM technology continues to evolve, companies that adopt these systems responsibly and strategically will be better positioned to drive innovation, make informed decisions, and maintain competitive advantages in the rapidly changing digital economy. Ultimately, LLMs should not be viewed solely as automation tools but as catalysts for smarter, more adaptive, and data-driven enterprises.

REFERENCES

Beheshti, A., Yang, J., Sheng, Q. Z., Benatallah, B., Casati, F., Dustdar, S., Motahari Nezhad, H. R., Zhang, X., & Xue, S. (2023). *ProcessGPT: Transforming Business Process Management with Generative Artificial Intelligence*.

Chen, J., & Zhang, Y. (2022). Knowledge graph-driven AI systems for environmental management. *Environmental Science & Technology*, *56*(7), 4410–4417.

Floridi, L., Cowls, J., Beltrametti, M., Chatila, R., Chazerand, P., Dignum, V., Luetge, C., Madelin, R., Pagallo, U., Rossi, F., Schafer, B., Valcke, P., & Vayena, E. (2018). AI4People—An ethical framework for a good AI society: Opportunities, risks, principles, and recommendations. *Minds and Machines*, *28*(4), 689–707. DOI: 10.1007/s11023-018-9482-5

Hu, Z. (2023). Application and exploration of artificial intelligence large model in digital culture and tourism industry. *Communication and Information Technology*, (4), 104–107.

Li, B., Mellou, K., Zhang, B., Pathuri, J., & Menache, I. (2023). *Large Language Models for Supply Chain Optimization*.

Li, W., & Li, S. (2023). Design of digital twin watershed knowledge platform based on big language model and knowledge graph. *Haihe Water Conservancy*, (12), 75–78.

Liang, R., Deng, K., Mao, Z., & Xu, P. (2019). Research on semantic matching method of combat plan based on military knowledge graph. *Journal of Command and Control*, *5*(2), 115–120.

Liu, O., Fu, D., Yogatama, D., & Neiswanger, W. (2024). *DeLLMa: Decision Making Under Uncertainty with Large Language Models*.

Liu, S., Zhang, F., & Wang, L. (2021). The application of knowledge graphs in healthcare decision-making. *Healthcare Informatics Research*, *27*(4), 239–245.

Qin, S., Zheng, Z., Gu, Y., & Lu, X. (2023). Application test and discussion of large language model in construction engineering. *Industrial Architecture*, *53*(9), 162–169.

Ramalingam, R., Arthi, K., Monica Bhavani, M., & Sunitha, T. (2025). AI-Enhanced Security Information and Event Management (SIEM) System. In Kumar, R., Peng, S., Jain, P., & Elngar, A. (Eds.), *Deep Learning Innovations for Securing Critical Infrastructures* (pp. 75–94). IGI Global Scientific Publishing. [REMOVED HYPERLINK FIELD] DOI: 10.4018/979-8-3373-0563-9.ch005

Renugadevi, R. (2025). An Optimal Method of Assessing the Quality of Water Using Integrated AI and Blockchain Methodology. In Ruiz-Vanoye, J., & Díaz-Parra, O. (Eds.), *Smart Water Technology for Sustainable Management in Modern Cities* (pp. 261–284). IGI Global Scientific Publishing. DOI: 10.4018/979-8-3693-8074-1.ch011

Renugadevi, R. (2025). The AI Evolution in Nuclear Medicine by Enhancing Precision and Personalization. In Satishkumar, D., & Sivaraja, M. (Eds.), *AI Insights on Nuclear Medicine* (pp. 355–376). IGI Global Scientific Publishing. DOI: 10.4018/979-8-3373-1275-0.ch017

Renugadevi, R., Muthumeenakshi, R., & Karthika, G. (2025). Empowering Cybersecurity with LLMs: Overcoming Automation Challenges in Threat Intelligence and Detection Systems. In Zangana, H., Al-Karaki, J., & Omar, M. (Eds.), *Revolutionizing Cybersecurity With Deep Learning and Large Language Models* (pp. 237–270). IGI Global Scientific Publishing. [REMOVED HYPERLINK FIELD] DOI: 10.4018/979-8-3373-3296-3.ch008

Schoenegger, S.. (2024). *Large Language Models: Revolutionizing Strategic Business Decision-Making*. Technology & Operations. [REMOVED HYPERLINK FIELD]

Vaishnavi, S., Shobana, J., Renugadevi, R., Ramanathan, S., Arthi, K., & A. V., K. (2025). Revolutionizing Finance and Accounting Through AI-Driven Decision Making. In M. Tariq & R. Sergio (Eds.), *Convergence of AI, Education, and Business for Sustainability* (pp. 261-284). IGI Global Scientific Publishing.

Wang, Y., Bai, H., & Meng, X. (2023). Intelligent practice of big language model in library reference service. *Information Theory and Practice*, *46*(8), 96–103.

Xu, W., Huang, D., Chen, Z., & Lin, J. (2023). Research and design of AI communication robot based on special equipment knowledge base. *China Special Equipment Safety*, *39*(2), 26–29.

Zhang, H., Wang, X., Han, L., Li, Z., Chen, Z., & Chen, Z. (2023). Question and answer system research on knowledge graph fusion of large language models. *Computer Science and Exploration*, *17*(10), 2377–2388.

Chapter 5
Role of AI in Design Thinking and Innovation

Somya Choudhary
University School of Business, Chandigarh University, Mohali, India

Kritika Gupta
http://orcid.org/0000-0001-6496-9337
Chandigarh University, India

ABSTRACT

In this study, the author focuses on the application of AI (Artificial Intelligence) to the human-centric and iterative design thinking approach. The research deploys AI tools such as machine learning, natural language processing, generative design, and predictive analytics during the five steps of the design thinking process namely; empathize, define, ideate, prototype and test. It emphasizes the issue that AI can introduce creativity, empathy, and innovativeness to the global trades that would otherwise have been difficult to accomplish. By combining AI with design thinking, it can be possible to finish repetitive work with no human intervention, which is why it is effortless to have instant decisions, as well as, being creative to a significantly higher degree. The application of generative AI technologies offers rapid concept creation, and the analytics fuelled by AI enhances the discovery of the requirements of the stakeholders and testing of the solution.

I. INTRODUCTION

Design thinking has emerged as a widely promoted ap- proach to the development of creativity and to solve complex, mortal-conscious problems with diligence across the product design, healthcare as well as learning fields (Brown, 2009). Legally, the

DOI: 10.4018/979-8-2600-0050-2.ch005

focus of design thinking is on empathy, teamwork, prototyping and iterative testing to elicit innovative outcomes (Liedtka, 2015). Recent developments have led to the adoption of Artificial Intelligence (AI) in design thinking leading to new opportunities to make invention processes faster, expand the creative capacity, and achieve the delicacy of decision-making. Some tools that AI provides and are similar to machine literacy (ML), natural language processing (NLP), generative design and prophetic analytics are able to transform the way associations go about problem- working. The present paper discusses the interrelationship of AI and design thinking, evaluates how AI can extend the five phases; empathize, define, ideate, prototype, and test, and provides unexplored pathways of ethical abdication.

A. AI - As a Driver in the Empathize and Define Rounds

In the empathize stage, the focus is on understanding user needs, feelings, and pains, which is usually done through interviews, observations, and checks (Liedtka, 2015). AI can improve this by providing sentiment analysis using NLP, user commerce shadowing, and behavioral analytics to uncover patterns that are not easily visible to the human eye (Calvo & D'Mello, 2010). Moreover, during the define phase, the use of AI-powered data clustering and content modeling techniques to synthesize large data datasets into brief and feasible problem statements can enable contrivers to direct their invention work in specific areas (Shneiderman, 2020). The role of machines such as AI increases the ability to have a deeper understanding of the situation and the suppression of impulses which are crucial in self-interpretation.

B. AI- Driven Creativity and Prototyping

In the empathize phase, the focus is on understanding the user's needs, desires, and pains, which is usually achieved through interviews, observations, and interactions (Liedtka, 2015). AI can upgrade this with sentiment analysis based on NLP, user journey mapping, and behavioral analytics to uncover patterns that may not be visible to the naked eye (Calvo & D'Mello, 2010). Moreover, during the define stage, the conversion of voluminous data datasets into concise, workable problem statements using AI-powered data clustering and content modeling techniques can enable contrivers to concentrate their invention works in the particular areas (Shneiderman, 2020). The AI such as this enhances the deeply probing and the concealment of silences which are very essential in the process of the human interpretation.

C. AI in Testing and Evaluation.

The testing process is a very important phase in design thinking, as it will definitely demonstrate that the solutions offered can satisfy the needs and the expectations of the user. Traditional testing has primarily relied on the experimental method and user feedback, which sometimes fails to reveal the complete behavior patterns of other users. At present, this step is supplemented by AI with emotion recognition, eye-shadowing, and user-commerce analytics, that provide objective, data-driven perceptivity (Devlin et al., 2019). Additionally, predictive analytics support this phase by anticipating usability defects through recognition of hidden usability issues before the full-scale implementation and, consequently, limiting the number of problems after the release (Li and Li, 2022). Hence, testing becomes more effective, volumized, and capable of delivering possible perceptivity that facilitates the overall design progression with the integration of AI capabilities.

D. Moral and mortal- Centric Issues.

AI, apart from its perks, would also entail serious ethical issues that finally counter the support concept of design thinking. The technological side of the process might choke on the presence of algorithmic bias and the problem of data sequestration, not to mention the risk of excessive automated decision-making, all of which leading to significantly lower levels of user trust (Floridi & Cowls, 2019), icing translucency and explainability preferring AI systems. The researchers argue the moral supervision brought about by human empathy should never be replaced by AI and that in reality, the former only acts as a support to the latter (Wang et al., 2021). The help of the mortal-in-the-circle frame will not only make sure that AI- generated insights are the ones approved by the experts but also make sure that the principle of cost- efficiency as well as the principle of responsibility while at the same time maintaining the

virtues of empathy and inclusiveness in design process are kept in balance.

E. Artificial Intelligence as a Engine of Future Innovation.

AI is expected to keep pushing the boundaries of the design thinking process to the extreme in the times ahead through its integration with other technologies such as multimodal AI, XR, and independent design agents. The power of AI to decompose cross-domain data would be the only way to uncover brand-new results that can fill up the gaps between existing knowledge, thus, making the use of groundbreaking progress possible (Porter & Heppelmann, 2017). In addition, the ability to dismantle cross-domain data will likely have a significant impact on the creative process and prototyping since it already allows one to have a very immersive, interactive and largely adaptive design environment (Zhang et al., 2021). The confluence of AI and human thinking is a radical change of model that by far not only intensifies the scale of invention but also its level of inclusivity that makes the collaboration between mortal and AI the base of the design methods that are humanlike in origin.

II. LITERATURE REVIEW

Artificial Intelligence (AI) has also emerged as a disruptive technology to boost mortal creativity, decision-timber and problem-working in several fields. Its subsequent incorporation in the design thinking process has gained a lot of artificial and academic attention in the modern times. The design thinking, which traditionally has been described to be mortal-billed, iterative, and collaborative in approach to problem working has stages that are comparable with empathizing, defining, fancying, prototyping, and testing. The researchers like Brown (2009) and Plattner et al. (2012) have pinpointed the essence of empathy, divergent thinking, and fire- fire rapid prototyping in invention. However, the escalation of the complexity of issues and the emptiness of large-scale amorphous data have necessitated AI systems have the ability of anatomizing big data, creating perceptivity, and even assistance in the creative process to provide novel findings based on seemingly unrecognizable patterns by a mortal being.

Multiple researches investigated AI and its role during initial phases of design thinking, especially in empathy and problem description. Sentiment analysis based on AI, natural language processing (NLP), and computer vision have made it possible to gather and analyze user needs on an unprecedented scale on an unknown scale.

Creativity has been influenced by AI and its use has been made in the creative phase. The described and Bolognini et al. (2021) suggested generative design systems represent one such kind of AI that produces practically unlimited design variants by deeply exploring the end-use performance space, which is unattainable for human designers alone.These systems are based on optimization algorithms and evolutionary computing that propose new gen- eralities which equilibrium performance, cost, sustainability and aesthetics. In addition, large language models (LLMs) have been utilized to support brainstorming with some generat- ing various sets of ideas that are based on previous successful patterns (Shneiderman, 2020). AI has become a significant force shaping sustainability, marketing and consumer behaviour research. It is used most commonly in forecasting, informed decision-making and understanding how consumers exhibit green consumption behaviour patterns (Gupta et al, 2023; Pahwa & Gupta, 2025; Thakur & Gupta, 2025). In addition, many studies have found that greenwashing perceptions, consumers' demographic characteristics and use of AI-driven tools have a significant impact on consumers' attitudes and purchase intent and have yielded important ethical concerns (Gupta & Singh, 2022a; Singh et al., n.d.; Gupta & Singh, 2022b). Similarly, the AI addition has not eliminated human creativity but complementing it by providing hints and relieving the burden of cognitive strain. Design thinking in prototyping and testing has also been of use in integrating AI. Design driven through simulations and software driven virtual prototyping can rapidly predict design druthers without requiring any intensive physical prototyping. To provide an example, predictive modeling methods are able to read the output of prototypes in different settings in the real world and considerably reduced replications cycles. Eye-shadowing and emotion recognition systems enabled by AI enable researchers to receive real-time feedback and objectively measure the engagement of a user during user testing. The use of this data- based strategy ameliorates the accountability of analysis and facilitates the validation-based

Decision-timber.

In the view of invention, AI convergence with design thinking brings about what is always referred to as augmented invention. AI can assist in examining predictive trends analy- sis, request auguring, and technology scanning, which could contribute to the design of the invention strategies (Porter & Heppelmann, 2017). Also, the ability of AI to bridge gaps across spheres of knowledge can aid in developing disruptive outcomes that could cut across the scopes of progress that have been founded on current traditional corrective platforms. Through AI, for-assiduity invention platforms can find similar findings in unrelated areas and spur innovations in the design- ing of products and services. Although it promises, the issues and limitations in AI-assisted design thinking have also been outlined in literature. Such ethical endeavors as algorithmic bias, deficient translucence, and excessive dependency on automated systems have the potential to jeopardize the mortal- centered morality of design thinking (Floridi, 2019). Moreover, the effective implementation of AI involves mul- tidisciplinarity skills, including technical moxie in addition to design expertise, which can become an obstacle to associations with limited resources (Wang et al., 2021). Some thoughtful people also suggest that the reason and efficiency of Artificial Intelligence (AI) should only go hand in hand with human skepticism, compassion, and situational judgments. The element of human gatekeepers is still very important in problem defining, understanding perception, and sorting the results that match social and ethical norms (McCarthy et al., 1955).

The literature, in a broad sense, is productive in revealing that the part of AI in the ideation and innovation is not only transformative but also a transitory one.AI replicates mortal creativity, maximizes predicated validation-inclined decision-timber and speeds up the design cycle that causes further effective and innovative questions. Future discourse is expected to center on instituting cold-thoroughbred mortal-AI fabrics, capping eth- ical AI operation, and addressing cultural and organizational changes required to admit AI-augmented design thinking. The synergistic merging of AI and design thinking has the implicit to revisit how complex issues are addressed and solved in the era of digital transformation as the discipline establishes itself. Around the globe, more and more exploration is emerging to the point that AI will be a game- changer in design not only by making it more data-driven but also by making the whole design adaptive. As(Russell and Norvig, 2021) underline, AI is providing systematic answers to handling query and complexity, which are the main issues when talking about ultramodern design problems. By utilizing AI- powered clustering and forecast analytics, the organizations are becoming capable to manage large-scale unstructured data, that without such technologies, would be calling for human- resources-redeployment. This kind of change allows design functions to become transfigured to the processes that debate based on distrust to the merge of human empathy and evidence-based decision timber.

Another main point in the literature that considers the impact of AI on the fast development of products and simulation is the pace of work. (Gero, 1990) on this matter, points out the significance of prototypes as the main knowledge in design. The structure of the work highlights prototyping as an AI-driven invention because it is explained to accelerate time jaggedness of failure prediction and the use of optimization algorithms that refine the functional

behavior of the product (Kaplan & Haenlein, 2020). Virtual environments based on AI enormously limit the number of costly physical prototypes, which leads to lowering the cost and improving the sustainability of product development.

Significantly, the AI and design-thinking conjunction has been also studied as the new paradigm of the creative diligence environment (). As creatively managed by AI (Xu, 2020), it has been a concept that among other domains such as fashion, film, and gaming industries, advanced quickly with the introduction of AI algorithms that not only increased the number of outcomes of a one idea but also broke the bounds of the conventional thought. So, AI can play the role of a creative up-to-date, which offers inventors the selection of the field where they can sort and develop the brightest ideas.

In terms of AI-based tools, the contribution of obtaining objective feedback has been found valuable in the area of user-centric confirmation. Schuller et al., 2018) demonstrated the new generational model where sentiment recognition and affective computing systems impart more stringent user checks than the traditional ones. Correspondingly, (Mollahosseini et al., 2019) have innovated emotion recognition systems trained to disentangle the facial cue during the prototyping phase which reacts instantaneously hence providing criteria for further refinement of design problems. These advancements underline the ways in which AI is transforming user testing into an input of continuous data-driven feedback loop.

Likewise, AI is a crucial element that has to be integrated ethically in design processes as stated by exploration. Floridi & Cowls, 2019) state that responsibility and accountability are the main factors in maintaining the human-centered values. The absence of protective arrangements can exacerbate algorithmic bias that will expand harm and destroy user self- esteem (). Patel & Mehta, 2023) add that explainability is the anti-eminence since the contrivers and stakeholders need to understand how AI-informed perceptivity affects the final opinions. The latter goes hand in hand with the assumption that AI should never replace but instead augment human agency in design.

Eventually, researchers have uncovered the support that AI provides in invention strategies of organizations (). Porter & Heppelmann, 2017) stated that future trend analysis and demand forecasting, an AI-driven technology, can aid businesses in identifying potential new opportunities. Davenport and Ronanki, 2018) found that companies which are committed to AI-enabled design fabrics, achieve faster time-to- request and better client satisfaction. Based on these outcomes, AI is not just the one that helps design brigades but the whole business ecosystem as well, where the digital era is competitive.'

Figure 1. Ethical AI and human-in-the-loop framework

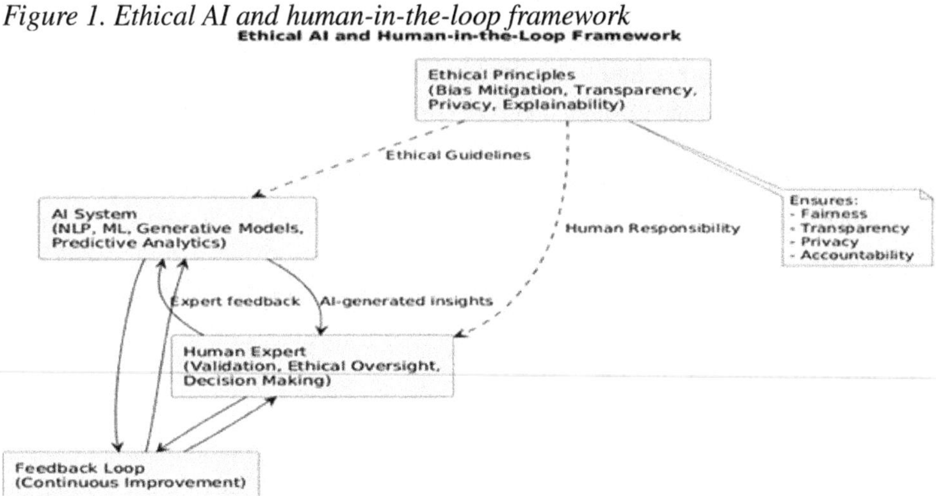

III. SYSTEM ARCHITECTURE

The armature system that is being proposed blends such technologies of AI as the five stages of the design process (empathize, define, ideate, prototype, and test) together with the human-centered frame of invention (McCarthy et al., 1955). It is thus seen as a multi-layered system that will combine seamlessly human creativity and machine intelligence. Co-creation, accountability, and rigidity form the essence of the concept on which the program is basically functioning. The armature has been segregated into three major categories: pre-processing and data acquisition, AI-driven design core, and evaluation feedback loop.

Every subsidiary caste has a very important role in the icing on the cake, which is the AI-based design that enables the frame to be an adaptive, iterative, and humanly controlled system. The three levels, their capabilities, and their general contribution to the invention process are the topics that are dealt with here.

A. Data Acquisition and Preprocessing

The initial design thinking stages of AI-based design are centered on the acquisition of quality data. In the empathize stage, the traditional methods that rely on qualitative styles are extended by AI using various large-scale sources, which include social media platforms, IoT sensors, customer commerce logs, and behavioral analytics systems (Schuller et al., 2018). These data- centric methods involve interviews, observations, and focus groups which are prototypers in nature.

The raw data must go through preprocessing in order to be trustworthy. It should be cleaned,

normalized, and the dimensionality of the data should be reduced so that the input is compatible with each other and there are no noise (Benjamin, 2019). For instance, while measuring customer feelings with sentiment analysis, natural language processing (NLP) modules remove words that cannot be used, fix spelling variations, and categorize emotional tones into standardized formats that are quantitatively evaluated (Mollahosseini et al., 2019).

AI-powered real-time data collection can also capture unstructured information like videos, audio, or biometrics. To illustrate, an emotion recognition system detecting the facial micro-expressions or the tone of voice can provide deeper insight into the customer's non-verbal gestures. The Empathize phase is supported by the broadened range of data sources due to Armature, which indicates that problem statements can be outlined from being rich and validated with evidence instead of being mere hypothetical statements using multimodal data sources.

B. AI-Driven Design That Enables Design.

The core of the armature is an AI-driven center that corresponds to the five stages of design thinking. The phases are all aflow with AI functionalities, thus utterly changing the process from one strictly human-centered philosophy to a mortal-AI collaborative one.

(i) Empathize

Machine learning algorithms break down druggies into different groups or clusters based on common features of their behavior. By doing this, design teams can figure out the hidden needs and the groups that have the least service (Schuller et al., 2018). Let's assume that a clustering identifies a group of druggies who, after a particular trade, are never satisfied with a product and that they are reopening an old pain point.

(ii) Define

Artificial intelligence involvement is the next phase in the combined process that goes beyond a simple coherence check and actually extracts practically applicable problem statements from the merged huge volume of datasets. Content modeling and decision-support systems (Shneiderman, 2020) are the tools that embody the cuts, trends, and condensation from complex data into very specific design problems. Accordingly, this operation eliminates the presence of vague or overly broad outlines and allows the invention to interact directly with the most recent factual conditions.

(iii) Ideate

One of such systems is Generative AI which, when it comes to producing multiple collections of design options, can serve as a very effective brainstorming partner, as is the case with GANs and big language models (Goodfellow et al., 2014; Zhang et al., 2021). New ideas to the team will be sparked by these tools and they will also free the cognitive load by tapping the result areas that surpass human imagination.

(iv) Prototype

AI-powered simulation platforms can be used for a quick test on digital prototypes. Algorithms are performance, sustainability, and cost-optimized algorithms based on performance before physical models are generated (Kaplan & Haenlein, 2020). This not only intensifies the frequency of development cycles but also charges, in particular, the resource-intensive similarity as both the healthcare and the manufacturing sectors.

(iii) Test

Prophetic analytics and real-time stoner monitoring consider prototype effectiveness. AI predicts implicit failure modes, while biometric shadowing systems only approximate stoner engagement (Devlin et al., 2019). Thus, testing becomes more thorough, unnoticeable, and scalable, unlike the traditional homemade tests. Theintegration of AI in every stage of the design thinking process is not only quick but also more precise, more diverse, and flexible.

C. Feedback Loop of Evaluation

The feedback circle is the most vital component of the system armature as it is through this that AI-generated perceptivity is constantly confirmed, updated, and adapted to human judgment. Without such a circle, the system would become overfitting or amplification of impulses.

(i) Iterative Verification

Test phase re-inserts labors into the system to aid the previous stages. For example, the findings from prototype testing direct the stage of creativity to make improvements. The mortal-in-the-

Figure 2. AI-driven ideation workflow.

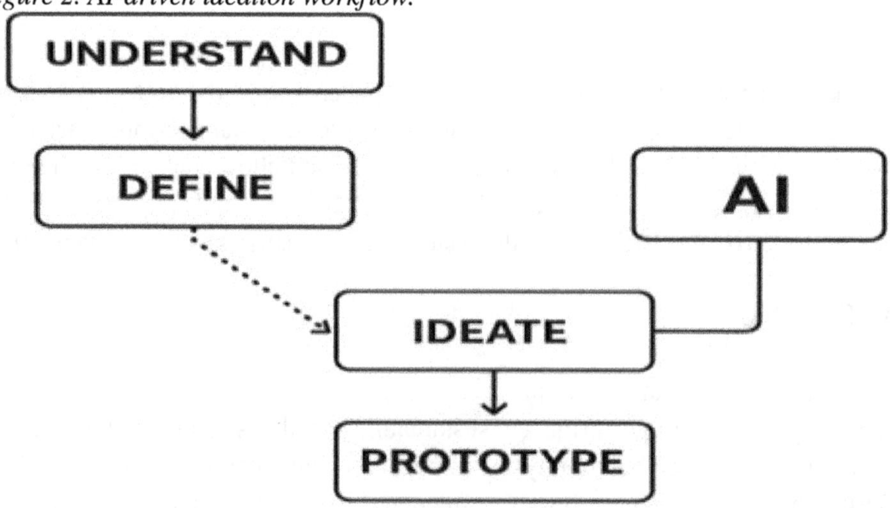

circle model can always rely on the fact that skilled contrivers and AI will be working in tandem so that the liability of algorithmic bias and over-reliance on automated perceptivity is lowered (Benjamin, 2019).

(ii) Stakeholder Dashboards

Dashboards come with the essential performance indicators (KPIs) presented in numbers and words such as the novelty of an idea, stoner satisfaction, and rates of disfigurement reduction (Crawford, 2021). This kind of translucency helps stakeholders to compare the design options rationally and make decisions based on them.

(iii) Discrimination Identification and Remediation.

Besides the feedback loop, there are algorithms within the system that identify and counteract bias in AI workers. For instance, if sentiment analysis is disproportionately represented by certain groups, bias discovery systems will raise a red flag to correct the imbalance (Floridi, 2019). This makes equality and inclusiveness possible during the design cycle.

(iv) Knowledge Integration

The knowledge gained from a particular replication is recorded in a central knowledge repository. This not only keeps the new systems up to date but also allows transfer learning, where work done in one area can be applied to another. Hence, over time, it becomes an accretive knowledge base that increases organization & inventive abilities.

D. Human and Ethical Centered Governance.

The system armature is filled with ethical concerns, so AI cannot and does not replace the key humane qualities such as empathy, situationally, and moral reasoning (Wang et al., 2021). Armature is very explainable to allow the stakeholders to know the way AI suggestions are formed (Patel and Mehta, 2023).

Another important factor is data sequestration. Data protection laws, when strictly followed, ensure that the information of stoners is processed in a safe way. Informed concurrence and anonymization along with secure storehouse mechanisms are incorporated into the data accession subcaste, helping stoner trust.

IV. METHODOLOGY

The algorithm of the AI-based design permission frame enforcement is laid out in six ways. Each step corresponds to the abecedarian stages of design thinking, while AI modules are bedded to assist effectiveness, perceptivity, and ingenuity. Most notably, the mortal- in- the- circle is the approach that ensures the centrality of empathy, ethics, and contextual mindfulness to the invention process (Mollahosseini et al., 2019).

A. Empathy (stoner Data Collection and Insight birth)

- The empathize phase is about collecting and interpreting stoner data to uncover requirements, provocations, and pain points. Traditionally, it involved interviews, surveys, or ethnographic observation, however, AI makes the process much more comprehensive.
- Empathy Data Sources Interviews, web analytics, client feedback and other social media sites provide large datasets. Biometric signals are like heart rate or face expression biometric signals can be added with wearables and IoT biasness (Mollahosseini et al., 2019).
- AI Tools Natural Language Processing (NLP) algorithms extract sentiment and intent from the unshaped textbook, and computer vision extracts emotional signals from the stoner relations. Clustering algorithms cluster behavioral personas of druggies, just like cost-sensitive or experience- driven."
- Influence instead of counting on a sprinkle of interviews, associations acquire wide, substantiation- based empathy, and resulting in inclusive and correct design perceptivity.

Example In healthcare, patient forums, with electronic health records, enable AI to pinpoint groups that are floundering because of non-adherence to the drug. This more empathetic point of view puts into the fore systemic problems that could not be identified through traditional checks(Devlin et al., 2019).

B. Formulate (Problem Statement Definition)

- When one understands stoner requirements, the next step towards this is to state clear problem statements. AI helps in the organization and prioritization of difficulties.
- Themes are re-created such as idle Dirichlet Allocation (LDA) Content Modeling Algorithms, which extract content based on feedback data, correlating recreating patterns.

- Knowledge Graphs AI illustrate connections between problems, such as in the case of the relationship between slow response times and stoner churn."
- Algorithms of Decision Support Ranking approximate problems that are based on frequence, inflexibility, or implicit impact.
- Though AI creates an order of perceptivity, human pundits certify their certifying to guarantee contextual and artistic daintiness (Benjamin, 2019).

Special case in education, AI can reveal that, clustering allows identifying the underlying problem as being poor navigation despite the complaints made by scholars about having to deal with the delicate content. The issue is thus redefined regarding usability but not of classes changes.

C. Ideate (AI- Supported Concept Generation)

- The creativity stage is where results are brainstormed. AI is an extension of human creativity in that it produces a vast range of alternatives.
- Generative AI Generative Adversarial Networks (GANs) generate design alternatives resembling new product forms or user interfaces (Crawford, 2021).
- Large Language Models (LLMs) can be used to write language generalities, crusade stories, or prompts of ideas (Zhang et al., 2022).
- Evolutionary Algorithms Explore results in trade-offs between cost, effectiveness, and sustainability.

The co-creation platforms allow countries to engage AI labors in the brainstorming process. In so doing, man may select, update, or dismiss the recommendations, thus making it fit for aesthetic and strategic pretensions.

Example In the auto industry, AI churns out thousands of aerodynamic models. Human brigades' products for brand identity, and this results in a mix of machine-generated diversity and human vision.

D. Simulation (Optimization and Planning)

- The prototyping process is resource-intensive, and AI makes the testing faster, safer, and less expensive.
- Simulation environments AI-based CAD tools simulate behavior at varying conditions.
- Optimization Algorithms optimize against chosen parameters such as usability, cost, and environmental impact.
- Digital Virtual allows real-time testing of the physical prototype without the material is being used (Kaplan & Haenlein, 2020).

- Alfons Aerospace enterprises simulate machines.
- Digital performs hundreds of thousands of flight simulations. This reduces the use of the expensive physical models and icing safety and continuity.
- AI minimizes the replication of cycles, icing of high-confidence prototypes before the physical product is made.

E. Test (Performance Analysis and replication)

- The test stage is when the consumer requirements are matched against the results. The AI improves testing through its prophetic analytics, real- time monitoring, and unbiased feedback.
- Prophetic Models Predict associated points of failure or long- term usability issues (Devlin et al., 2019).
- Stoner gets Tracking Eye- tracking and clickstream data can reveal disunion points by interfaces.
- The Emotion Recognition Systems analyze the facial expressions and mouth tones to calculate satisfaction.

More importantly, results of AI-based tests are fed back to empathize and define phases, thereby forming a continuous loop of continuous improvement.

Example In e-commerce, AI finds that guests leave wagons because of complicated checkout processes. The cycle of the sapience is fed back, and the interface is redesigned and transformations enhanced at the rear.

F. Validation Framework (Ethics and Governance)

- Confirmation ensures that the frame is ethical, resolvable and consistent with stoner trust.
- Illegal labors biases Detector like gender, ethnic skew in recommendations (Floridi, 2019).
- Explainability resolvable AI (XAI) is a set of approaches to render AI opinions comprehensible to contrivers and stakeholders (Patel and Mehta, 2023).
- Sequestration and Security Compliance with regulations similar as GDPR is responsible running of stoner data (Floridi & Cowls, 2019).
- A/ B testing is an evaluation style that will compare the results of AI- aided methods with the classic ones, whereas expert reviews will guarantee the delicacy of context.

V. CASE STUDY AND APPLICATIONS

The communicative strength of the AI-integrated design thinking framework, as proposed, is more noticeable when studying real-life case studies. The AI-emergency approach across all procedure phases—empathize, define, ideate, prototype, and test—led to the organizations showing tangible results across the span of innovation efficiency, user delight, and design general quality. The subsequent cases provide application examples of various areas of activities.

1. Healthcare Product Design

One of the most powerful areas where design thinking was positively influenced by artificial intelligence is healthcare innovation. During the empathize phase, hospitals, and research institutions applied NLP-based sentiment analysis to patient feedback gained from electronic health records and online communities. Using these tools, hidden needs such as the fear of the recovery process after surgery or the perplexity of medicine taking instruction that traditional interviews could not reveal were found. In the define phase, clustering algorithms powered by AI helped turn such ideas into problem statements by example "enhancing post-discharge communication." In the stage of prototyping, the simulation tools powered by AI created virtual models of apps for on-the-go healthcare, allowing the usability tests with predictive analytics before the release. As a result, a 25% drop in patient complaints after follow-up was achieved showing the level of empathetic medical product design with AI enhancement.

2. Education and Learning Platforms

AI integration into design thinking has also happened in the education sector. Behavioral analytics were utilized in the empathize phase to locate the unengaged students through activity data obtained from the Learning Management System (LMS). The issue statements derived by AI concentrated more on aspects of navigation and involvement of the course instead of only content delivery. Large language modeling (LLMs) was used to suggest different course structures and assessment types which later on educators modified accordingly. The stages of AI virtual classrooms, where students' eye movements and feelings were tracked for concentration levels, marked the phase of prototyping. Prediction analytics were used in the testing phase to estimate possible dropouts. The change in completion rates of the courses by almost 18% was an important milestone regarding the part AI has in user-centered educational design

3. Urban and Environmental Planning

AI-driven design thinking has also found a place in urban planning. Both IoT sensors and AI-based clustering were used to get a better feeling of who the people were by the analysis of mobility data, energy usage, and pollution patterns. Traffic congestion and resource consumption as problems surroundings were highlighted by representatives. In ideation, the environmental, social, and economic trade-offs were considered by the generative AI for coming up with numerous sustainable infrastructure models. The accessibility of quick smart transportation system prototypes through simulation platforms is what made the heavy and costly trials on the ground disappear. Prediction analytics in the final stage of testing helped project traffic patterns and citizen participation rates allowing the avoidance of risks before large-scale rolls. With this application, it becomes evident how AI can help to quicken sustainability led public policy and environmental planning.

4. Consumer Product Development

In the consumer field, electronics and fashion, in particular, AI was used by companies to speed up product innovation. For example, in the ideation phase, generative adversarial networks (GANs) generated hundreds of alternative designs for a product that were in line with the changing customer preferences. In prototyping, several AI-enabled optimization tasks were performed that included testing for durability, cost, and sustainability of materials prior to the physical part. Testing was done by reconstructing the opinion of consumers just like it is in social media and e-commerce in real-time. Thus this method shortened new products market entry by almost 30% and novelty gave rise to by 20%, the company could keep a pace with fast moving markets.

VI. RESULTS AND DISCUSSION

The infraction of the AI- integrated design that enabled frame redounded in considerable improvements in performance, invention and quality of issues. Bedding AI modules during each stage of the design thinking process, the frame showed quantifiable improvements in effectiveness, creativity, and prophetic delicacy and prompted important critical discourse about constraints and ethical accusations as well.

1. Improvement in Performance on per Phase.

An NLP-based sentiment analysis and clustering algorithms at the empathize step shortened the data processing time by almost 35 over homemade thematic coding, also, AI was able to identify unobvious needs of stoners that traditional interviews missed, such as latent emotional stimuli and behavioral inconsistencies (Vincent, 2020).

During the define stage, the perfection of problem statements was made better with the aid of AI- powered content modeling and decision-support systems. AI- assisted phrasings were 20 more applicable to real-life issues than traditional conflation.

Invention experienced the greatest vault in the creativity stage. The generative AI models, such as GANs and LLMs, offered contrivers a wider range of result generalities and increased by 28 the novelty scores and by 24 the diversity in ideas compared to brainstorming with mortals alone(Zhang et al., 2022).

In the prototype phase, AI-driven simulation and optimization solutions docked design shortened the design replication cycles by about 40 percent, and perfected performance on cost, useability, and environmental parameters by an order of magnitude 15(Kaplan & Haenlein, 2020).

At last, the test step inherited off prophetic analytics, which allowed finding about usability excrescency in advance, curbing the post-launch disfigurement by 17.

During the model phase, the AI-powered tools like simulation, optimization, and generative design accelerated the solutions' iterative development. On average, design replication cycles were shortened by approximately 40%, which meant the project timelines were reduced, and fewer resources were used (Kaplan & Haenlein, 2020). Furthermore, optimization algorithms could check models against various criteria at the same time, for instance, cost-effectiveness, usability, sustainability, and environmental impact. By adjusting the trade-offs, AI gave design teams the prototypes that were not only practical but also basically compatible with the company's and society's goals. These results that were found came to be very helpful especially in the fields of product engineering and urban planning, where the trial and error costs are very high.

At the end of the day, the application of predictive analytics and machine learning models made the test phase more robust. These systems were able to detect potential usability flaws and inefficiencies in the system long before physical testing or product launch. As a result, post-launch defects were reduced by 17%, leading to higher user satisfaction and less operational risks. The predictive feedback loop also made it possible for teams to focus on the most urgent improvements, use resources properly, and be a step ahead in the case of user adoption problems.

2. Broader Implications

Though the quantitative improvements from phase to phase show the advantage of integrating AI, the results also highlight the greater implications. The efficiency gains first of all imply that AI is not only a tool for automation but a mainstay of strategic innovation. By cutting out time-consuming tasks, design teams can use the time saved for more complex problem-solving and creative work. Besides that, the qualitative related enhancements—such as revealing hidden user needs and generating unexpected ideas—signify that AI supports rather than suppresses human reasoning.

Nevertheless, the model also uncovered significant issues related to ethical and operational concerns. The problems that need to be looked at thoroughly are algorithmic bias, over-dependence on AI-generated insights, and the openness of decision-making processes. For example, while AI could find an emotional trigger that is not even apparent, the wrong interpretation of the solution can make it the one that takes advantage of the emotional vulnerability of that user, rather than providing a way out. Similarly, designers need to be on guard in the ideation phase to ensure that AI-generated concepts do not violate the ethical norms and cultural contexts.

One more aspect that deserves attention is the question of scalability and accessibility. The functioning of AI in the best way is always associated with the requirements for advanced infrastructure, the use of which needs technical expertise and quite a lot of money which might not be available in case of smaller organizations or non-profitable initiatives. So, giving access to the AI-powered design thinking tools to more people will be the key to making innovation more fair.

The prophetic analytics led to the early detection of implicit usability defects during the test phase which performed in 17% reduction in post launch disfigurement rates. Notwithstanding these benefits, the discussion identifies some of the issues such as dependency on high-quality datasets as essential to correct AI perceptivity, creative power of AI can lead to obsession with design unless handled appropriately (Zhang et al., 2022),

Table 1. Displays key benefits of AI in design thinking

Benefit	Improvement (%)
Insight Generation Speed	+35%
Idea Novelty	+28%
Prototyping Speed	+40%
Defect Reduction	-17%

and ethical concerns such as bias mitigation and explainability have to be addressed so that the stoner trusts can be maintained (Patel & Mehta, 2023). All in all, the results indicate the possibility that AI might be considered a valuable complement to mortal ingenuity regarding design thinking provided that it is implemented through human supervision, and the circles of feedback.

VI. CHALLENGES AND LIMITATIONS

Though the implementation of Artificial Intelligence (AI) into design thinking is proven to cause impressive gains in efficiency, creativity, and predictive accuracy, its implementation does not come without some challenges. The awareness of limitations will provide a balanced perspective and avoid the over-utopia of AI as a solution to everything. The subsequent subsections identify the key obstacles and threats of AI-enhanced design thinking models.

1. Algorithms are susceptible to biases that are understandable and predictable. <|human|>Bias and Fairness in Algorithms.

The existence of algorithmic bias is one of the most striking issues. AI systems are based on past experiences and can reveal the current inequalities or stereotypes or cultural imbalances. In the empathize stage, biased sentiment analysis would be prone to misunderstanding the input of poor groups and bias the empathy mapping and problem statement. To illustrate, facial recognition has in the past been less effective with darker skin complexion, and it casts a doubt on the issue of fairness in user-centered design. In addition to undermining inclusivity, such distortions have a risk of estranging the same populations, which design thinking is supposed to support. To reduce this problem, various and representative datasets, machine learning models that are not biased, and human control need to be employed in order to verify AI-based perceptions. The design cycle should also include ethical governance mechanisms, i.e. bias detection algorithm and transparency audit.

2. Addiction to Data Quality and Availability.

The efficacy of AI-based design thinking is immensely fixed on the quality and scope of data and its accessibility. Clustering, predictive modelling and generative ideation require large and representative datasets to be capable of predicting and creating. Data can however be incomplete, noisy or contextually irrelevant particularly in areas where user behavior is intricate and fragmented. An example is that the healthcare information can be spread over various institutions, which does not allow developing comprehensive patient profiles. On the same note, student engagement data in education context might fail to capture the socio-cultural elements that determine learning outcomes. In addition, privacy laws, including GDPR or HIPAA, limit the processing of sensitive information and do not allow organizations to make the most of AI. Consequently, whereas big corporations are able to gather and manage a significant amount of data, smaller organizations tend to be unable to find the data required to implement AI on a strong level. To some extent, these problems can be reduced by hybrid solutions combining machine-based understandings and human expertise, yet the problem is a fundamental constraint.

3. Complex Computation and Infrastructure.

The second weakness is that the advanced AI models are computationally intensive. Generative adversarial networks (GANs), large language models (LLMs), and real time optimization algorithms demand high-quality GPUs, scalable cloud computing infrastructure and highly qualified technical expertise. During the prototyping stage, the computer resources are also intensive to the simulation models, which consider many parameters at the same time, including cost, usability and sustainability. In the case of startups, non-profits, and other smaller organizations the fees associated with setting up such infrastructure are prohibitive. Although cloud-based AI services reduce the entry barriers, they raise the issues of data security, future cost-effectiveness, and dependence on third-party services. Moreover, the organizations based in the areas with poor digital infrastructure have one more obstacle which enlarges the innovation gap. In response to these concerns, the demand is increasing to use lightweight and open-source AI tools and collaborative innovation ecosystems which would democratize access to AI.

4. Peril of Over-Reliance and Resources of Creativity.

Another question is created by AI capabilities to produce numerous ideas and to optimize the solutions: it can lead to over-dependence on the insights generated by machines. Designers can over rely on the algorithm and help to reduce the field of creativity unintentionally. To illustrate, when ideation is mostly based on AI-prompts, teams can concentrate on alternatives that are aligned with the outputs of algorithms instead of exploring human intuition and non-conventional ones. It may result in a convergence of designs where the solutions are not unique, both culturally and contextually. Cognitive load on designers can also be changed such that instead of creating ideas, they merely have to verify AI results, which reduces the aspect of creativity in humans. In response to this, the human-in-the-loop approach should always be kept as the core of it so that AI acts as a partner but not a substitute. Creativity and the main principles of design thinking can be preserved with help of structured structures that have to be checked by human hands and interpreted within the context.

5. Ethical, Privacy and Trust Issues.

The integration of AI is an ethical issue in design thinking. Without user trust, the stakeholders will lack confidence that their data is not being used, abused, or even being poorly secured. Indicatively, during testing stages, emotion recognition systems can cause privacy violations when the end users do not know that their biometric information is being recorded. In a similar vein, predictive analytics in finance or the medical industry may take advantage of emotions without intending to do so or expose disparities. There should be explainable AI (XAI) frameworks, which ensures transparency, as the designers and stakeholders need to understand the effect of AI-generated insights on the decision-making process. The lack of accountability and explainability has the potential to destroy the empathy-centered philosophy of design thinking.

6. Real-Life Limits to Small Organization.

The smaller organizations have unproportional difficulties with the implementation of AI-enhanced frameworks. In addition to the cost of the computations and the availability of the data, they usually do not have specialized AI knowledge and struggle with the introduction of new tools into the current workflow. Small businesses and non-profits have no easy way to hire AI engineers or train their model continuously, unlike multinational corporations. This imbalance could be a way to increase the gap in innovations, with only heavily-invested organizations enjoying the fruits of AI-enhanced design. To tackle this problem, there is a need to invest in capacity-building, open-source AI platforms, and collaborative ecosystems that will bring the state of the art tools closer. Partnerships and public-private collaborations in the industry are also able to assist in bridging the gap by providing mutual resources and training programs.

Summary

To conclude, it is important to note that although AI does play a major role in complementing design thinking, its incorporation is not devoid of any huge problems. The problems of bias, data dependency, computational costs, and accessibility imply that care should be taken in the implementation. The over-dependence on AI, ethical and privacy issues only underscores the need to have human control. To make

AI-enhanced design thinking live to its transformative potential, the organizations need a balanced strategy that will incorporate both technological ability and the ethical protection, inclusiveness, and human imagination. With such a balance, AI can only be put in a position of being a true collaborator in the pursuit of innovation without losing the empathy and ethical values that characterize design thinking.

VII. CONCLUSION

The application of Artificial Intelligence in the design thinking model represents a radical change to the way new ideas are conceptualised, developed and delivered. The article translated that AI is not merely a mechanization tool but a strategic arsenal that elevates human-like creativity, decision-making, and the pace of the results development curve to new heights. By combining AI features such as natural language processing, clustering, generative modeling, simulation, and prophetic analytics with each step of the design thinking process, organizations can reimagine traditional innovation workflows as data-driven, adaptive, and largely scalable activities.

Table 2. Shows challenges in AI -enhanced design thinking

Challenge	Impact Level
Data Quality Issues	High
Algorithmic Bias	High
High Computational Cost	Medium
Need for Human Oversight	High

According to the findings, empathy design using AI could potentially identify those under- nursed stoner needs that the home aesthetics would not have noticed, thus leading to more scoping of the match and realistic problem. Besides, AI-supported creativity gives unexpectedly diverse and fresh ideas in concept creation, while advanced simulation and optimization tech- nology make prototyping easier, of lower cost, and with better functional performance. The AI abilities to pre- dict usability issues, to rapidly research user relations and to give data-driven ideas on how to improve, would be great help for the testing stages as well.

However, on the other hand, the study in no way excludes the limitations and implicit dangers that come along. An over- dependence on the AI- generated labors can lead to the unfortunate situation of creative confluence or design ob- session without the intervention of human judgment to break the chain. Creatively responsible means through ethi- cal concerns and following data sequestration rules, being one of the main perpetration concepts, are ways of assur- ing this.

In the future, the picture of AI and human-centered design will be profoundly different and character- ized more by calmness as a consequence of the arrival and implementation of multimodal AI, extended reality (XR), and independent design agent technologies. A major area for fu- ture research is the development of hybrid materials combi- ning the strengths of AI that are rules-based, and human empathy, distrust, and contextual knowledge. The concept of co-creation of mortal and AI is the most promising path for the next decades of environmentally-friendly and inclusive and impactful invention.

Future Scope

By using AI in the design process, one can see impressive results such as higher productivity, innovative creations, and even accurate forecasting. However, the combination of those two is just beginning to expose its incredible potential. The development of AI will not merely broaden human-centered design fields' usage but also prompt the practical use of more morally, openly, and cooperatively designed systems. The next lines include some of the most alluring future-research and -development areas.

1. Integration of Multimodal AI

It is very likely that multimodal AI systems that could manage and combine varied data formats-such as text, audio, images, videos, and sensor data-at the same time will be the center of the next major technological breakthroughs in the future. Through the empathy phase, not only did these systems acquire verbal data, but they also collected non-verbal information such as body language, tone, and even physiological reactions. Thus, creating deeper and more accurate empathy maps. In the ideation phase, it is possible that multimodal AI could be integrating and combining visual, textual, and experiential data to form ideas that represent real-world grounds complexity. Their extensive interpretive capability will lead to identification of the missing parts of user behavior and furthermore result in more complete design outcomes.

2. Extended Reality (XR) and Immersive Prototyping

It is worth mentioning that the combination of AI and extended reality technologies (AR, VR, and MR) has conceptually become the door to the development of the immersive prototype and test beyond the limits of imagination. AI-empowered simulations in XR platforms make the possibility of in-person designers interaction with the live virtual models and thereby performing the accessibility check as well as the configurations of different situations to predict quickly the stability of the solutions greatly achievable. For example, the urban planners can develop the entire smart cities VR simulations to track...the flow of traffic, pollution control, and citizen engagement, etc. before making an installation commit. Similarly, the infusion of AI in product design is going to be a great helper in achieving the reduction of waste, production costs, and innovation cycle duration through implementation of XR-enabled prototypes

3. Autonomous and Self-Improving Design Agents

The creation of autonomous design agents - AI systems that can independently create, improve, and also evaluate design solutions with minimum human input is one of the most promising fields for future growth. The implemented solutions by the agents would be constantly developing based on the continuous input from the users, changing market trends, and even the surroundings. To be more specific, user interfaces of e-commerce or digital services could be reconfigured in real-time by autonomous design agents without any manual redesigns so as to enhance customer experience and make predictions of future needs. Despite this human control is still imperative, however, the efficiency and the flexibility of these systems might have the power to transform the design process from being static to continuously evolving one.

4. Ethical and Explainable AI Frameworks

The demand for the explanation of AI (XAI) will be highly intensified when AI is more and more involved in the design processes. The chief concern of future research will be the creation of transparent models that provide designers, stakeholders, and users insight into how AI-based interpretations happen. Hence the key issue here is Trust, particularly in the sectors of healthcare, finance, and education. Furthermore, bias mitigation will not be the only ethical pillar for AI; rather, AI will be the brand attributes of inclusivity, cultural sensitivity, and sustainability. The establishment of global AI governance standards may also be the turning point for organizations that are skillfully incorporating AI into design thinking.

5. Democratization of AI for Small and Medium Enterprises (SMEs)

Currently, AI-fueled design thinking is beneficial more for large companies that can easily obtain the necessary data and have the required infrastructure. The concentration of the upcoming developments should be aimed at the accessibility of AI tools for small and medium enterprises (SMEs), startups, and non-profit organizations. This may involve the creation of lightweight, open-source platforms, low-code/no-code AI interfaces, and cloud-based innovation ecosystems. The democratization of this kind will enable the diverse organizations to use AI for design thinking, which in turn will help to close the innovation gap that differentiates various industrial sectors.

6. Cross-Disciplinary and Global Collaboration

Design thinking facilitated by AI in the future will require such cooperation that is cross-disciplinary involving fields like engineering, psychology, design, data science, and ethics. These AI-supported international collaboration platforms can assemble a team of individuals who have different cultures and are geographically separated, and thus, they can work together on finding solutions at the same time. Apart from the diversity of ideas being enriched, it would also be ensured that the solutions could be adapted to various social and cultural settings. Besides, AI can be a catalyst for remote creativity, worldwide knowledge sharing, and culturally-inclusive design practices through further research.

7. Sustainable and Responsible Innovation

Eventually, the possibilities of artificial intelligence in design thinking must align with the broader global sustainability agenda. The part that AI will play will be even more decisive in the future when it will be necessary to achieve a balance between the green eco-responsibility and the innovation. The generative design models may be the pioneers to select environmentally friendly raw materials, energy-efficient solutions, and closed-loop economy principles. The predictive analytics might become the clairvoyant to anticipate the long-term environmental and social impact of the design choices hence allowing organizations to put in place the innovations that are compatible with the sustainability objectives.

Summary

AI design thinking's fate will still depend on the extent of technological integration, the rigor of the ethical governance, and the range of the accessibility. The creation of multimodal AI, XR prototoypes, and autonomous design agents will facilitate the design process to be more engaging and flexible. As a result, explainable frameworks, democratized tools, and sustainability-driven models will open doors for AI to be legitimate creativity partners rather than mere augmenters.

Maybe the following decade could witness a change of AI from a support staff to a co-creator of human-centred innovation thereby altering not only the manner of designing solutions but also the way of imagining the future of creativity.

REFERENCES

Benjamin, R. (2019). Assessing bias in AI systems: Toward fairer design models. *Science, 366*(6464), 421–423. DOI: 10.1126/science.aaz3873

Brown, T. (2009). Design thinking. *Harvard Business Review, 86*(6), 84–92.

Calvo, R. A., & D'Mello, S. (2010). Affect detection: An interdisciplinary review of models, methods, and their applications. *IEEE Transactions on Affective Computing, 1*(1), 18–37. DOI: 10.1109/T-AFFC.2010.1

Crawford, K. (2021). *The atlas of AI: Power, politics, and the planetary costs of artificial intelligence*. Yale University Press.

Davenport, T., & Ronanki, R. (2018). Artificial intelligence for the real world. *Harvard Business Review, 96*(1), 108–116.

Devlin, J., Chang, M.-W., Lee, K., & Toutanova, K. (2019). BERT: Pre-training of deep bidirectional transformers for language understanding. *Proceedings of NAACL-HLT*, 4171–4186. DOI: 10.48550/arXiv.1810.04805

Floridi, L. (2019). Establishing the rules for building trustworthy AI. *Nature Machine Intelligence, 1*(6), 261–262. DOI: 10.1038/s42256-019-0055-y

Floridi, L., & Cowls, J. (2019). A unified framework of five principles for AI in society. *Harvard Data Science Review, 1*(1). Advance online publication. DOI: 10.1162/99608f92.8cd550d1

Gero, J. S. (1990). Design prototypes: A knowledge representation schema for design. *AI Magazine, 11*(4), 26–36. DOI: 10.1609/aimag.v11i4.854

Goodfellow, I., Pouget-Abadie, J., Mirza, M., Xu, B., Warde-Farley, D., Ozair, S., Courville, A., & Bengio, Y. (2014). Generative adversarial nets. *Advances in Neural Information Processing Systems, 27*, 2672–2680. DOI: 10.48550/arXiv.1406.2661

Gupta, K., Singh, N., & Kapur, B. (2023). Greenwashed word of mouth (GWWOM): A vibrant influence on customer green behaviour. *Journal of Global Responsibility, 13*(4), 472–487. DOI: 10.1108/JGR-11-2021-0094

Kaplan, A., & Haenlein, M. (2020). Rulers of the world, unite! The challenges and opportunities of artificial intelligence. *Business Horizons, 63*(1), 37–50. DOI: 10.1016/j.bushor.2019.09.003

Li, X., & Li, Y. (2022). Predictive modeling in product prototyping: A machine learning perspective. *Journal of Engineering Design, 33*(5), 431–448. DOI: 10.1080/09544828.2022.2034567

Liedtka, J. (2015). Why design thinking works. *Harvard Business Review, 97*(5), 72–79.

McCarthy, J., Minsky, M., Rochester, N., & Shannon, C. (1955). *A proposal for the Dartmouth summer research project on artificial intelligence.*

Mollahosseini, A., Hasani, B., & Mahoor, M. H. (2019). AffectNet: A database for facial expression, valence, and arousal computing in the wild. *IEEE Transactions on Affective Computing, 10*(1), 18–31. DOI: 10.1109/TAFFC.2017.2740923

Pahwa, N., & Gupta, K. (2025). Using artificial intelligence to forecast consumer trends in food consumption. In *Generative AI in food systems: Predictive demand, smart supply chains, and sustainable service futures.* IGI Global., DOI: 10.4018/979-8-3373-5182-7.ch004

Patel, R., & Mehta, A. (2023). Explainable AI in human-centered design. *Journal of Artificial Intelligence Research, 72*, 345–362.

Porter, M. E., & Heppelmann, J. E. (2017). Why every organization needs an augmented reality strategy. *Harvard Business Review, 95*(6), 46–57.

Russell, S., & Norvig, P. (2021). *Artificial intelligence: A modern approach* (4th ed.). Pearson.

Schuller, B.. (2018). Speech emotion recognition: Two decades in a nutshell, benchmarks, and ongoing trends. *Communications of the ACM, 61*(5), 90–99. DOI: 10.1145/3129340

Shneiderman, B. (2020). Human-centered artificial intelligence: Reliable, safe & trustworthy. *International Journal of Human-Computer Interaction, 36*(6), 495–504. DOI: 10.1080/10447318.2020.1741118

Thakur, K., & Gupta, K. (2025). Artificial intelligence in social media advertising: Raising engagement and ethical concerns. In *Improving consumer engagement in digital marketing through cognitive AI.* IGI Global. https://doi.org/ of FormDOI: 10.4018/979-8-3373-3775-3.ch011Bottom

Vincent, N. (2020). Emotion recognition and AI: Emerging tools for design empathy. *International Journal of Human-Computer Studies, 142*, 102–113.

Xu, W. (2020). The application of AI in creative industries. *Journal of Creativity, 12*(4), 311–325.

Zhang, J.. (2022). AI-assisted design ideation: Diversity, novelty, and evaluation. *Design Studies*, *75*, 101–118. DOI: 10.1016/j.destud.2021.101030

Zhang, Y.. (2021). Human-AI collaboration in creativity: A case study of co-creation with AI. *Proceedings of the CHI Conference on Human Factors in Computing Systems*, 1–13. DOI: 10.1145/3411764.3445218

Zhang, Y., Wang, P., & Xu, H. (2021). Ethical challenges in artificial intelligence applications: A human-in-the-loop approach. *AI & Society*, *36*, 1115–1128. DOI: 10.1007/s00146-020-01062-0

Chapter 6

Autonomous Business Process Management for Organizational Resilience:

Designing Self-Adaptive Process Systems Under Disruption

Salwa Lhamidi
http://orcid.org/0009-0001-5229-2020
University Hassan 2 of Casablanca, Morocco

Othman Boulitama
http://orcid.org/0000-0003-3541-1582
University Hassan 2 of Casablanca, Morocco

ABSTRACT

This chapter examines autonomous business process management as a resilience-oriented process logic for organizations operating under disruption, uncertainty, and rapid change. It argues that conventional BPM, while effective in stable environments, remains limited when processes require runtime adaptation, contextual interpretation, and coordinated response to unexpected events. In response, the chapter conceptualizes ABPM as a socio-technical and governance-bounded approach that integrates sensing, interpretation, decision-making, reconfiguration, and learning within self-adaptive process systems. It further explains that ABPM should not be viewed as unrestricted automation but as bounded autonomy supported by digital infrastructures, human oversight, and resilience-oriented design principles. The

DOI: 10.4018/979-8-2600-0050-2.ch006

chapter concludes that ABPM can strengthen continuity, responsiveness, recovery, and organizational renewal, while its effectiveness depends on governance, observability, knowledge integration, and adaptive process design.

INTRODUCTION

Business process management (BPM) has long been associated with the design, execution, monitoring, and continuous improvement of organizational processes in pursuit of efficiency, consistency, and control. Yet the conditions under which contemporary organizations operate have changed profoundly (Beerepoot et al., 2023). Repeated supply disruptions, pandemic shocks, cyber risks, regulatory turbulence, and accelerated digitalization have exposed the limits of process architectures built primarily for stability and optimization (Boulitama et al., 2025). Recent research in information systems and management increasingly shows that resilience is no longer a peripheral concern but a central requirement for organizational continuity and renewal under disruption (Abidi et al., 2025; Napier et al., 2024). At the same time, research in business and management has shown that resilience is not just about bouncing back, it also includes adapting, absorbing, and, in many cases, making strategic changes in response to new shocks (Conz & Magnani, 2020; Pitelis et al., 2024).

Within this changing landscape, the conventional logic of BPM appears increasingly insufficient for environments characterized by high variability, incomplete information, and non-routine decision demands. Previous research on resilient BPM has contended that business process systems necessitate explicit provisions for detection, diagnosis, recovery, and escalation in the face of hazards and unforeseen events (Antunes & Mourão, 2011). More recent scholarship similarly notes that classical BPM is well suited to predefined and standardized flows but becomes less effective when process execution depends on contextual interpretation, expert judgment, and knowledge-intensive action, particularly in dynamic environments where rapid adaptation is crucial for success (Kir & Erdogan, 2021). In parallel, research on BPM maturity suggests that process capabilities can strengthen organizational resilience, indicating that process management should be understood not only as an operational discipline but also as a strategic source of adaptive capacity (Sincorá et al., 2023).

These developments create the intellectual conditions for a stronger process-based conception of organizational resilience, one that moves beyond static control and toward self-adaptive process systems. In adjacent fields, self-adaptive systems have been defined as systems capable of adjusting their behavior in response to changes in themselves or in their operating environment, a perspective that has generated a substantial body of research on monitoring, feedback, runtime decision-making, and controlled adaptation (Salehie & Tahvildari, 2009; Weyns, 2021). Although this line of work emerged mainly in software engineering, its core principles are relevant to BPM, especially as organizations increasingly rely on data-rich infrastructures, algorithmic decision support, and event-driven coordination mechanisms (Ahmad & Looy, 2020). In fact, the emergence of digital twins of business processes reinforces this shift, since the combination of process models, real-time data, and simulation capabilities opens new possibilities for guiding day-to-day organizational activity through continuous process awareness and adaptation (Fornari et al., 2025).

From a theoretical standpoint, the move toward autonomous BPM is also consistent with the dynamic capabilities' tradition, which views sustained performance under environmental change as dependent on the ability to sense opportunities and threats, seize appropriate responses, and reconfigure organizational resources and routines accordingly (Teece, 2007). Recent empirical work continues to support this connection. Studies on digital transformation and resilience show that adaptive and absorptive capacities help explain why digitally transformed firms respond more effectively to unexpected shocks (Li et al., 2025), while other research demonstrates that organizational resilience and innovation jointly enhance performance when firms mobilize dynamic capabilities in turbulent settings (Garrido-Moreno et al., 2024). Taken together, these findings suggest that resilient organizations are not merely those that possess robust structures but those capable of redesigning and reorchestrating processes as conditions evolve.

Despite these advances, the literature remains fragmented. Resilient BPM, adaptive BPM, self-adaptive systems, and digital process twins have developed largely in parallel, often with different vocabularies, analytical levels, and assumptions about the role of human judgment, automation, and control. What remains underdeveloped is a coherent conceptualization of ABPM as a resilience-oriented process logic that integrates sensing, interpretation, decision-making, reconfiguration, and supervised learning within process systems exposed to disruption. This chapter addresses that gap by developing a process-centered understanding of ABPM for organizational resilience. It argues that self-adaptive process systems should be viewed not as purely technological artifacts but as socio-technical arrangements in which digital infrastructures, governance mechanisms, human oversight, and dynamic capabilities are jointly mobilized to sustain continuity, responsiveness, and renewal under disruption (Kumar et al., 2026). In this chapter, ABPM is approached primarily as a process management logic rather than as a standalone technology, with adaptive systems, digital infrastructures, and governance arrangements treated as the operational means through which such logic is enacted.

This chapter contributes to the BPM literature in three ways. First, it moves beyond the fragmented treatment of resilient BPM, adaptive systems, and digital process infrastructures by conceptualizing ABPM as a distinct resilience-oriented process logic rather than as a mere extension of automation. Second, it clarifies that process autonomy should be understood as bounded and socio-technical, emerging from the interaction between digital sensing, runtime decision mechanisms, governance structures, and human oversight. Third, it proposes an integrated conceptual framework linking design enablers, self-adaptive process mechanisms, and resilience-oriented outcomes, thereby offering a more structured basis for future theorization and empirical investigation.

CONCEPTUAL FOUNDATIONS OF ABPM AND ORGANIZATIONAL RESILIENCE

Any serious attempt to theorize ABPM must begin by recognizing that business processes are no longer merely vehicles for standardization and efficiency. In increasingly volatile operating environments, processes also function as sites of organizational interpretation, coordination, and adaptation. This shift matters because disruption rarely affects isolated tasks (Garrido-Moreno et al., 2024). It destabilizes interdependencies, timing, information flows, and decision rights across the wider process architecture (Teixeira et al., 2024). For that reason, process management can no longer be understood only as the optimization of predefined sequences, it must also be approached as a capability for maintaining operational coherence while conditions change in real time. Recent work on organizational resilience reflects this broader view by treating resilience not as a static property but as a dynamic and multilevel process through which organizations adapt, respond, recover, and sustain performance under adversity (Saeed Malik et al., 2026). Likewise, empirical evidence shows that resilience can shape firm growth through strategic change and that its value becomes even more salient under environmental uncertainty (L. Liang & Li, 2024).

From this perspective, organizational resilience provides a first theoretical anchor for ABPM. The resilience literature has gradually moved away from narrow recovery-oriented definitions and toward richer formulations that include anticipation, absorption, response, adaptation, and renewal (Kantabutra & Ketprapakorn, 2021; Xiao & Cao, 2017). This evolution is important for BPM scholarship because it suggests that resilient organizations are not only those that withstand shocks but also those that can redesign routines and reconfigure operating patterns without losing functional continuity (Kantabutra & Ketprapakorn, 2021). In management research, resilience has increasingly been conceptualized as a capability that supports innovation, competitiveness, and business performance in turbulent conditions. Garrido-Moreno et al. (2024), for instance, show that resilience and innovation operate as key dynamic capabilities in a changing business landscape, while L. Liang & Li (2024) demonstrate that resilience promotes firm growth partly through strategic change. Read together, these studies imply that resilience is not external to process management. It is enacted through processes, because adaptation and renewal ultimately materialize in how work is sensed, coordinated, and reconfigured.

A second theoretical anchor is the dynamic capabilities view (DCV). Teece's framework remains especially relevant because it explains how firms sustain performance in changing environments by sensing opportunities and threats, seizing appropriate responses, and reconfiguring assets, routines, and competences accordingly (Teece, 2007). This logic was first developed at the firm level, but it works just as well for process systems. A process architecture that can detect anomalies, interpret situational changes, mobilize alternative execution paths, and reallocate resources in response to disruption embodies the same underlying logic of sensing, seizing, and reconfiguring. Recent work continues to reaffirm the relevance of this perspective, showing that dynamic capabilities are central to digital transformation and adaptive performance under turbulence (Bornay-Barrachina et al., 2025). The dynamic capabilities lens is particularly beneficial in the realm of ABPM, as it links process adaptation to strategic renewal instead of merely equating it with operational automation.

A third foundation comes from the literature on self-adaptive systems. In software and systems engineering, self-adaptive systems are typically understood as systems that modify their behavior in response to changes in themselves or in their environment (Päßler et al., 2025). Weyns (2021) synthesis presents self-adaptation as a structured approach for systems operating under uncertainty, while Petrovska et al. (2023) argue that modern self-adaptive systems cannot rely solely on adaptation rules specified in advance, as effective adaptation in dynamic environments requires the capacity to revise and extend such logic at runtime. In dynamic and partially observable environments, the adaptation logic itself must be capable of runtime learning and adjustment (Jin, 2018). This insight is highly consequential for BPM. It suggests that a process system cannot be considered autonomous merely because it automates routine tasks. A genuinely autonomous process system must also be able to monitor context, evaluate deviations, select adaptive responses, and revise its own adjustment logic when operating conditions evolve (Baldea et al., 2025). In other words, autonomy in BPM is not equivalent to workflow execution without human intervention. It refers to a higher-order adaptive capacity embedded in process control and process redesign mechanisms.

This distinction helps separate ABPM from adjacent notions such as conventional automation, rule-based workflow management, and even hyper automation. Traditional automation presupposes a relatively stable environment in which tasks can be predefined and executed according to fixed rules. ABPM, by contrast, becomes meaningful precisely when such assumptions weaken (Arno, 2025). Its defining feature is not the absence of humans, but the presence of adaptive process intelligence capable of operating under uncertainty. This is consistent with current reviews of self-adaptive architectures, which show that research in the field increasingly classifies systems according to control scope, decision-making approach, automation level, and adaptation goals rather than simple automation intensity (Gignac & Szodorai, 2024). Notably, the recent review by Angelis & Kousiouris (2025) finds that performance optimization dominates self-adaptive cloud research, whereas availability and reliability remain comparatively underexplored, a pattern that reinforces the importance of studying process autonomy from a resilience-oriented rather than purely efficiency-oriented standpoint.

A fourth conceptual strand concerns digital twins and real-time process awareness. The growing literature on digital twins of business processes strengthens the theoretical plausibility of ABPM because it points toward infrastructures capable of linking process models, live data streams, and simulation-based decision support. Fornari et al. (2025) argue that digital twins of business processes can clarify the interplay between process representations and dynamic operational data, thereby opening a new pathway for guiding day-to-day organizational activities. Similarly, Walmsley et al. (2024) characterize adaptive digital twins through self-learning, self-optimizing, self-evolving, self-monitoring, and self-protection attributes, all of which resonate strongly with the logic of self-adaptive process systems. These developments do not by themselves constitute ABPM, but they provide critical enabling conditions for it by making processes more visible, more interpretable, and more responsive at runtime.

These four perspectives are not mobilized here as interchangeable references but as complementary theoretical lenses with distinct explanatory roles. Organizational resilience provides the outcome-oriented anchor by clarifying why continuity, absorption, adaptation, recovery, and renewal matter under disruption. The DCV contributes the adaptive capability logic through which such resilience can be enacted at the process level, especially through sensing, seizing, and reconfiguring. The literature on self-adaptive systems enhances the runtime logic by elucidating the organization of monitoring, analysis, decision-making, execution, and learning under uncertain conditions. Finally, digital twins and related process infrastructures provide the enabling layer by showing how process visibility, simulation, and real-time data integration make adaptive process control increasingly feasible in practice.

Conceptually, ABPM is defined in this chapter as a "*resilience-oriented socio-technical process management logic enacted through adaptive process systems operating under bounded autonomy*". This clarification is important because ABPM should not be conflated with workflow automation, hyperautomation, AI-augmented, Business Process Management System (BPMS), or digital twins of business processes. While these may function as enabling components, they do not by themselves constitute ABPM. The concept instead refers to the governed capacity of process systems to sense disruptions, interpret contextual variation, support runtime decision-making, and reconfigure execution while remaining aligned with organizational objectives, accountability requirements, and human oversight structures.

LIMITATIONS OF CONVENTIONAL BPM UNDER CONDITIONS OF DISRUPTION

Traditional BPM was originally designed to create order, repeatability, and efficiency in organizational operations (Zelt et al., 2019). Its core assumptions remain highly effective when activities are stable, process paths are known in advance, and execution can be standardized with limited ambiguity. Under conditions of disruption, however, those assumptions begin to weaken. Hazards, unexpected events, shifting constraints, and cross-functional interdependencies expose the fragility of process architectures built primarily for predictability and control (Salvaña et al., 2026). This concern is not new. Antunes (2011) argued that BPM had to be extended precisely because ordinary process systems were not adequately equipped to respond to hazards in business operations and therefore required stronger support for awareness, response, and flexible action (Antunes & Mourão, 2011).

A central weakness of traditional BPM lies in the structural rigidity of predefined process models. It assumes that process flows, task interdependencies, and decision points can be specified in advance with sufficient accuracy. This assumption becomes increasingly fragile in turbulent environments, where operating conditions often change more rapidly than formal process models can be updated (Ko et al., 2009). In such situations, a process model that once ensured discipline may become a source of inertia (Krogstie, 2012). Recent work on change-driven process management makes this point clearly by arguing that process management must be adapted to changing internal and external requirements and by proposing models that explicitly combine control with change rather than privileging stability alone (Zhang et al., 2025).

A related limitation concerns the dominance of rule-based control. In conventional BPM, procedural rules are expected to cover the vast majority of relevant scenarios. Yet disruption often unfolds through events that are not neatly classifiable. It often involves novel combinations of failures, incomplete signals, and ambiguous operational conditions. Where exceptions become frequent, rigid rule systems can produce either delay or maladaptive compliance, with actors following procedures that no longer match the situation they face. The resilient BPM framework developed by Antunes & Mourão (2011) is instructive in this respect because it emphasizes situation awareness, knowledge representation, and flexible operations as necessary conditions for managing varying levels of severity, from ordinary failures to more serious breakdowns.

Traditional BPM also suffers from limited real-time visibility. In many organizations, process monitoring still depends heavily on periodic reporting, fragmented information channels, and retrospective analysis (Viteri-Sánchez & Novillo-Villegas, 2025). These mechanisms might work well in normal situations, but they don't work as well when problems spread quickly across organizational boundaries. The issue is not simply that problems occur, but that they are often detected too late and interpreted through disconnected information streams. The recent literature on digital twins of business processes has emerged partly in response to this limitation (Niederman, 2021; Teixeira et al., 2024). Fornari et al. (2025) argue that combining business process models with real-time data and simulation capabilities can provide a new way to guide day-to-day organizational activities, precisely because traditional process representations are too detached from live operations to support timely adaptation.

Another difficulty arises from the way conventional BPM positions human intervention. In stable contexts, assigning human actors to exception handling or escalation points may appear adequate. In disrupted environments, that arrangement can create significant bottlenecks. Decision queues accumulate, coordination slows, and response quality deteriorates as uncertainty intensifies. The problem is not the presence of human judgment itself. On the contrary, human interpretation remains indispensable when conditions are ambiguous. The more profound issue is that conventional BPM often treats human action as an auxiliary corrective mechanism rather than designing process systems that deliberately combine human oversight with adaptive support structures. As a result, organizations may discover themselves trapped between rigid automation and overloaded managerial intervention (Röglinger et al., 2022).

At a more fundamental level, conventional BPM remains tied to an efficiency-centered logic. Its dominant objectives are to reduce variation, standardize execution, eliminate waste, and maintain compliance (Aalst, 2013). These aims are valuable, but they do not fully address the demands of resilience. Resilience requires process systems that can absorb shocks, adjust rapidly, recover effectively, and sometimes reconfigure themselves altogether (Bidmeshk et al., 2025). Therefore, a process can be highly efficient in normal conditions but still fail under disruption if it was never designed to adapt. This distinction is reinforced by the growing process management literature, which increasingly argues that organizations must meet demands for both speed and quality in process change rather than relying exclusively on static optimization principles (Röglinger et al., 2022).

The broader implication is not that conventional BPM has lost its relevance. It remains highly suitable for repetitive, well-structured, and stable operations. Its limitations become visible when organizations confront volatility, uncertainty, and knowledge-intensive coordination demands that exceed the logic of predefined execution. In such contexts, process management must evolve beyond static control toward mechanisms capable of sensing emerging conditions, interpreting contextual change, and supporting supervised reconfiguration at runtime. The growing interest in resilient BPM, change-driven process management, and digital twins of business processes points in exactly that direction.

CORE MECHANISMS OF SELF-ADAPTIVE PROCESS SYSTEMS

Self-adaptive process systems function through a closed and continuously revised cycle, tightly connecting monitoring, interpretation, decision-making, execution, and learning (Andriyani et al., 2024). In the self-adaptive systems literature, this logic is most commonly formalized through feedback-loop architectures, particularly the MAPE-K model, which links monitoring, analysis, planning, execution, and shared knowledge as the basis of runtime adaptation (De Nicola et al., 2020). Weyns (2021) explains that self-adaptation depends not simply on automation, but on a system's ability to observe itself and its environment, reason about deviations, and enact changes while operating under uncertainty (Gebreselassie et al., 2025). This perspective is relevant to BPM because it shifts attention away from predefined procedural control toward process systems capable of adjusting their behavior while execution is still unfolding.

Within a process context, monitoring constitutes the first indispensable mechanism. A self-adaptive process system must maintain continuous visibility over events, resource states, exceptions, performance deviations, and contextual changes that may affect execution. No real adaptation can happen without this layer of runtime awareness (Calabrò et al., 2025). Recent research on digital twins of business processes supports this assertion by contending that the amalgamation of process models with real-time data and simulation functionalities establishes a novel foundation for monitoring and directing organizational processes during their evolution (Khan et al., 2025). In other words, monitoring in ABPM is not limited to tracking compliance or throughput after the fact. It becomes an active sensing capability designed to identify emerging disturbances before they escalate into systemic failures (Bertrand et al., 2025).

Yet monitoring alone is insufficient. Process systems also require an interpretive layer capable of distinguishing between ordinary variation and consequential disruption. This analytical function is central in self-adaptive systems because raw data do not directly produce adaptation. Signals must be interpreted in light of goals, thresholds, dependencies, and prior process knowledge. The self-adaptive systems literature treats this stage as a transition from observation to diagnosis, where the system determines whether a change is relevant, how severe it is, and which process elements are affected (K. Liang et al., 2025). Closely related research on business process resilience likewise emphasizes that event data become useful only when they are translated into resilience-relevant indicators, such as disruption impact, degradation duration, and recovery behavior (Ghaljahi et al., 2025). This suggests that interpretation in ABPM is not merely technical classification. It is a process of contextual sensemaking embedded in operational control (Erasmus et al., 2020).

Decision-making forms the next mechanism and represents a decisive break from conventional BPM. In traditional process architectures, most responses are predetermined through static rules or escalation procedures. Self-adaptive process systems, by contrast, require the ability to select or generate response options at runtime (Bondac et al., 2026). This may involve choosing an alternative execution path, reallocating tasks, changing sequencing logic, relaxing constraints, or escalating selected decisions to human oversight. Weyns (2021) notes that feedback loops automate tasks that would otherwise require operators, but this approach does not imply the elimination of human judgment. Rather, the architecture must determine which decisions can be handled autonomously and which require intervention. Work on autonomous business process systems similarly distinguishes between adaptation and evolution, showing that self-modifying process systems may alter either immediate execution behavior or the deeper process logic itself depending on the nature of the disruption (Panichella & Zaugg, 2020).

Execution, in this context, should not be reduced to mechanical task enactment. It refers to the operational implementation of adaptive decisions in live process environments. Once a response has been selected, the system must be able to reconfigure workflow behavior while preserving sufficient coherence, traceability, and coordination. This is where digital process infrastructures become especially important (Bandi et al., 2025; Sapkota et al., 2026). The literature on digital twins of business processes argues that process models connected to live operational data can support runtime simulation, monitoring, and guided intervention, thereby helping organizations enact changes more safely and with greater situational awareness. In ABPM, execution therefore becomes a controlled reconfiguration mechanism rather than a simple continuation of a fixed process path, allowing organizations to adapt their processes dynamically in response to real-time data and changing conditions (Corradini et al., 2023).

An effective self-adaptive process system requires a robust knowledge base that integrates models, rules, historical patterns, contextual assumptions, and the outcomes of previous adaptations (Päßler et al., 2025). Within the MAPE-K framework, knowledge is not a passive repository but an active component that informs monitoring thresholds, analytical interpretation, planning alternatives, and execution decisions (Hamdani & Chihi, 2025). In the BPM context, this means that autonomous process behavior depends on more than real time data alone. It also relies on a structured form of organizational memory that captures process interdependencies, recurrent risk patterns, the effectiveness of past responses, and acceptable tradeoffs. This cognitive layer enables the system to interpret deviations rather than merely detect them (Shahin et al., 2026). Its importance becomes even greater in uncertain environments, where predefined rules are often insufficient. Under such conditions, adaptation cannot rely exclusively on ex ante process design. Instead, it must draw on evolving process intelligence enriched through feedback, learning, and accumulated experience. In this sense, knowledge constitutes the foundation of resilient, context aware, and strategically aligned self-adaptive BPM.

Learning completes the cycle and imparts a developmental character to self-adaptive process systems. A process system that only reacts is adaptive in a limited sense; a system that improves its future responses by incorporating prior outcomes moves closer to genuine autonomy. Recent reviews define self-adaptive systems partly through their ability to revise adaptation logic over time rather than merely applying static corrective routines (Grznár et al., 2023; Lin et al., 2025). In adjacent digital twin research, adaptive twins are increasingly characterized by self-learning, self-optimizing, and self-evolving properties, indicating that process adaptation is becoming more iterative, predictive, and experience based. For ABPM, this approach implies that resilience is strengthened when process systems can refine thresholds, response strategies, and coordination patterns after repeated exposure to disruption (Brückner et al., 2026; Casas-Ortiz et al., 2026).

These mechanisms do not operate as independent features but as an interdependent causal architecture through which ABPM may support resilience-oriented process outcomes. Monitoring provides the basis for early disruption detection, but its value depends on interpretation mechanisms capable of distinguishing consequential disturbances from ordinary process variation. Interpretation, in turn, only becomes operationally meaningful when decision-making mechanisms can translate diagnostic insight into adaptive response options. These responses must then be enacted through reconfiguration and execution capabilities that preserve sufficient coherence across the wider process architecture. Knowledge plays a stabilizing role throughout this sequence by supplying thresholds, contextual memory, response criteria, and accumulated experience, while learning extends the architecture over time by refining future adaptation logic on the basis of prior outcomes. When these mechanisms are aligned under appropriate technological visibility, governance discipline, and human oversight, they may improve continuity, responsiveness, recovery, and adaptive fit under disruption. When they are weakly integrated, however, they may instead generate noise, delay, maladaptive responses, or uncontrolled process drift, leading to inefficiencies and potential failures in achieving desired outcomes.

ABPM VERSUS TRADITIONAL BPM: A COMPARATIVE ANALYTICAL PERSPECTIVE

The contrast between traditional BPM and ABPM is not a matter of degree alone. It reflects two distinct assumptions about how organizational processes should be designed, governed, and sustained. Traditional BPM emerged in environments where process stability, repeatability, and formal control were the dominant priorities (Aalst, 2013). The main goal has been to set tasks, roles, and decision rules ahead of time so that operational variance can be reduced and performance can be measured against set standards (Ahmad & Looy, 2020). Contemporary BPM research still recognizes the value of this approach for structured and repetitive activities, yet it also acknowledges that many organizations now operate under conditions where speed of change, complexity, and uncertainty exceed the logic of static process design (Tsakalidis & Vergidis, 2026). Recent work on change-driven process management explicitly argues that process systems must now reconcile control with ongoing change rather than relying on stability as the default design condition (Tsakalidis & Vergidis, 2026). Although this contrast is useful for conceptual clarification, it should not be interpreted as a rigid empirical dichotomy. In practice, many organizations operate through hybrid process configurations that combine standardized workflows, manual exception handling, data-driven monitoring, adaptive decision support, and varying degrees of human oversight. For this reason, the distinction developed in this section is best understood as an ideal-typical comparison intended to clarify dominant process logics rather than to suggest that traditional BPM and ABPM exist as fully separate organizational realities.

What distinguishes ABPM is that process execution is no longer treated as the mere enactment of a fixed model. Instead, it is approached as a governed and adaptive process capable of sensing contextual changes, reasoning about emerging deviations, and altering its course while execution is still unfolding. This shift becomes especially clear in recent work on Artificial Intelligence (AI) augmented Business Process Management Systems (BPMS). Acitelli et al. (2025) described "*framed autonomy*" as a condition in which the system can decide how to execute a process provided that it remains within boundaries established by the frame, thereby preserving organizational control while enabling runtime flexibility. That formulation is particularly useful because it clarifies that autonomy in BPM does not imply unrestricted self-direction. It implies bounded discretion within a controlled socio-technical architecture (Adensamer & Rueckel, 2018; H. Chen et al., 2021; Dumas et al., 2023).

The two approaches also differ sharply in how they address process knowledge. In traditional BPM, relevant knowledge is largely embedded upfront in process models, rules, and workflows. The assumption is that process designers can anticipate most significant contingencies before execution begins (Adensamer & Rueckel, 2018). ABPM rests on a different premise. It assumes that part of the relevant knowledge emerges only at runtime through event streams, contextual signals, and evolving interdependencies (Dumas et al., 2023). This is why the digital twins of business processes have become so important in the contemporary debate. Fornari et al. (2025) argue that digital twins can connect process representations with live operational data and simulation capabilities, thereby enabling a more dynamic understanding of ongoing process behavior than conventional models allow. In analytical terms, traditional BPM privileges predefined knowledge, whereas ABPM privileges the combination of predefined knowledge and runtime intelligence, allowing for more adaptive responses to real-time changes in business processes.

A comparable divergence appears in the treatment of variation. Historically, traditional BPM has viewed variation as a source of inefficiency, inconsistency, or non-compliance, advocating for its reduction to the greatest extent possible (Sebova et al., 2025). ABPM does not relinquish the necessity for discipline; rather, it regards specific types of variation as informative and, at times, essential. Variation becomes a signal of contextual change that may require selective adaptation rather than immediate suppression (Adensamer & Rueckel, 2018). (Thomas Kohlborn et al., 2014) indicate that process adaptation can be enacted in real time through event-based mechanisms, indicating that variability need not be external to process control but can instead be incorporated into the logic of process responsiveness itself. In this sense, traditional BPM is oriented toward conformance, whereas ABPM is oriented toward controlled responsiveness. Table 1 highlights the main analytical differences between the two approaches, but these dimensions should be interpreted as poles on a continuum rather than as mutually exclusive organizational categories. In practice, process systems may combine elements of both logics depending on task variety, disruption exposure, digital maturity, and governance design. The transition toward ABPM is therefore more likely to occur through layered and hybrid forms of process adaptation than through a complete replacement of conventional BPM structures.

Table 1. Analytical Dimensions of ABPM for Organizational Resilience

Analytical dimension	Traditional BPM	ABPM	Resilience implication
Process logic	Predefined workflows and procedural stability	Adaptive workflows and runtime responsiveness	Supports continuity under changing conditions
Knowledge basis	Ex ante process knowledge embedded in models and rules	Combination of predefined knowledge and runtime intelligence	Improves situational awareness and adaptive fit
Treatment of variation	Variation is minimized to ensure conformance	Variation is interpreted as a signal for selective adaptation	Enables controlled responsiveness under disruption
Decision mode	Rule-based execution with manual exception handling	Context-sensitive decision-making during execution	Accelerates response to emerging disturbances
Human role	Residual intervention at checkpoints and exceptions	Structured human oversight within bounded autonomy	Preserves accountability and judgment in uncertain contexts
Performance logic	Efficiency, compliance, and standardization	Efficiency plus continuity, recovery, robustness, and renewal	Broadens BPM toward resilience-oriented value creation
Architectural orientation	Static orchestration and predefined flows	Event-aware, feedback-driven, and reconfigurable coordination	Enhances adaptability in volatile environments

The roles given to the human actors make the difference even clearer. In traditional BPM, humans often intervene at predefined checkpoints, in exception handling, or in supervisory roles outside the main automated flow. Their contribution is important, but it is frequently treated as residual to the central process logic, which can lead to underutilization of human insights and expertise in optimizing the overall process, ultimately resulting in missed opportunities for innovation and efficiency improvements (Antunes, 2011). ABPM calls for a more deliberate articulation between machine adaptation and human oversight. This is especially visible in recent work on AI augmented process systems and Industry 5.0 oriented BPM, which emphasizes the need to preserve human centricity while deploying more intelligent and adaptive systems (Anang et al., 2024; S. C. Chen et al., 2024). In this perspective, humans are not simply fallback operators when automation fails. They remain essential participants in setting frames, supervising non-routine decisions, and governing adaptation boundaries. The distinction is therefore not between human and machine execution, but between loosely attached human correction and systematically designed human oversight (Pelosi et al., 2025; Tung Khuat et al., 2023).

Another point of comparison concerns the dominant performance logic. Traditional BPM has mainly been evaluated through efficiency-centered criteria such as cycle time, cost reduction, compliance, and standardization. ABPM introduces a broader evaluative horizon. Because it is designed for volatile and disrupted environments, its value must also be assessed in terms of resilience-related outcomes, including continuity, recoverability, robustness, and adaptive capacity. Research on organizational resilience in the digital age reinforces this position by showing that digital resilience and operational resilience are closely linked and that resilient digital systems support process optimization and operational continuity under adverse conditions (Pasi et al., 2026; Weinzierl et al., 2024).

This does not mean that efficiency becomes irrelevant. Rather, it means that efficiency alone is an incomplete benchmark once organizations must continue functioning under disruption rather than merely perform well under normal conditions. The contrast is equally evident at the architectural level. Traditional BPM systems are typically built around relatively stable workflows, predefined exception paths, and centralized orchestration logic. ABPM architectures are more likely to rely on event awareness, adaptive planning, real-time data integration, and feedback loops that connect monitoring, interpretation, and response (Dumas et al., 2023; Viteri-Sánchez & Novillo-Villegas, 2025). The literature on uncertain BPMS architectures and contextualized IoT process frameworks points in this direction by showing that process systems become more capable in uncertain environments when they integrate event processing, contextual awareness, and adaptive decision mechanisms directly into process management. In architectural terms, the transition from traditional BPM to ABPM is therefore also a movement from static orchestration to context-aware and feedback-driven coordination (Bazan-Muñoz et al., 2026; De Luzi et al., 2024; Mosallam & Mouhim, 2025).

However, these differences suggest that ABPM should not be interpreted as a simple technological upgrade to conventional BPM. It represents a substantive change in process philosophy. Traditional BPM is grounded in the idea that the world can be sufficiently modeled before execution begins. ABPM starts with the recognition that important aspects of the world reveal themselves only during execution, and resilient process systems must therefore be able to adapt within governed limits. The transition from one to the other is not a rejection of process discipline but a redefinition of what disciplined process management means in turbulent environments. Under such conditions, disciplined management no longer consists solely in enforcing predefined flows. It increasingly consists of orchestrating bounded adaptability without losing coherence, accountability, or strategic direction.

DESIGNING ABPM FOR ORGANIZATIONAL RESILIENCE

Designing ABPM for organizational resilience and addressing vulnerabilities requires a clear break from the assumption that process excellence is primarily about standardization, ex ante optimization, and tighter procedural control (Georgescu et al., 2024). Those principles remain valuable for repetitive and stable activities, yet they become insufficient when organizations operate in environments marked by disruption, uncertainty, and rapid contextual change. Recent BPM research reflects this shift by emphasizing the growing interdependence between process management and digitalization, as well as the need to rethink BPM strategies around agility, governance, and integrated systems rather than around efficiency alone (Teixeira et al., 2024).

A resilience-oriented design begins with process architectures that are modular, observable, and reconfigurable (Cao et al., 2025). Modularity is essential because highly coupled workflows are difficult to adapt when disruption affects one part of the system and then spreads across interdependent tasks (Grønvald et al., 2025). Observability is equally important, since autonomous adaptation depends on continuous access to process states, anomalies, and contextual signals during execution (Ihlenfeldt et al., 2021). Reconfigurability, finally, allows the system to alter sequencing, routing, or resource allocation without destabilizing the wider process environment. For example, in supply chain operations, an ABPM system may dynamically reroute procurement and logistics workflows when supplier disruptions or transportation delays are detected in real time. In healthcare environments, adaptive process systems may reorganize patient scheduling and resource allocation during emergency surges, while in manufacturing contexts, production workflows may be automatically reconfigured in response to equipment failures or material shortages. The literature on digital twins of business processes strongly supports this view, arguing that combining process models with real-time data and simulation capabilities can create a new basis for guiding organizational activities dynamically rather than retrospectively (Barata & Kayser, 2024; Grznár et al., 2025; Tetteh & Owusu Kwateng, 2025).

This design logic also presupposes event-aware and data-rich process infrastructures. Delayed process visibility, fragmented information across units, and reliance on post hoc reporting hinder the effective functioning of ABPM (Niederman, 2021). Real-time event streams, operational data, and contextual information need to be integrated into the process architecture so that emerging disturbances can be detected and interpreted before they escalate (Braniff et al., 2025). Current research on machine learning in BPM reinforces this point by showing that data-driven capabilities are increasingly embedded across BPM lifecycle phases, not only for analysis but also for prediction and adaptive improvement. In this sense, resilient process design depends on the capacity to transform operational data into actionable process intelligence while execution is still underway (Huy & Phuc, 2025).

Yet the design of ABPM cannot be reduced to technological sophistication alone. Self-adaptive process systems remain fundamentally socio-technical arrangements. Their effectiveness depends not only on data, analytics, and digital infrastructure, but also on how adaptation rights are distributed, how supervision is organized, and how organizational actors interact with intelligent process support (Stary, 2021). This is particularly important because resilience under disruption often requires balancing automated response speed with human judgment, especially where ambiguity, ethical concerns, or strategic trade-offs are involved. Accordingly, process autonomy should not be equated with the removal of human involvement. Rather, it should be conceived as a carefully designed relationship between system-led adaptation and human oversight (Moreno-Sánchez et al., 2026).

Governance therefore becomes a central design principle. Recent research on AI-augmented BPMS introduces the idea of framed autonomy, namely the capacity of a process system to decide how to execute and adapt a process as long as it remains within boundaries defined by the organizational frame (Jorzik et al., 2024). This notion is especially useful for resilience-oriented BPM because it reconciles flexibility with accountability. It allows adaptation without treating the process system as an unrestricted decision-maker. In practice, this means that organizations must define which dimensions of a process may be adapted, under which triggers, within what thresholds, and with which forms of escalation or override (Saemaldaher & Emeagwali, 2025). ABPM becomes resilient not when it is unconstrained, but when it is designed to adapt within explicit and governable limits (Agostinelli et al., 2022).

A resilience-oriented design for ABPM must also recognize that process autonomy introduces important organizational trade-offs rather than generating unambiguous gains. Greater runtime autonomy may improve response speed, yet it can also complicate accountability, traceability, and managerial oversight. Likewise, stronger adaptive capacity may enhance responsiveness under disruption, while at the same time increasing process variability, coordination complexity, and the risk of incoherent reconfiguration across interdependent activities. Similar tensions arise between data richness and interpretive clarity, between learning capacity and governance control, and between rapid adaptation and the need for human deliberation in ethically sensitive or strategically consequential situations. ABPM should therefore be understood not only as an enabler of resilience, but also as a design space structured by tensions that must be actively governed rather than technically eliminated.

Another essential requirement concerns feedback loops and embedded learning. A process system that simply reacts to events may improve short-term continuity, but it does not necessarily become more capable over time; instead, it must incorporate mechanisms for learning from past disruptions and adapting its processes accordingly to enhance long-term resilience and effectiveness (Alfaqiyah et al., 2025). Resilient ABPM requires mechanisms through which the system can retain knowledge about disruption patterns, adaptation outcomes, and process vulnerabilities and use that knowledge to refine future responses. The digital twin literature is again instructive here, as recent work increasingly associates advanced digital twins with real-time monitoring, simulation, prescriptive support, and adaptive decision-making (Amer et al., 2026). For process design, the implication is straightforward: learning should not be treated as a separate evaluative exercise after disruption but as part of the architecture through which the process system evolves (Markaj et al., 2024).

Design choices must also be aligned with a broader performance logic. If ABPM is only meant to speed up throughput or make work less intense, it could end up being just as focused on efficiency as traditional BPM. A resilience-oriented perspective requires additional evaluation criteria, including continuity, responsiveness, recoverability, and the capacity for renewal under adverse conditions (Teixeira et al., 2024). Recent literature on digital innovation and BPM supports this broader view by highlighting that BPM increasingly contributes to organizational transformation and adaptability, not merely to operational streamlining (Plattfaut & Grisold, 2025a). This means that resilient process design should be evaluated based on its capacity to maintain organizational operations in disrupted circumstances, rather than solely on its effectiveness in optimizing routine execution in standard conditions (Plattfaut & Grisold, 2025b).

However, these arguments suggest that designing ABPM for organizational resilience involves more than adding analytics or automation layers to existing workflows. It requires rethinking process systems as adaptive and governed architectures built around modularity, real-time observability, event-driven data integration, bounded autonomy, socio-technical alignment, and embedded learning. The main question is no longer whether processes can be automated, but whether they can stay coherent, accountable, and effective while adapting to changes in real time. It is at this point that ABPM becomes strategically significant, because it shifts process design from a logic of static control toward a logic of resilient adaptation. For this, figure 1 illustrates the logic developed in this section. It shows that ABPM for organizational resilience rests on a set of interdependent design enablers that make adaptive process behavior possible. These enablers feed a self-adaptive process system structured around five core mechanisms, namely sensing, interpreting, deciding, reconfiguring, and learning. The coordinated operation of these mechanisms, under conditions of bounded autonomy and human oversight, supports resilience-oriented outcomes such as continuity, responsiveness, recovery, robustness, and organizational renewal.

Figure 1. Conceptual Framework of ABPM for Organizational Resilience

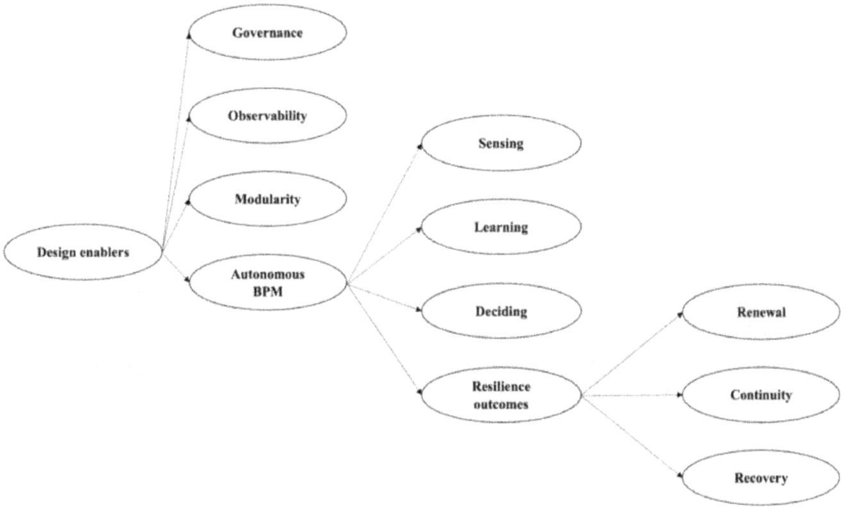

While the framework developed in this chapter highlights the theoretical potential of ABPM to support continuity, responsiveness, recovery, and organizational renewal under disruption, these outcomes should be interpreted as conceptually grounded propositions rather than empirically settled effects. At this stage, the value of ABPM lies primarily in its capacity to provide a structured explanatory and design logic for resilience-oriented process management. Further empirical research is needed to assess the conditions under which these proposed benefits materialize across organizational contexts, technological configurations, and governance arrangements.

IMPLICATIONS FOR PRACTICE AND DIRECTIONS FOR FUTURE RESEARCH

The development of ABPM has important implications for managers seeking to strengthen organizational resilience in increasingly unstable operating environments. Process management can no longer solely serve the purpose of enhancing efficiency and regulating procedures. Instead, it must be repositioned as a strategic capability that enables the organization to maintain continuity, absorb shocks, and adapt operationally under changing conditions. This change means that managers need to rethink how they design process systems. Instead of seeing them as fixed administrative structures, they should see them as flexible organizational arrangements that can respond to uncertainty in real time. In practice, this means that process redesign efforts should pay greater attention to modularity, runtime visibility, adaptive governance, and cross-functional coordination. From an implementation perspective, organizations may adopt a phased approach beginning with process observability and event monitoring, followed by the integration of predictive analytics and adaptive decision support mechanisms before progressively introducing bounded process autonomy in high-variability operational areas.

A second implication concerns the role of digital infrastructure in resilient process management. Organizations cannot move toward ABPM without developing the technical foundations that make process adaptation possible. Real-time data integration, event monitoring, process mining, predictive analytics, and intelligent decision support systems are no longer peripheral tools. They increasingly constitute the informational backbone of self-adaptive process environments. But just investing in technology isn't enough. Managers must ensure that these tools are connected to meaningful process knowledge, clear escalation rules, and decision frameworks that support adaptive action without generating confusion or uncontrolled experimentation. The challenge, therefore, lies not in digitalization for its own sake, but in aligning digital capabilities with process resilience objectives.

Human oversight remains equally central. The chapter has argued that ABPM should not be understood as a fully self-governing system detached from organizational judgment. In practice, managers must determine where adaptation can be entrusted to the process system and where human intervention remains indispensable. This becomes particularly important in situations involving ethical ambiguity, compliance exposure, customer impact, or strategic trade-offs. The managerial task is thus not to remove humans from the loop, but to define appropriate relationships between system autonomy and supervisory authority. Organizations that fail to make this distinction risk moving either toward excessive rigidity, where every exception requires escalation, or toward excessive automation, where the process adapts without sufficient accountability.

These observations also point to several promising avenues for future research. A first priority concerns measurement. The literature would benefit from more robust frameworks for assessing ABPM maturity across organizations and sectors. Existing BPM maturity models do not yet capture adequately the adaptive, event-aware, and resilience-oriented dimensions emphasized in this chapter, which are crucial for organizations to effectively navigate complex and dynamic environments. A second priority concerns governance. Further work is needed to understand how framed autonomy can be operationalized in environments where process failures carry significant legal, social, or reputational consequences. This issue is especially important in fields like healthcare, finance, logistics, and public administration, where adaptive process behavior can make systems more resilient but also create new risks if not managed properly.

The relevance and effectiveness of ABPM are also likely to depend on important boundary conditions. Its value may be greater in volatile environments characterized by process interdependence, real-time data availability, and recurrent disruptions, where adaptive coordination can meaningfully improve continuity and responsiveness. By contrast, its applicability may be more limited in highly stable process settings, in organizations with low digital maturity, or in contexts where data quality, governance capacity, or regulatory constraints restrict the safe exercise of bounded autonomy. For this reason, ABPM should not be treated as a universally superior process model, but as a context-dependent resilience logic whose effectiveness depends on technological, organizational, and institutional conditions.

Another important avenue concerns the relationship between human judgment and system adaptation. The evaluation of ABPM performance should also extend beyond traditional efficiency indicators. Relevant metrics may include disruption detection speed, adaptation response time, process recovery duration, continuity maintenance under disruption, adaptive decision accuracy, process reconfiguration stability, and the effectiveness of human oversight during runtime adaptation. Together, these indicators provide a broader assessment framework aligned with resilience-oriented process performance. While current research increasingly recognizes the importance of socio-technical alignment, there is still limited empirical understanding of how managers and employees interact with self-adaptive process systems in practice. Future studies could examine how trust in adaptive systems develops, how decision rights are renegotiated, and how human actors respond when process systems begin to recommend or enact changes autonomously. Longitudinal research would be especially valuable here, as it could capture the gradual co-evolution of organizational routines, governance structures, and digital process capabilities over time.

Finally, Future studies should meticulously investigate the performance implications of ABPM. Understanding the conditions under which process autonomy improves not only responsiveness but also continuity, recovery, and long-term organizational renewal remains a significant challenge. Comparative studies across industries, disruption types, and digital maturity levels would be particularly useful in clarifying whether the benefits of ABPM are universal or contingent on specific organizational and environmental configurations. Such research would help move the field beyond conceptual promise toward a more grounded understanding of how ABPM contributes to resilience in practice.

CONCLUSION

This chapter has argued that ABPM represents a meaningful evolution of process management in environments where disruption is no longer exceptional but structural. Conventional BPM remains relevant for stable and repetitive activities, yet its emphasis on predefined flows, standardized control, and ex ante optimization offers only a partial response to contexts characterized by uncertainty, interdependence, and rapid change. Under such conditions, process effectiveness depends not only on efficiency but also on the capacity to sense deviations, interpret contextual shifts, decide adaptively, and reconfigure execution while maintaining organizational coherence. ABPM becomes important precisely because it addresses this expanded requirement.

The discussion has also shown that organizational resilience can be understood through a process lens. Resilience is not merely a strategic aspiration or an ex-post recovery outcome. It is enacted through process systems capable of sustaining continuity, absorbing disruption, supporting timely response, and enabling renewal when established routines no longer fit the environment. From this perspective, ABPM is best interpreted as a resilience-oriented process logic in which adaptive mechanisms are embedded directly into the design and governance of business processes. Its value lies not in replacing human actors with technology, but in structuring a more effective relationship between runtime adaptation, organizational knowledge, and bounded oversight.

A central implication of the chapter is that designing ABPM requires a socio-technical approach. Data integration, real-time observability, digital twins, and adaptive decision mechanisms are critical enablers, but they are not sufficient on their own. Their contribution depends on governance boundaries, human supervision, accountability structures, and learning routines that shape how adaptation is enacted in practice. For that reason, ABPM should not be reduced to a technical upgrade of workflow automation. It is a broader organizational design challenge involving architecture, decision rights, feedback loops, and resilience-oriented evaluation criteria.

The chapter further suggests that the future relevance of BPM will increasingly depend on its ability to move beyond a narrow efficiency logic. In environments that are not stable, process systems can't be judged just by cycle time, compliance, or cost savings. They must also be assessed by their ability to preserve continuity, support recovery, maintain robustness, and enable organizational renewal. This wider evaluative frame doesn't make efficiency any less important; it just puts it in a more realistic context for understanding process value in today's businesses.

Several avenues remain open for future research. More empirical work is needed to examine how ABPM capabilities develop across sectors, how bounded autonomy can be governed in high-risk environments, and how human actors interact with self-adaptive process systems over time. Longitudinal and comparative studies would be particularly beneficial in assessing whether ABPM enhances resilience in a lasting manner or if its advantages are dependent on particular technological and organizational contexts.

Ultimately, the strategic significance of ABPM lies in its capacity to reposition process management for a world defined by ongoing disruption. When uncertainty lasts for a long time, resilient organizations are not just those with well-documented procedures. They are also those whose process systems can change without losing control, accountability, or direction. In that sense, ABPM marks a shift from managing processes as fixed execution structures toward designing them as governed systems of continuous adaptation.

REFERENCES

Abidi, O., Richet, J. L., & Vitari, C. (2025). Digital Transformation and Resilience: Dimensions and Interactions. Https://Services.Igi-Global.Com/Resolvedoi/Resolve.Aspx?Doi=10.4018/JGIM.367873, *33*(1), 1–64. DOI: 10.4018/JGIM.367873

Acitelli, G., Alman, A., Maggi, F. M., & Marrella, A. (2025). Achieving framed autonomy in AI-augmented business process management systems through automated planning. *Information Systems*, *133*, 102573. DOI: 10.1016/j.is.2025.102573

Adensamer, A., & Rueckel, D. (2018). Differences Between BPM and ACM Models for Process Execution. *Lecture Notes in Business Information Processing*, *319*, 270–279. DOI: 10.1007/978-3-319-94214-8_18

Agostinelli, S., De Luzi, F., di Canito, U., Ferraro, J., Marrella, A., & Mecella, M. (2022). A Data-Centric Approach to Design Resilient-Aware Process Models in BPMN. *Lecture Notes in Business Information Processing, 458 LNBIP*, 38–54. DOI: 10.1007/978-3-031-16171-1_3

Ahmad, T., & Looy, A. Van. (2020). Business Process Management and Digital Innovations: A Systematic Literature Review. *Sustainability 2020, Vol. 12, Page 6827, 12*(17), 6827. DOI: 10.3390/su12176827

Alfaqiyah, E., Alzubi, A., Aljuhmani, H. Y., & Öz, T. (2025). How Industry 4.0 Technologies Enhance Supply Chain Resilience: The Interplay of Agility, Adaptability, and Customer Integration in Manufacturing Firms. *Sustainability 2025, Vol. 17, Page 7922, 17*(17), 7922. DOI: 10.3390/su17177922

Amer, Y., Soufali, A., & Zaghwan, A. (2026). A digital twin-based framework for predictive quality assurance and supply chain resilience in the automotive industry. *Advanced Engineering Informatics*, *69*, 103969. DOI: 10.1016/j.aei.2025.103969

Anang, A. N., Obidi, P. O., Mesogboriwon, A. O., & Obidi, J. O. (2024). THE role of Artificial Intelligence in industry 5.0: Enhancing human-machine collaboration. *World Journal of Advanced Research and Reviews*, *24*(2), 380–400. DOI: 10.30574/wjarr.2024.24.2.3369

Andriyani, Y., Suripto, , Yohanitas, W. A., Kartika, R. S., & Marsono, . (2024). Adaptive innovation model design: Integrating agile and open innovation in regional areas innovation. *Journal of Open Innovation*, *10*(1), 100197. DOI: 10.1016/j.joitmc.2023.100197

Angelis, A., & Kousiouris, G. (2025). An Overview on the Landscape of Self-Adaptive Cloud Design and Operation Patterns: Goals, Strategies, Tooling, Evaluation, and Dataset Perspectives. *Future Internet 2025, Vol. 17, Page 434, 17*(10), 434. DOI: 10.3390/fi17100434

Antunes, P. (2011). BPM and exception handling: Focus on organizational resilience. *IEEE Transactions on Systems, Man, and Cybernetics. Part C, Applications and Reviews, 41*(3), 383–392. DOI: 10.1109/TSMCC.2010.2062504

Antunes, P., & Mourão, H. (2011). Resilient Business Process Management: Framework and services. *Expert Systems with Applications, 38*(2), 1241–1254. DOI: 10.1016/j.eswa.2010.05.017

Arno, P. (2025). From Theory to Practice: Empirical Insights on BPM Lifecycle Adoption and Low-Code Utilization in SMEs. *IEEE Access : Practical Innovations, Open Solutions, 13*, 117385–117403. DOI: 10.1109/ACCESS.2025.3585261

Baldea, M., Georgiou, A. T., Gopaluni, B., Mercangöz, M., Pantelides, C. C., Sheth, K., Zavala, V. M., & Georgakis, C. (2025). From automated to autonomous process operations. *Computers & Chemical Engineering, 196*, 109064. DOI: 10.1016/j.compchemeng.2025.109064

Bandi, A., Kongari, B., Naguru, R., Pasnoor, S., & Vilipala, S. V. (2025). The Rise of Agentic AI: A Review of Definitions, Frameworks, Architectures, Applications, Evaluation Metrics, and Challenges. *Future Internet 2025, Vol. 17, Page 404, 17*(9), 404. DOI: 10.3390/fi17090404

Barata, J., & Kayser, I. (2024). How will the digital twin shape the future of industry 5.0? *Technovation, 134*, 103025. DOI: 10.1016/j.technovation.2024.103025

Bazan-Muñoz, A., Pautasso, C., Ortiz, G., & Garcia-de-Prado, A. (2026). BPSmart-CARE: A framework for managing contextualized actions in IoT systems through the integration of business process modelling and complex event processing. *Internet of Things : Engineering Cyber Physical Human Systems, 36*, 101887. DOI: 10.1016/j.iot.2026.101887

Beerepoot, I., Di Ciccio, C., Reijers, H. A., Rinderle-Ma, S., Bandara, W., Burattin, A., Calvanese, D., Chen, T., Cohen, I., Depaire, B., Di Federico, G., Dumas, M., van Dun, C., Fehrer, T., Fischer, D. A., Gal, A., Indulska, M., Isahagian, V., Klinkmüller, C., & Zerbato, F. (2023). The biggest business process management problems to solve before we die. *Computers in Industry, 146*, 103837. DOI: 10.1016/j.compind.2022.103837

Bertrand, Y., Stevens, A., Deforce, B., De Smedt, J., De Weerdt, J., & Serral, E. (2025). Approaches for IoT-enhanced predictive process monitoring. *Process Science 2025 2:1, 2*(1), 7-. DOI: 10.1007/s44311-025-00011-x

Bidmeshk, O. G., Woo, C., & Mehraeen, M. (2025). Managing operational alignment complexity: A recommender system approach. *Journal of Open Innovation, 11*(4), 100627. DOI: 10.1016/j.joitmc.2025.100627

Bondac, G.-T., Stanescu, S.-G., Ionescu, C. A., Duica, A., & Uzlău, M. C. (2026). Decision-Making in Complex Systems Using AI-Based Decision Support: The Role of Trust, Transparency, and Data Quality. *Electronics 2026, Vol. 15, Page 372, 15*(2), 372. DOI: 10.3390/electronics15020372

Bornay-Barrachina, M., López-Cabrales, Á., & Salas-Vallina, A. (2025). Sensing, seizing, and reconfiguring dynamic capabilities in innovative firms: Why does strategic leadership make a difference? *Business Research Quarterly, 28*(2), 399–420. DOI: 10.1177/23409444231185790;PAGEGROUP:STRING:PUBLICATION

Boulitama, O., Lhamidi, S., Taifi, R., & Brahim, S. (2025). Digital transformation and supply chain capabilities: A sectoral maturity perspective in morocco. *Polish Journal of Management Studies, 32*(1), 61–80. DOI: 10.17512/pjms.2025.32.1.04

Braniff, A., Akundi, S. S., Liu, Y., Dantas, B., Niknezhad, S. S., Khan, F., Pistikopoulos, E. N., & Tian, Y. (2025). Real-time process safety and systems decision-making toward safe and smart chemical manufacturing. *Digital Chemical Engineering, 15*, 100227. DOI: 10.1016/j.dche.2025.100227

Brückner, M., Thyssen, C., & Huwer, J. (2026). AI for adaptive science teaching: Strengthening teacher self-efficacy and perceived usefulness. *Frontiers in Education, 11*, 1759274. DOI: 10.3389/feduc.2026.1759274

Calabrò, A., Marchetti, E., Skrzypek, P., & Marchel, J. (2025). CONCERN: A model-based monitoring infrastructure. *Internet of Things : Engineering Cyber Physical Human Systems, 31*, 101557. DOI: 10.1016/j.iot.2025.101557

Cao, Z., Zhao, H., Wang, Y., He, C., Zhou, D., & Han, X. (2025). A Resilience Quantitative Assessment Framework for Cyber–Physical Systems: Mathematical Modeling and Simulation. *Applied Sciences 2025, Vol. 15, Page 8285, 15*(15), 8285. DOI: 10.3390/app15158285

Casas-Ortiz, A., Jimenez-Tellez, N., Portaz, M., & Santos, O. C. (2026). A holistic framework to create intelligent systems for psychomotor learning: a systematic review and insights from the virtual reality Kenpo learning simulator. *Virtual Reality 2026 30:2, 30*(2), 72-. DOI: 10.1007/s10055-026-01326-1

Chen, H., Wen, Y., Zhu, M., Huang, Y., Xiao, C., Wei, T., & Hahn, A. (2021). From Automation System to Autonomous System: An Architecture Perspective. *Journal of Marine Science and Engineering 2021, Vol. 9, Page 645, 9*(6), 645. DOI: 10.3390/jmse9060645

Chen, S. C., Chen, H. M., Chen, H. K., & Li, C. L. (2024). Multi-Objective Optimization in Industry 5.0: Human-Centric AI Integration for Sustainable and Intelligent Manufacturing. *Processes 2024, Vol. 12, Page 2723, 12*(12), 2723. DOI: 10.3390/pr12122723

Conz, E., & Magnani, G. (2020). A dynamic perspective on the resilience of firms: A systematic literature review and a framework for future research. *European Management Journal, 38*(3), 400–412. DOI: 10.1016/j.emj.2019.12.004

Corradini, F., Pettinari, S., Re, B., Rossi, L., & Tiezzi, F. (2023). Executable Digital Process Twins: Towards the Enhancement of Process-Driven Systems. *Big Data and Cognitive Computing 2023, Vol. 7, Page 139, 7*(3), 139. DOI: 10.3390/bdcc7030139

De Luzi, F., Leotta, F., Marrella, A., & Mecella, M. (2024). On the Interplay Between Business Process Management and Internet-of-Things. *Business & Information Systems Engineering 2024 67:2, 67*(2), 265–288. DOI: 10.1007/s12599-024-00859-6

De Nicola, R., Jähnichen, S., & Wirsing, M. (2020). Rigorous engineering of collective adaptive systems: Special section. *International Journal of Software Tools for Technology Transfer, 22*(4), 389–397. DOI: 10.1007/s10009-020-00565-0

Dumas, M., Fournier, F., Limonad, L., Marrella, A., Montali, M., Rehse, J. R., Accorsi, R., Calvanese, D., De Giacomo, G., Fahland, D., Gal, A., La Rosa, M., Völzer, H., & Weber, I. (2023). AI-augmented Business Process Management Systems: A Research Manifesto. *ACM Transactions on Management Information Systems, 14*(1), 1–19. Advance online publication. DOI: 10.1145/3576047

Erasmus, J., Vanderfeesten, I., Traganos, K., & Grefen, P. (2020). Using business process models for the specification of manufacturing operations. *Computers in Industry, 123*, 103297. DOI: 10.1016/j.compind.2020.103297

Fornari, F., Compagnucci, I., Callisto De Donato, M., Bertrand, Y., Beyel, H. H., Carrión, E., Franceschetti, M., Groher, W., Grüger, J., Kilic, E., Koschmider, A., Leotta, F., Li, C. Y., Lugaresi, G., Malburg, L., Mangler, J., Mecella, M., Pastor, O., Riss, U., & Valderas, P. (2025). Digital Twins of Business Processes: A Research Manifesto. *Internet of Things : Engineering Cyber Physical Human Systems, 30*, 101477. DOI: 10.1016/j.iot.2024.101477

Garrido-Moreno, A., Martín-Rojas, R., & García-Morales, V. J. (2024). The key role of innovation and organizational resilience in improving business performance: A mixed-methods approach. *International Journal of Information Management, 77*, 102777. DOI: 10.1016/j.ijinfomgt.2024.102777

Gebreselassie, B. A., Garcia, N. M., & Midekso, D. (2025). A Systematic Review of Self-Adaptive Mobile Applications with Cooperative Dimension. *IoT 2026, Vol. 7, Page 6, 7*(1), 6. DOI: 10.3390/iot7010006

Georgescu, I., Bocean, C. G., Vărzaru, A. A., Rotea, C. C., Mangra, M. G., & Mangra, G. I. (2024). Enhancing Organizational Resilience: The Transformative Influence of Strategic Human Resource Management Practices and Organizational Culture. *Sustainability 2024, Vol. 16, Page 4315, 16*(10), 4315. DOI: 10.3390/su16104315

Ghaljahi, M., Omidi, L., & Karimi, A. (2025). Resilience assessment in process industries: A review of literature. *Heliyon, 11*(4), e42498. DOI: 10.1016/j.heliyon.2025.e42498

Gignac, G. E., & Szodorai, E. T. (2024). Defining intelligence: Bridging the gap between human and artificial perspectives. *Intelligence, 104*, 101832. DOI: 10.1016/j.intell.2024.101832

Grønvald, J. M., Nørgaard, M., Christensen, C. K. F., & Mortensen, N. H. (2025). Quantifying the Effects of Modular Product Architectures: A Data-Driven Framework for Evaluating Product Variety and Complexity. *Applied Sciences 2025, Vol. 15, Page 12284, 15*(22), 12284. DOI: 10.3390/app152212284

Grznár, P., Burganová, N., Mozol, Š., & Mozolová, L. (2023). A Comprehensive Digital Model Approach for Adaptive Manufacturing Systems. *Applied Sciences 2023, Vol. 13, Page 10706, 13*(19), 10706. DOI: 10.3390/app131910706

Grznár, P., Papánek, L., Marčan, M., Krajčovič, M., Antoniuk, I., Mozol, Š., & Mozolová, L. (2025). Enhancing Production Efficiency Through Digital Twin Simulation Scheduling. *Applied Sciences 2025, Vol. 15, Page 3637, 15*(7), 3637. DOI: 10.3390/app15073637

Hamdani, R., & Chihi, I. (2025). Adaptive human-computer interaction for industry 5.0: A novel concept, with comprehensive review and empirical validation. *Computers in Industry, 168*, 104268. DOI: 10.1016/j.compind.2025.104268

Huy, P. Q., & Phuc, V. K. (2025). Unveiling how business process management capabilities foster dynamic decision-making for effectiveness of sustainable digital transformation. *Business Process Management Journal, 31*(8), 67–103. DOI: 10.1108/BPMJ-06-2024-0467

Ihlenfeldt, S., Wunderlich, T., Süße, M., Hellmich, A., Schenke, C. C., Wenzel, K., & Mater, S. (2021). Increasing Resilience of Production Systems by Integrated Design. *Applied Sciences 2021, Vol. 11, Page 8457, 11*(18), 8457. DOI: 10.3390/app11188457

Jin, Z. (2018). The System Dynamic Adaptability Concern. *Environment Modeling-Based Requirements Engineering for Software Intensive Systems*, 217–239. DOI: 10.1016/B978-0-12-801954-2.00012-1

Jorzik, P., Klein, S. P., Kanbach, D. K., & Kraus, S. (2024). AI-driven business model innovation: A systematic review and research agenda. *Journal of Business Research, 182*, 114764. DOI: 10.1016/j.jbusres.2024.114764

Kantabutra, S., & Ketprapakorn, N. (2021). Toward an Organizational Theory of Resilience: An Interim Struggle. *Sustainability 2021, Vol. 13, Page 13137, 13*(23), 13137. DOI: 10.3390/su132313137

Khan, M. Q., Alvi, M. A. H., Nawaz, H. H., & Umar, M. (2025). Impact of Digital Twins on Real Practices in Manufacturing Industries. *Inventions 2025, Vol. 10, Page 106, 10*(6), 106. DOI: 10.3390/inventions10060106

Kir, H., & Erdogan, N. (2021). A knowledge-intensive adaptive business process management framework. *Information Systems, 95*, 101639. DOI: 10.1016/j.is.2020.101639

Ko, R. K. L., Lee, S. S. G., & Lee, E. W. (2009). Business process management (BPM) standards: A survey. *Business Process Management Journal, 15*(5), 744–791. DOI: 10.1108/14637150910987937

Krogstie, J. (2012). Model-based development and evolution of information systems: A quality approach. *Model-Based Development and Evolution of Information Systems: A Quality Approach*, 1–439. https://doi.org/DOI: 10.1007/978-1-4471-2936-3/COVER

Kumar, R., Pillai, S., O'connell, E., & Denny, P. (2026). Role of Unified Namespace (UNS) and Digital Twins in Predictive and Adaptive Industrial Systems. *Machines 2026, Vol. 14, Page 252, 14*(2), 252. DOI: 10.3390/machines14020252

Li, P., Chen, Y., & Guo, X. (2025). Digital transformation and supply chain resilience. *International Review of Economics & Finance, 99*, 104033. DOI: 10.1016/j.iref.2025.104033

Liang, K., Ren, X., Yin, H., Jiang, J., Xiang, L., & Dai, W. (2025). An Adaptive Process Monitoring Method Based on an Improved ADALINE Neural Network. *Applied Sciences 2025, Vol. 15, Page 2090, 15*(4), 2090. DOI: 10.3390/app15042090

Liang, L., & Li, Y. (2024). How does organizational resilience promote firm growth? The mediating role of strategic change and managerial myopia. *Journal of Business Research*, *177*, 114636. Advance online publication. DOI: 10.1016/j.jbusres.2024.114636

Lin, K. Y., Li, M. H., Lo, F., Huang, H.-C., Matsuno, K., & Watanabe, R. (2025). Adaptive learning with human factors and Artificial Intelligence: Associations with training effectiveness in programming education. *International Journal of Industrial Ergonomics*, *110*, 103834. DOI: 10.1016/j.ergon.2025.103834

Markaj, A., Mercangöz, M., & Fay, A. (2024). Design and implementation of an Autonomous Systems Training Environment framework for control algorithm evaluation in autonomous plant operation. *Computers & Chemical Engineering*, *189*, 108798. DOI: 10.1016/j.compchemeng.2024.108798

Moreno-Sánchez, P. A., Del Ser, J., van Gils, M., & Hernesniemi, J. (2026). A design framework for operationalizing trustworthy artificial intelligence in healthcare: Requirements, tradeoffs and challenges for its clinical adoption. *Information Fusion*, *127*, 103812. DOI: 10.1016/j.inffus.2025.103812

Mosallam, A. S., & Mouhim, S. (2025). Contextual Modeling and Intelligent Decision-Making for IoT Systems: A Combined Ontology and Machine Learning Approach. *Engineering Proceedings 2025, Vol. 112, Page 71, 112*(1), 71. DOI: 10.3390/engproc2025112071

Napier, E., Liu, S. Y. H., & Liu, J. (2024). Adaptive strength: Unveiling a multilevel dynamic process model for organizational resilience. *Journal of Business Research*, *171*, 114334. DOI: 10.1016/j.jbusres.2023.114334

Niederman, F. (2021). Process Theory: Background, Opportunity, and Challenges. *Foundations and Trends in Information Systems*, *5*(1), 1–230. DOI: 10.1561/2900000017

Panichella, S., & Zaugg, N. (2020). An Empirical Investigation of Relevant Changes and Automation Needs in Modern Code Review. *Empirical Software Engineering*, *25*(6), 4833–4872. DOI: 10.1007/s10664-020-09870-3

Pasi, B. N., Dhamak, P. S., Rane, S. B., Todkari, V. C., & Mishra, A. R. (2026). Dual impact of AI on business processes: Fuzzy TOPSIS prioritization and theoretical framework for strategic adoption. *Business Process Management Journal*, •••, 1–33. DOI: 10.1108/BPMJ-11-2025-1896

Päßler, J., ter Beek, M. H., Damiani, F., Johnsen, E. B., & Tapia Tarifa, S. L. (2025). Analysing Self-Adaptive Systems as Software Product Lines. *Journal of Systems and Software*, *222*, 112324. DOI: 10.1016/j.jss.2024.112324

Pelosi, D., Cacciagrano, D., & Piangerelli, M. (2025). Explainability and Interpretability in Concept and Data Drift: A Systematic Literature Review. *Algorithms 2025, Vol. 18, Page 443, 18*(7), 443. DOI: 10.3390/a18070443

Petrovska, A., Hutzelmann, T., & Kugele, S. (2023). A Theoretical Framework for Self-Adaptive Systems: Specifications, Formalisation, and Architectural Implications. *Proceedings of the ACM Symposium on Applied Computing, 23*, 1440–1449. DOI: 10.1145/3555776.3577665

Pitelis, C. N., Teece, D. J., & Yang, H. (2024). Dynamic Capabilities and MNE Global Strategy: A Systematic Literature Review-Based Novel Conceptual Framework. *Journal of Management Studies*, *61*(7), 3295–3326. DOI: 10.1111/JOMS.13021;-JOURNAL:JOURNAL:14676486;WGROUP:STRING:PUBLICATION

Plattfaut, R., & Grisold, T. (2025a). Business process management as a continuous enabler for digital innovation – insights from top executives. *Business Process Management Journal*, *31*(8), 244–266. DOI: 10.1108/BPMJ-05-2025-0759

Plattfaut, R., & Grisold, T. (2025b). Business process management as a continuous enabler for digital innovation – insights from top executives. *Business Process Management Journal*, *31*(8), 244–266. DOI: 10.1108/BPMJ-05-2025-0759

Röglinger, M., Plattfaut, R., Borghoff, V., Kerpedzhiev, G., Becker, J., Beverungen, D., vom Brocke, J., Van Looy, A., del-Río-Ortega, A., Rinderle-Ma, S., Rosemann, M., Santoro, F. M., & Trkman, P. (2022). Exogenous Shocks and Business Process Management: A Scholars' Perspective on Challenges and Opportunities. *Business & Information Systems Engineering*, *64*(5), 669–687. DOI: 10.1007/s12599-021-00740-w

Saeed Malik, F., Terzidis, O., Malik, F. S., & Terzidis, O. (2026). The metamorphosis of organizational resilience: A strategic roadmap for SMEs in the digital age. *Review of Managerial Science*, *2026*, 1–35. DOI: 10.1007/s11846-026-01002-3

Saemaldaher, K., & Emeagwali, O. L. (2025). The Role of Open Innovation in Enhancing Organizational Resilience and Sustainability Performance Through Organizational Adaptability. *Sustainability 2025, Vol. 17, Page 5846, 17*(13), 5846. DOI: 10.3390/su17135846

Salehie, M., & Tahvildari, L. (2009). Self-adaptive software. [TAAS]. *ACM Transactions on Autonomous and Adaptive Systems*, *4*(2), 1–42. Advance online publication. DOI: 10.1145/1516533.1516538

Salvaña, M. L. O., Bolingot, H. J. M., & Tangonan, G. L. (2026). A self-organized criticality model of extreme events and cascading disasters of hub-and-spoke air traffic networks. *International Journal of Disaster Risk Reduction*, *133*, 106009. DOI: 10.1016/j.ijdrr.2026.106009

Sapkota, R., Roumeliotis, K. I., & Karkee, M. (2026). AI Agents vs. Agentic AI: A Conceptual taxonomy, applications and challenges. *Information Fusion*, *126*, 103599. DOI: 10.1016/j.inffus.2025.103599

Sebova, L., Svec, A., & Chovanova, K. (2025). The Role of Business Process Management in Hotel Innovation and Competitiveness. *Tourism and Hospitality 2025, Vol. 6, Page 122, 6*(3), 122. DOI: 10.3390/tourhosp6030122

Shahin, M., Maghanaki, M., & Chen, F. F. (2026). Integration of Lean Analytics and Industry 6.0: A Novel Meta-Theoretical Framework for Antifragile, Generative AI-Orchestrated, Circular–Regenerative, and Hyper-Connected Manufacturing Ecosystems. *Big Data and Cognitive Computing 2026, Vol. 10, Page 65, 10*(2), 65. DOI: 10.3390/bdcc10020065

Sincorá, L. A., de Oliveira, M. P. V., Zanquetto-Filho, H., & Alvarenga, M. Z. (2023). Developing organizational resilience from business process management maturity. *Innovation & Management Review*, *20*(2), 147–161. DOI: 10.1108/INMR-11-2021-0219

Stary, C. (2021). Digital Twin Generation: Re-Conceptualizing Agent Systems for Behavior-Centered Cyber-Physical System Development. *Sensors 2021, Vol. 21, Page 1096, 21*(4), 1096. DOI: 10.3390/s21041096

Teece, D. J. (2007). Explicating dynamic capabilities: The nature and microfoundations of (sustainable) enterprise performance. *Strategic Management Journal*, *28*(13), 1319–1350. DOI: 10.1002/smj.640

Teixeira, A. R., Ferreira, J. V., & Ramos, A. L. (2024). Optimization of Business Processes Through BPM Methodology: A Case Study on Data Analysis and Performance Improvement. *Information 2024, Vol. 15, Page 724, 15*(11), 724. DOI: 10.3390/info15110724

Tetteh, F. K., & Owusu Kwateng, K. (2025). Digital pathways to supply chain viability: Exploring nice and must-to have practices in dynamic environment –insight from SEM and NCA. *Journal of Enterprise Information Management*, •••, 1–38. DOI: 10.1108/JEIM-05-2025-0367/1308502

Thomas Kohlborn, D. O. M. P. J. P. (2014). Event-Driven Business Process Management: Where are we now? A comprehensive synthesis and analysis of literature. *Business Process Management Journal*, *20*(4), 615–633. DOI: 10.1108/BPMJ-07-2013-0092

Tsakalidis, G., & Vergidis, K. (2026). Systematicity and generalizability in business process redesign methodologies: A systematic literature review. *Science of Computer Programming*, *248*, 103392. DOI: 10.1016/j.scico.2025.103392

Tung Khuat, T., Jacob Kedziora, D., & Gabrys, B. (2023). The Roles and Modes of Human Interactions with Automated Machine Learning Systems: A Critical Review and Perspectives. *Foundations and Trends in Human-Computer Interaction*, *17*(3–4), 195–387. DOI: 10.1561/1100000091

van der Aalst, W. M. P. (2013). Business Process Management: A Comprehensive Survey. *International Scholarly Research Notices*, *2013*(1), 507984. DOI: 10.1155/2013/507984

Viteri-Sánchez, C., & Novillo-Villegas, S. (2025). A Framework for a Sustainable Adoption of Business Process Management. *Sustainability 2025, Vol. 17, Page 9827*, *17*(21), 9827. DOI: 10.3390/su17219827

Walmsley, T. G., Patros, P., Yu, W., Young, B. R., Burroughs, S., Apperley, M., Carson, J. K., Udugama, I. A., Aeowjaroenlap, H., Atkins, M. J., & Walmsley, M. R. W. (2024). Adaptive digital twins for energy-intensive industries and their local communities. *Digital Chemical Engineering*, *10*, 100139. DOI: 10.1016/j.dche.2024.100139

Weinzierl, S., Zilker, S., Dunzer, S., & Matzner, M. (2024). Machine learning in business process management: A systematic literature review. *Expert Systems with Applications*, *253*, 124181. DOI: 10.1016/j.eswa.2024.124181

Weyns, D. (2021). *An introduction to self-adaptive systems : a contemporary software engineering perspective*. https://www.wiley.com/en-ie/An+Introduction+to+Self-adaptive+Systems%3A+A+Contemporary+Software+Engineering+Perspective-p-9781119574910

Xiao, L., & Cao, H. (2017). Organizational Resilience: The Theoretical Model and Research Implication. *ITM Web of Conferences, 12*, 04021. DOI: 10.1051/itmconf/20171204021

Zelt, S., Recker, J., Schmiedel, T., & vom Brocke, J. (2019). A theory of contingent business process management. *Business Process Management Journal*, *25*(6), 1291–1316. DOI: 10.1108/BPMJ-05-2018-0129

Zhang, Z., Kang, Y., Lu, Y., & Li, P. (2025). The Role of Artificial Intelligence in Business Model Innovation of Digital Platform Enterprises. *Systems 2025, Vol. 13, Page 507, 13*(7), 507. DOI: 10.3390/systems13070507

Chapter 7
Autonomy Gradient Architecture:
Designing Self-Evolving and Governance-Aware Business Process Systems in the Age of Intelligent Enterprises

Ismailov Azizbek
Alfraganus University, Tashkent, Uzbekistan

Aarthi Selvakumar
http://orcid.org/0009-0006-9064-2091
Marwadi University, Rajkot, India

Ravikumar Natarajan
http://orcid.org/0009-0009-3705-1681
Marwadi University, Rajkot, India

Dilmurod Turaev
Termez University of Economics and Service, Termez, Uzbekistan

Karimov Rashid Salayevich
Mamun University, Khiva, Uzbekistan

Jumayev Akbar
Tashkent State University of Economics, Tashkent, Uzbekistan

ABSTRACT

The chapter introduces the Autonomy Gradient Architecture (AGA) that is a governance aware platform applied in the creation of self-evolving business process systems in intelligent business firms. Beyond binary automation conceptualizations, AGA conceptualizes autonomy as a continuum with structured components, that is, a continuum between assisted decision support and adaptive orchestration and self-learning processes. Embedded governance layers that bring transparency, accountability, and regulatory alignment are integrated within the structure, using analytics, orchestration, sensing, feedback and governance. Evidence and data in

DOI: 10.4018/979-8-2600-0050-2.ch007

the financial services and healthcare cases demonstrate that resilience, efficiency and accuracy were improved and in measurable ways. The chapter entails autonomy and co-development of the governmental strategies as a strategic foundation of sustainable competitive advantage in Industry 5.0 settings.

1. INTRODUCTION

Business Process Management (BPM) has also developed to a large extent beyond the rule-based workflow automation into smart and dynamic process ecosystems. The adoption of Artificial Intelligence, robotic process automation, predictive analytics, and large language models have facilitated the transition of processes to dynamic decision-making and responsiveness to the context instead of merely executing something. This has led to the emergence of Autonomous Business Process Management (ABPM) in which the processes are able to detect environmental changes, discern data trends and restructure execution routes with little human intervention (Vidgen & Wang, 2006). However, technological abilities have developed at a high pace whereas conceptual clarity in terms of autonomy degrees is still in pieces. Companies often implement automation systems without a systematic system to direct the transformation of the assisted workflows into the fully adaptive orchestration. There is no harmonized autonomy design paradigm, which poses governance, ethical and operational issues. This chapter proposes the Autonomy Gradient Architecture (AGA) as an organised continuum framework that combines autonomy advancement and governance wisdom that enterprises can use to develop self-adapting, transparent, and tactical business process systems.

1.1 Evolution from BPM to ABPM

The traditional BPM was preoccupied with process modeling, process standardization and optimization according to predefined workflow and enterprise system. The earlier attempts at automation were made towards the implementation of rules, compliance, and digitalisation of tasks. With AI-driven analytics and real time data infrastructure seen and heard, BPM has shifted to intelligent systems which are able to make predictive insights and adaptive responses. The next step in this development is Autonomous Business Process Management because workflows can no longer be understood as a fixed sequence, but rather as dynamic systems which can learn both according to past tendencies and other contextual information. Reinforcement Learning, computer process twins, and AIs can allow the processes to make changes in execution paths, dynamically divide resources, and anticipate risks (Szelągowski & Lupeikiene, 2020). The change also means a renegotiation of the ownership and the forms of governance of the process with human roles being more of a strategic control than a manual implementation. The transformation of more autonomy in the working process is not simply a sign of technological advancement but also a change in paradigm in the business perception of the process of control, flexibility, and performance control. Figure 1 demonstrating the way of an evolution to the AI-based and purely autonomous business process systems with adaptive intelligence and governance integration is based on the traditional rule-based BPM.

Figure 1. Evolution from BPM to ABPM Framework

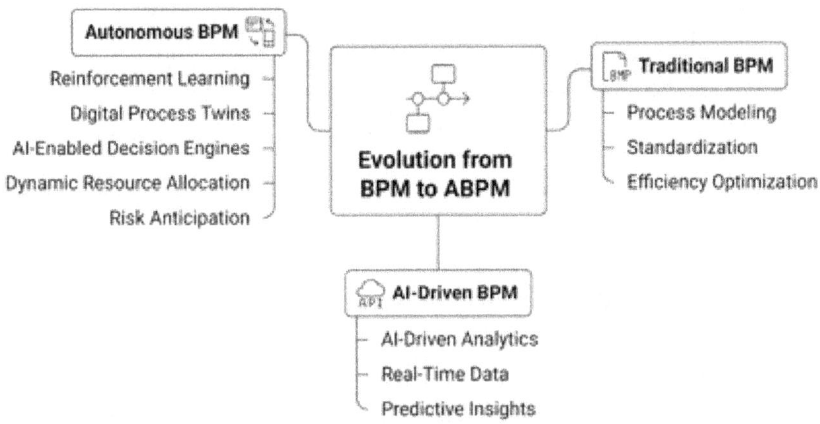

1.2 Problem of Unstructured Autonomy

Although there is a rapid advancement in technologies, the organizational strategies of autonomy are conceptually loose in many cases. Most companies implement AI, RPA, or analytics software in isolation, which leads to disjointed automation programs as opposed to combined autonomy plans. Autonomy is often considered to be in two forms: manual and automated without acknowledging having intermediate forms of assisted intelligence, semi-autonomous execution, or adaptive orchestration. Such unorganized development presents such risks as the lack of algorithmic transparency, non-regulatory compliance, inconsistency in decision responsibility, and degradation of human agency. Besides, the lack of formal measures of autonomy makes it hard to measure maturity levels or correspondence to governance. Systems can undergo technical advancements whilst being strategically off course with enterprise purpose. Members of society may lose trust in greater autonomy without any inherent safety systems like audit trails, explainability modules and human-in-the-loop validation (Gudipudi, 2025). But the issue does not lie in the capability of technology but in the design. It needs to have a well-organized system to handle autonomy advancement in a systematic manner without compromising on transparency, accountability, and integrity. Table 1 Comparative overview The shift to Autonomous Business Process Management vs. traditional BPM in terms of the control logic, level of intelligence, flexibility and governance integration.

Table 1. Evolution from BPM to ABPM

Dimension	Traditional BPM	Intelligent BPM	Autonomous BPM
Decision Logic	Rule-based	Data-assisted	AI-driven adaptive
Human Role	Manual execution	Supervisory	Strategic oversight
Adaptability	Static workflows	Conditional adjustments	Dynamic orchestration
Governance	External audit	Compliance monitoring	Embedded governance intelligence
Learning Capability	None	Predictive	Self-evolving

1.3 Need for Autonomy Gradient Architecture

Autonomy Gradient Architecture Autonomy Gradient Architecture is a formal architecture design proposal to overcome the drawbacks of fragmented autonomy deployment. AGA grounds its conceptualization of autonomy as a dichotomy, instead of a continuum of structured levels between assisted decision support and completely adaptive, self-evolving orchestration. This gradient deployment facilitates companies to plan strategically the deployment of autonomy based on the risk tolerance, regulatory limits and operating complexity. AGA combines sensing, intelligence, orchestration, feedback and governance layers into a seamless architecture that has accountability and flexibility (Villamil et al., 2025). Governance intelligence is internalized as a part of it and not an external control process. Gradient based deployment planning and maturity of autonomy matrices give objective ways to move along. The combination of performance measurement and governance alignment signals helps AGA to make sure that the autonomy of the process develops in a responsible manner (Ravikumar & Aarthi, 2025a). The architecture hence assists businesses to develop systems that are not only smart and flexible but also clear, conforming and bound to long-term strategic goals.

2. BACKGROUND TO THE STUDY

The digital transformation environment has radically changed the way businesses conduct their operations, spurring the move towards adopting intelligent technologies in organizational operations. Cloud platforms, infrastructures, real-time data streams, sophisticated analytics, and ecosystems, have helped companies to work in highly dynamic environments that are volatile and complex. Process intelligence has, in this regard, become an important asset to retain agility and resilience. The following technologies Artificial Intelligence, robotic process automation, and process mining have increased the functional range of BPM, enabling the continuous monitoring, predictive forecasting, and automatic adaptation (Sang et al., 2025). At the same time, enterprise architecture frameworks have also been modified to include governance, compliance, and ethical oversight in digital infrastructures. The coming together of these developments brings opportunity and threat. Although integration of technology boosts the performance of operations, it also brings with it issues of transparency, bias, accountability, and regulation alignment. These technological and governmental underpinnings are critical to comprehend in order to have a structured autonomy model that can facilitate intelligent and accountable change in the enterprise.

2.1 Digital Transformation and Process Intelligence

Digital transformation has changed organizational focus that was directed at efficient optimization to dynamic capability development. Present day businesses are in ecosystems that have fast changing markets, changing demands of customers requiring personalization, and unending innovation cycles. Management of this complexity has been made the focus of process intelligence (Salameh & Bass, 2022). Event log analytics, real-time data ingestion, and predictive modelling help organizations to implement solutions ahead of time, detect anomalies and predict demand trends. Machine learning algorithms improve the quality of decisions made by finding the correlation and deviation that are not within the human analysis limits. Digital process twins also allow the simulation of a different execution pathway prior to implementation (Ravikumar, 2025b). The capabilities make processes transform into anticipatory systems that can be continuously refined. But additional smartness also needs to be controlled to avoid the unintended consequences. The technology substrate that can be built on is the combination of sensing, analytics and feedback mechanisms that can be used to build autonomous process architectures. Process intelligence thus is a facilitator and a condition of the systematic design of autonomy.

2.2 AI, RPA, and Process Mining Convergence

The combination and integration of AI, robotic process automation, and process mining technologies have enabled the transition to autonomous operations to occur faster. RPA can accomplish task-level automation based on the rule-based implementation, whereas AI puts forward predictive reasoning and contextual interpretation. Process mining gives insights into actual execution trends, with the help of event logs, and performance bottlenecks. These technologies can form adaptive loops with known inefficiencies that are discovered and fed into an adjustment mechanism. Predictive process mining is able to predict changes in direction and propose remedial steps before that service outage (Ayoubi et al., 2019). The orchestration engines powered by AI will be able to dynamically distribute resources depending on the changes in workloads and risk indicators. This merging decreases manual control and enhances flexibility of the process. The complexity of integration is, however, linked to siloed deployments with no single governance. In the absence of architectural fit, autonomous capabilities can be local instead of global. An organized system is thus required to align technological convergence in enterprise-wide strategies of autonomy, to provide consistency, scalability and responsibility.

2.3 Governance and Enterprise Architecture Foundations

Enterprise architecture ensures the underlying framework in the integration of technology innovation with strategy and regulation of an organization. The governance systems such as policy engines, audit trails, risk management systems, and compliance monitoring systems are used to guarantee operational integrity of digital infrastructures. With greater autonomy, governance will have to be different than reactive control, but should be proactive intelligence, built in the system design (Lawal & Oduleye, 2020). Human control mechanisms and explainability requirements are being integrated with ethical AI principles into the enterprise frameworks to ensure that the issue of bias and transparency is addressed. The layers of architecture should maintain traceability of the decision and assignment of accountability and overrides during high-risk situations. The compatibility of the ability to autonomy and the protection of the governance defines the sustainability in the long term (Aarthi et al., 2025). The integration of governance in architectural layers makes the autonomy progression not to jeopardize trust or compliance. Thus, the enabling environment in which Autonomy Gradient Architecture can perform well balancing adaptive performance and institutional responsibility and strategic coherence is found in enterprise architecture foundations. Figure 2 framework shows how adaptive and responsible process autonomy is supported by the combination of enterprise architecture, governance mechanisms, ethical AI, traceability and accountability.

Figure 2. Autonomy Gradient Architecture Governance

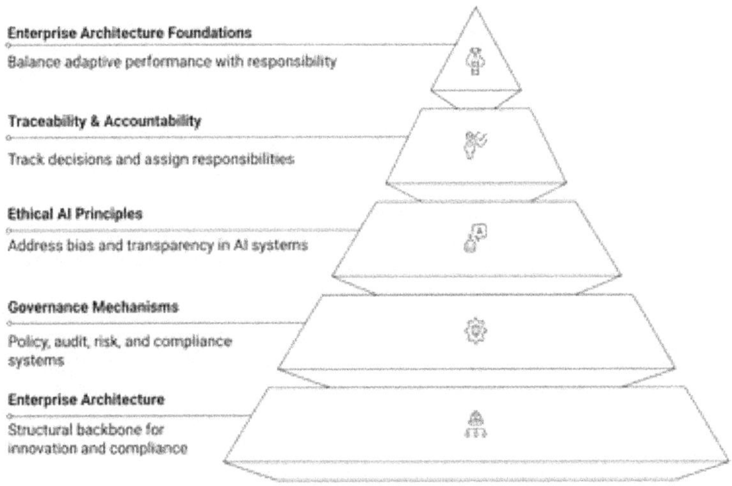

3. RESEARCH OBJECTIVES

The research objectives in this chapter are aimed at systematizing the concept of Autonomy Gradient Architecture (AGA) in Autonomous Business Process Management. The aims shift towards descriptive analysis and are oriented towards the building of theory, integration and governance, design and measurability. A strong focus is placed on the creation of a formal continuum of autonomy and the introduction of accountability mechanisms into intelligent systems and the empowerment of self-evolving process ecosystems in accordance with strategic enterprise objectives. The conceptual and methodological development of the chapter will be guided by the following objectives:

- To conceptualize and formalize the Autonomy Gradient as a structured continuum of process intelligence levels ranging from assisted decision support to fully adaptive orchestration.
- To develop a governance-aware architectural framework that integrates explainability, auditability, compliance monitoring, and human-in-the-loop controls within autonomous process systems.

- To design a layered Autonomy Gradient Architecture (AGA) that connects sensing, intelligence, orchestration, feedback, and governance modules into a cohesive enterprise model.
- To propose measurable evaluation criteria, including autonomy index, governance alignment coefficient, adaptive cycle ratio, transparency score, and self-evolution efficiency indicators.
- To examine how digital process twins, reinforcement learning, and AI-driven reasoning enable the creation of self-evolving, context-sensitive, and strategically aligned process ecosystems.

4. CONCEPTUAL MODEL

Autonomy Gradient Architecture (AGA) conceptual model offers an organized guide to the development of governance cognizant and self-evolving business process systems. Instead of taking autonomy as a binary state, the model characterizes autonomy as a scaled and stratified construct as part of enterprise architecture. The model combines a spectrum of autonomy scales with a stack of technological elements and intrinsic governance systems. It defines the manner in which sensing, analytics, decision intelligence, orchestration, feedback, and supervision functions work together in a dynamically evolving manner to facilitate performance that is adaptive and yet maintain accountability (Hetman et al., 2025). The model also comes with indicators that can be measured to connect the autonomy progression with transparency, compliance, and alignment with the strategic direction. The conceptual model also places autonomy as a controlled ability and not as uncontrolled automation by matching the socio-technical system theory with the AI governance and enterprise architecture concepts. The architecture is such that intelligence development has been coupled with ethical protection and risk management, such that organizations can gradually progress over time in a systematic way out of stagnant automation in favor of robust and accountable autonomous process ecosystems (Halliday, 2023). Figure 3 that shows the capacity of sensing, analytics, decision intelligence, orchestration, feedback, and oversight to orchestrate to form resilient autonomous process ecosystems.

Figure 3. Building Blocks of Autonomous Process Ecosystems

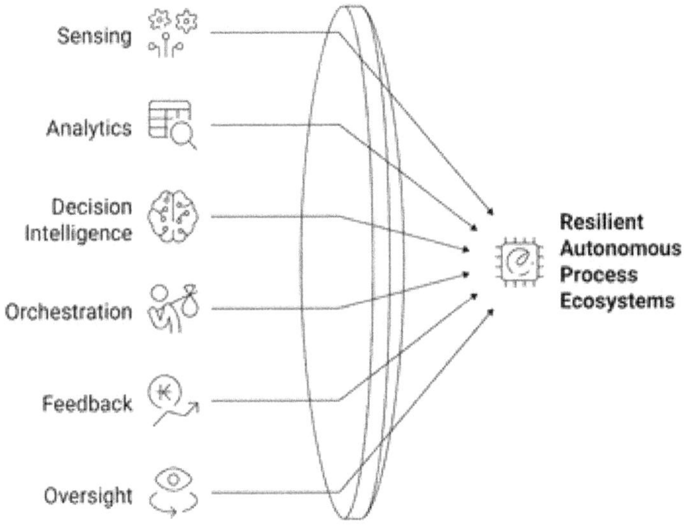

4.1 Autonomy Gradient Continuum Levels

The Autonomy Gradient Continuum sets the graded degrees of process intelligence in which the human-assisted implementation develops to full orchestration of adaptability. The analytics dashboards and automated rules can help in decision making on the bottom, although the human being will be the primary decision makers. The intermediate level includes the semi-autonomous performance of operations with the help of predictive models and an AI-driven recommendation, which offers the possibility to change the workflow conditionally. The next-order levels are adaptive orchestration, where the processes have the ability to dynamically alter the execution paths based on the contextual input, performance criteria and environmental turbulence. Self-evolving systems at the topmost level operate on the application of reinforcement learning and digital process twins and optimize the operations continuously under minimal human intervention (Adesanya et al., 2021). Each tier is associated with certain governance requirements, the level of risk and extent of control. The continuum does not experience sudden changes and the autonomy scales are controlled. Making the degree of autonomy a formal one, makes businesses measure maturity, risk exposure and consistent smart powers to strategic objectives and regulatory limits. Table 2 Systematised information regarding the degree of autonomy and the attributes and level of control.

Table 2. Autonomy Gradient Continuum Levels

Level	Autonomy Type	System Capability	Human Involvement	Governance Intensity
Level 1	Assisted	Decision support	High	Standard oversight
Level 2	Semi-Autonomous	Conditional execution	Moderate	Policy monitoring
Level 3	Adaptive	Dynamic orchestration	Limited	Embedded checkpoints
Level 4	Self-Evolving	Reinforcement learning	Minimal	Continuous audit & override

4.2 Layered AGA Architecture Design

The Layered AGA Architecture is premised on the configuration of autonomous capabilities into layers of functions that are interconnected. Data streams, events logs and unstructured records of the internal and external environment are recorded by the sensing layer in real time. The intelligence layer applies the machine learning models, predictive process mining, and AI-driven analytics in order to decode patterns and foresee deviations. The orchestration layer dynamically recreates workflows, resources and executions sequences based on the provided performance metric and risk indicators (Munnangi, 2018). The feedback layer assesses performance and compares performance against the strategic benchmarks and refines systems parameters in order to support continuous improvement. The last but not the least, the explainability modules, audit trails, compliance engines and the human oversight interfaces are incorporated in the governance layer so as to ensure transparency and accountability. These layers are a complete ecosystem instead of individual modules. The structured layering with its modular scalability, internal consistency, and the containment of risks make the enterprises have the power to proceed with autonomy in bits without destabilizing operations and the company strategy. Table 3 Operational functions of AGA layers and functional breakdowns.

Table 3. Layered AGA Architecture Components

Layer	Core Function	Technologies Involved	Output
Sensing	Data capture	Event logs, APIs	Contextual signals
Intelligence	Pattern analysis	ML, Predictive mining	Forecasts & insights
Orchestration	Workflow adaptation	AI engines, RPA	Dynamic execution
Feedback	Performance tracking	KPI dashboards	Adaptive refinement
Governance	Compliance control	Policy engines, XAI	Transparency & audit

4.3 Governance-Aware Control Mechanisms

This is due to governance-sensitive control mechanisms entrenched in AGA so as to prevent that enhanced autonomy does not affect ethical integrity or regulatory adherence. Transparency and trust Explainability frameworks give automated decisions traceable reasoning, improving their explanation. The audit trails will document process amendments, rationale behind decisions as well as interventions, which are made to the system to verify compliance. Policy engines are dynamically implemented limits of regulations and organizational rules within execution settings. Human-in-the-loop validation systems permit control in situations that are too risky or unpredictable and maintain the human agency and responsibility (Debbadi & Boateng, 2025). Risk-sensitive override systems allow a quick response in the event of anomalies or breaches of threshold. There are also bias detecting modules that track the outputs of the algorithms to counter the discriminatory behavior. Governance controls are not set as supervisory instruments but rather as system building blocks. This integration makes sure the development of autonomy can be aligned with the ethical standards, legal regulations, and the enterprise values. Therefore, the governance conscious design has turned autonomy into an ability that is responsible and strategically sustainable in intelligent enterprises.

5. RESEARCH METHODOLOGY

The study design is a systematic conceptual theory-creating direction aimed at creating and advancing the Autonomy Gradient Architecture (AGA). Since Autonomous Business Process Management is still emerging, the paper focuses on model construction, architectural synthesis, and the formulation of metrics as opposed to massive empirical experimentation. Using the methodology, the experiences of Business Process Management, socio-technical systems theory, enterprise architecture, and AI governance are combined to build a unified model of autonomy

(Nwachukwu et al., 2025). Theoretical rigor is achieved by a systematic correspondence between the research purpose and objective, the conceptual constructs, and evaluation indicators. The structure ushers in quantifiable autonomy and governance parameters to facilitate empirical evaluation in subsequent application. In order to enhance the notion of practical relevance, the proposed architecture is analyzed in terms of case-based analytical validation in various spheres. This synthesis, metric operationalization, and validation approach of triangulation ensures a depth of concept, and relevancy of application. This methodology thus makes it possible to ensure that AGA is both theoretically based but also operationally explainable and applicable in different enterprise contexts.

5.1 Conceptual Theory-Building Approach

It is a conceptual theory-building approach used in the research to formalize autonomy in business process systems. As opposed to gathering primary empirical data, the methodology synthesizes interdisciplinary literature of BPM, Artificial Intelligence, governance systems, and enterprise architecture. Basic notions like levels of autonomy, alignment of governance, adaptive orchestration, and self-evolution are the systematically defined and put into place within a layered architectural model. The process is systematically based on the following steps: identification of theoretical gaps, formulation of constructs, architectural integration and development of relational connections between progression of autonomy and governance mechanisms (Zetta Software Tlc Shpk et al., 2025). The theory of socio-technical systems offers the basis of combining human control and machine intelligence, and the AI governance principles guide the development of accountability and transparency. The outcome is a normative framework, which is supposed to inform the ordered implementation. This is suitable in new areas where the clarity of concepts is the key to major empirical verification. The mechanism of building the theory would guarantee the internal consistency, rationality, and compliance with the ideas of established enterprise design.

5.2 Evaluation Metrics and Autonomy Index

In order to project the Autonomy Gradient Architecture on measurable evaluation indicators, it is proposed. The Autonomy Index measures the process intelligence degree, which consists in the amount of decision independence, orchestrating ability and learning ability. The Governance Alignment Coefficient measures the alignment between the autonomous activity and regulatory, ethical and organizational restraints. Other measures encompass Adaptive Cycle Time Ratio, which is the ratio of speed of reconfiguring the process, Decision Transparency Score, which evaluates explainability and traceability and Self-Evolution Index which is the continuous improvement performance. The aggregation of these metrics gives a multi-dimensional assessment model. Instead of determining automation volume, the indicators estimate the ratio between the growth of autonomy and the strength of governance. The metric framework assists comparative maturity evaluation among firms within the organization or industry (Ike et al., 2021). The evaluation system may be empirically validated, benchmarked, and planned in an organized way of deployment in intelligent enterprise setting by translating the conceptual constructs into quantifiable parameters. Table 4 Suggested quantitative measures of autonomy and alignment of governance.

Table 4. Evaluation Metrics for AGA

Metric	Definition	Measurement Focus	Strategic Value
Autonomy Index	Degree of decision independence	AI-driven actions	Maturity assessment
Governance Alignment Coefficient	Compliance synchronization	Policy adherence	Risk mitigation
Adaptive Cycle Ratio	Speed of process adjustment	Reconfiguration time	Operational agility
Transparency Score	Explainability level	Audit trace clarity	Trust building
Self-Evolution Efficiency	Continuous improvement rate	Learning optimization	Long-term sustainability

5.3 Validation through Case-Based Analysis

The suggested architecture is checked through the systematic analysis of the case that is based on the chosen industries such as finance and healthcare. The case-based method evaluates the action of autonomy gradients in the actual organizational cases, the workflow personalization analysis, regulation policies and the overall performance outcome. The proposed assessment indicators are implemented to each of the cases that enable to define the degree of autonomy, congruence in governing, and adaptive efficiency (Lyell et al., 2021). The comparative analysis creates abilities, drawbacks and regional conditions in the deployment of autonomy. It is a qualitative validation method which allows studying complex socio-technical interactions which cannot be acquired using quantitative methods exclusively. Practical information is also provided in analysis of cases on the issues of implementation like resistance to change, compliance constraints, and complexity of integration. Although the case-based validation is an exploratory type, the additional strength of the AGA framework applicability is offered. It demonstrates the ways in which the structured autonomy progression can be useful to the resilience, transparency, and strategic alignment within the various enterprise settings.

6. LITERATURE REVIEW

The Autonomous Business Process Management literature represents a progressive shift in the conventional approach of workflow automation to the concept of intelligent and adaptable process ecosystems that are self-optimizing. The initial research on BPM focused on process modelling, standardization and efficiency of processes via rule bases systems and enterprise integration platforms. As sophisticated analytics and Artificial Intelligence came into being, the focus of scholarship switched to the monitoring of predictive processes, detection of anomalies, and the allocation of resources dynamically. Modern discourse is gradually turning to process autonomy whereby the system is capable of perceiving the signals of the context, predicting disruptions, and reconfiguring the execution routes with the minimal human interventions. However, autonomy is described in the literature as the result of technological capability and not a systematic architectural development (BENTONVILLE USA & Natta, 2025). The notion of governance integration and accountability mechanisms is often talked about in isolation of process intelligence. The given fragmentation marks the necessity of having a single conceptual framework that would be able to balance autonomy scaling and governance design. The subsequent subsections provide the synthesis of streams of research that guide the development of the Autonomy Gradient Architecture.

6.1 Autonomous Business Process Management

ABPM research is based on the development of AI, process mining, and real-time analytics. Research focuses on predictive process monitoring, which uses event logs and machine learning models to predict delays or compliance violations or performance deviations before they happen. Proposals have been made to have adaptive workflow systems which dynamically reconfigure process paths when there are changes in the environment. The reinforcement learning methods also allow systems to acquire optimal decision policies with respect to reward feedback systems (Beigi & Franconi, 2025). Digital environments are also becoming more intelligent with AI-based decision engines combined with robot process automation that minimizes human involvement in routine processes. Although these progress have been achieved, majority of the literature portrays the concept of autonomy at an operational level as opposed to conceptualizing at the architectural or governance level. Limited systematic deployment is caused by there being no maturity continuum or organised autonomy gradient. The resultant implication is that a growing amount of appreciation exists that autonomy has to be contextualized within enterprise-wide governance frameworks in order to promote scalability, transparency, and strategic consistency.

6.2 AI Governance and Ethical Control Systems

As the field of smart automation has grown, much discussion about AI regulation and its moral containment has been generated. The literature on this field includes the transparency, explainability, fairness, accountability and regulatory compliance. Explainable AI systems are made to provide users trust and auditability, through the provision of them traceable lines of reasoning to gain awareness of why an algorithm decided to. The danger of ethical AI research is an algorithmic bias, misuse of data and non-transparent decision-making. The designs of governance processes are also adopting the principle of applying compliance checks and human controls within the framework of systems rather than apply external controls of oversight (Gabriel, 2025). The literature on enterprise risk management also indicates the need to have a dynamic compliance mechanism that would operate in real time. Despite the fact that the research in the field of governance is valid, it is not always associated with BPM architecture. The aspect of integration of governance intelligence in independent process design is not cultivated. This gap is what supports the need to directly introduce governance-reflective mechanisms into autonomy structures in order to create a balance between innovation and accountability and regulatory integrity.

6.3 Digital Twins and Reinforcement Learning in BPM

The recent research emphasizes the importance of digital process twins and reinforcement learning in the process of facilitating self-evolving ecosystems of processes. Digital twins are simulation models that recreate the conditions of actual processes in the real environment, and this means organizations can be provided with an opportunity to test other execution strategies, prior to implementation. This functionality is in aid of the scenario analysis, risk analysis and forecasting of performance (Aarthi, 2025). Reinforcement learning presents adaptive decision-making by use of continuous feedback where the system can improve the policies using rewards and penalties in the environment. When used together with predictive analytics, they form closed-loop optimization systems that can adjust themselves independently. However, it is also observed in the research that there are challenges with model drift, data quality, and computational complexity. Also, the autonomous learning systems need governance controls to avoid unplanned changes in behavior

(Gamboa-Cruzado et al., 2025). The intersection between digital twins and reinforcement learning is the beginning of dynamic and self-optimizing BPM, but they need to be designed in specific frameworks to keep such technologies consistent with the goals of the organization and ethical limits.

7. CASE STUDY

To illustrate the implementation of an Autonomy Gradient architecture to the practical case, two recent industry examples are generalised by experience of the on-scale execution of process intelligence + automation + governance controls made publicly available. The chosen cases demonstrate how autonomy was improved to go beyond the independent task automation to adaptive orchestration (dynamic routing, predicting exceptions, and autonomous monitoring) and workflow reconfiguration (understanding documents automatically, compressing the turnaround-time and more accuracy). All the cases can be described in terms of calculable outcomes (percentage improvements) and are balanced to autonomy-gradient behaviors: (i) sensing by consuming event logs and real-time indicators, (ii) intelligence by consuming predictive analytics or human-understood AI documents, (iii) orchestration by automated work allocation and reduced cycle time and (iv) governance by auditability, assurance of accuracy and human control of exceptions (Eshuis et al., 2019). This is followed by a highly rapid, cross-sector comparison, to establish some of the design principles that can be migrated and the seemingly resulting governance risk that arises where task automation no longer occurs and instead processes are customised.

7.1 Financial Services Adaptive Orchestration

Case (2025): Banking (payments / onboarding / reporting): Process Intelligence-coordination. Banking-specific process intelligence program introduced recently has demonstrated some measurable delivery in line with adaptive orchestration results: a 30% cycle time decrease on cross-border payments, 80% cycle time decrease on reporting SLA violations, and 18% increase in paying processing capacity on an end-to-end visibility and actionability were implemented on process variants. These results are suggestive of an autonomy development where an event log and an operational signal are continuously consumed (sensing), bottlenecks and potential violations are discovered early (intelligence) and work is diverted or increased in response to an SLA violation (orchestration) (Ravikumar, Aarthi, Pardaev, et al., 2025). The decrease in SLA violations means that exception handling became more proactive intervention than reactive firefighting where a fast response can be obtained and fewer delays can be observed at the downstream. Capacity increase means that the same amount of operational footprint was realized by reducing idle time and rework. That is akin to AGA terms of a shift between semi-autonomous execution (discrete task automation) to adaptive orchestration (system driven, KPI driven reconfiguring processes) without impairing governance needs through traceable reporting and constrained interventions (Bhardwaj, 2025). Percentages improvement in the reduction of the cycle time, decrease in the SLA breach and increase in the processing capacity attained in the banking operations by adaptive orchestration as presented in Figure 4 as percentages.

Figure 4. Financial Services Adaptive Orchestration Performance Gains

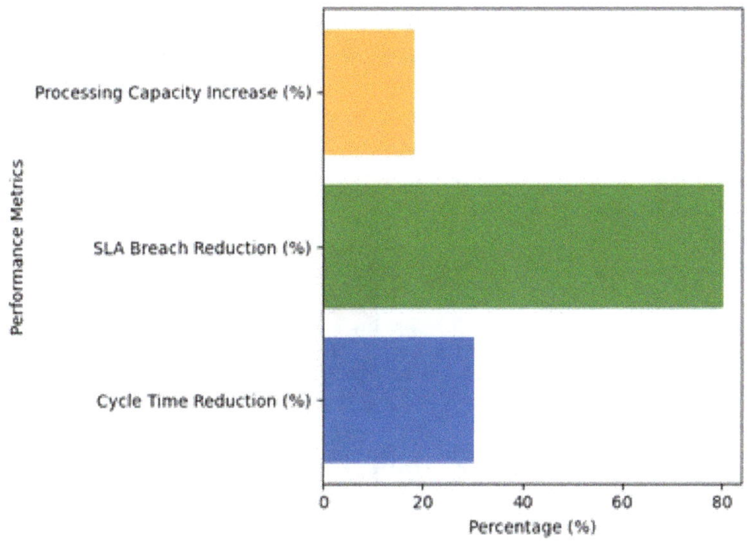

7.2 Healthcare Intelligent Workflow Reconfiguration

Case (reported 2025): Omni Healthcare UiPath AI revenue cycle document processing. Understanding and automation Documents with AI-assisted document-related workflows were introduced in the Omega Healthcare which maintains the revenue-cycle activities of hundreds of provider organizations. The reported results are 15,000 hours in saving of employees every month, 40% reduced the time in documentation, 50% reduced the time in document-processing, and 99.5% accuracy in the processes. It was mentioned that automation coverage was 60-70% of processes of clients whose ROI will be estimated at 30%. These results describe the reconfigurability of workflows as an intelligent one as compared to single bots:

unstructured documents (e.g., denial letters, medical records) are converted into structured signals (sensing), AI extraction and validation reduces the number of manual processes (intelligence), and pipelines of routing/processing are redesigned to reduce turnaround time (orchestration). It is also governed by the aid of the accuracy goals, auditable processing steps, and the ambiguous items that should be examined by humans under the exception, which is required in the regulated healthcare billing through in the regulated setting (Ravikumar, Aarthi, Dilshodbek, et al., 2025). As indicated in the terms of AGA, the case illustrates the transition to the upper strata of autonomy in which systems no longer perform the work in any way, but also transform the direction of the working process and improve the cycle-time performance without reducing the accountability. Figure 5 illustrating the percentage change in the documentation time reduction, turnaround time reduction, process accuracy, the automation coverage, and client ROI in the reconfiguring of a workflow involving AI.

Figure 5. Healthcare Workflow Performance Gains

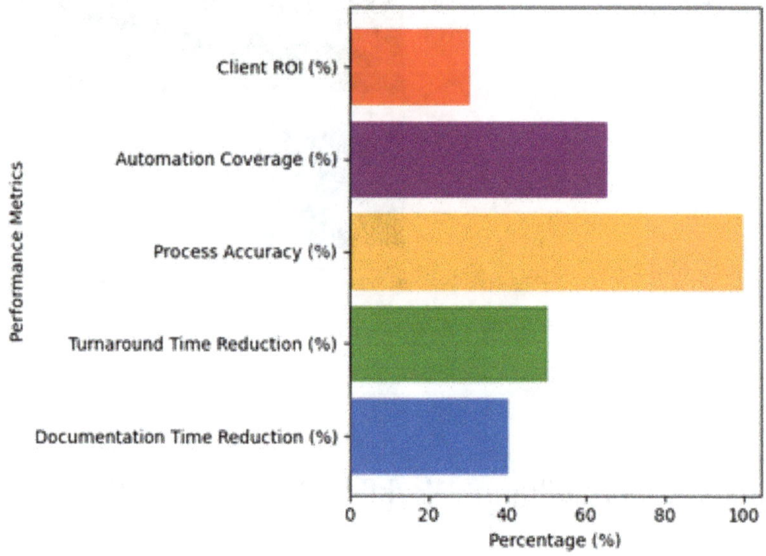

7.3 Cross-Sector Comparative Insights

The optimum returns are offered in both fields where flow end-to-end is implemented in autonomy and not independent jobs. Such a cycle-time compression (30%) and SLA reliability (80% breach is mitigated) is more akin with the consideration of the banking performance that the autonomy can be observed periodically and in a way which it is estimated that it would increase. Since the unstructured inputs are constantly transformed into structured decisions, the outcomes of healthcare are also realized such as throughput/time saving (15,000 hours/month) and quality of processing (99.5% accuracy) and autonomy, the latter, in its turn, is developed extensively. The architecture patterns of the AGA are the same: (i) process instrument with logs/signals, (ii) predict deviations, by having an intelligence, (iii) coordinate interventions, with KPI/risk signals, (iv) add governance (auditability, accuracy controls, human override) (Singh, 2024). It is only a more sophisticated form of governance: healthcare must be improved on the aspect of tracing the documentation, and determining whether it is correct; whereas the banking part is a lot to do with the SLA suitability and monitoring the operational risk. Table 5 Comparison of performance improvement of implementation of financial services and healthcare implementations.

Table 5. Case Study Performance Outcomes

Indicator	Financial Services	Healthcare
Cycle-Time Reduction	30%	50%
SLA Breach Reduction	80%	—
Capacity Increase	18%	—
Time Savings	—	15,000 hrs/month
Documentation Time Reduction	—	40%
Accuracy Rate	—	99.5%
Automation Coverage	—	60–70%

8. RESULTS AND DISCUSSION

The findings of the analyses of the financial services and healthcare cases can be used to indicate that organized autonomy advancement can produce quantifiable operational, governance, strategic results. In both industries, the autonomy of the tasks was further than automation into adaptive orchestration and smart workflow redesign. The banking case demonstrated the reduction of cycle-time, stabilization of SLA and throughput improvement, which showed the better systemic coordination and predictive responsiveness. The healthcare case has shown impressive time savings of documentation, compression of turnaround-time, and near perfect accuracy rates, which represents clever processing of unstructured information and real time reconfiguring of workflow (Oluwafemi et al., 2021). Notably, the performance gains were made together with internal checks and balances in support of the philosophy that autonomy should be modified to concur with the governance checks. However, the new threat dimension associated with algorithmic bias, explainability limitations, and human decision displacement also came with greater autonomy. This discussion thus compares performance advantages with transparency and accountability aspects and places emphasis on the fact that sustainable autonomy lies in ensuring that a balance between adaptive intelligence and responsible oversight is maintained.

8.1 Performance and Resilience Gains

The case of financial services showed that the cycle time was lowered by 30%, SLA violations were also lowered by 80% and processing capacity increased by 18%, that is, it is seen that operational resilience was improved. The enhancements of those demonstrate that both predictive monitoring and adaptive routing minimized the number of bottlenecks and exception escalation delays. The case scenario in healthcare provided an example of 15,000 hours per month saved on time through automation, cut down the time needed on documentation by 40%, cut down the turnaround time by half and accuracy in processing by 99.5%. These results demonstrate that the smart document perception and workflow re-design would have a potential to reduce the administrative latency by a significant margin without reducing the quality (Okoruwa et al., 2024). The concept of resilience was achieved in the two instances by averting errors in the form of proactive detection of deviation and multiplied dynamics of error adaptation rather than error redress. The findings also suggest that the autonomy gradients also allow the organizations to stay stable in-service provision (in a variably conditioned condition), augment throughput performance and be prone to environmental instability. Organized autonomy also leads to systemic stability and operational resilience besides efficiency. Figure 6 that shows that proportional or percentage contribution of reduction in cycle time, reduction in SLA breach along with increase in the capacity to the total performance gains in financial services adaptive orchestration.

Figure 6. Financial Services Performance Distribution

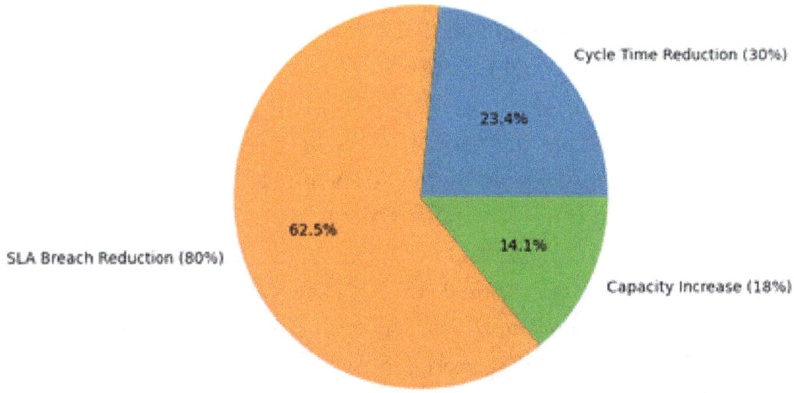

8.2 Governance Alignment and Transparency Outcomes

According to the evidence presented in the case, the mechanisms of governance were introduced along with autonomy deployment, which strengthens the principles of transparency and accountability. Within financial services, it can be seen that there is minimization of SLA breaches, this implies enhanced compliance monitoring and performance traceability. Auditable reporting systems supported adaptive orchestration through which the decision paths and exception handling could be monitored. Within the healthcare field, 99.5% accuracy of processes and formal systems of document validation demonstrate good alignment of governance within a controlled environment. The presence of the human-in-the-loop validation of ambiguous cases-maintained control over high-risks situations (Karras et al., 2025). These results show that autonomy scaling did not destroy governance but it only further architecturally integrated it. Transparency was ensured by steps in processing which were traceable, performance dashboards, and well-set escalation procedures. These findings confirm the theory that governance intelligence should be a structural layer and not a supervisory one. Proper development of decision intelligence and compliance assurance systems is thus needed to facilitate the development of effective autonomy.

8.3 Risks: Bias, Opacity, and Human Agency

Even though considerable improvements have been observed, the incidences point to the potential catastrophes of escalating the degree of autonomy. The algorithmic bias may be developed based on the AI-based orchestration in case the training data is founded on the historical inequities or biased representations. In automation of healthcare documents, there might be there is some form of opaqueness due to the use of AI extraction in the event that the decision rationale is not explicable well. Predictive routing is able to obscure decision criteria in the underlying decision making of financial services and this prevents such decisions being perceived to have been used in a compliance audit. Secondly, the ability to automate healthcare processes on the scale of 60-70% results in the problem of human agency loss and replacement of human skills as well. Over-dependence on adaptive systems may lead to a situation in which the operational staff is less aware of the situation (Babalola et al., 2022). The existence of these risks makes it clear why governance-conscious control systems like explainability modules, override systems, and bias monitoring systems are relevant. The discussion confirms the reality that autonomy must be gauged in a wise fashion. The process of transformation of sustainable enterprises is not founded only on the efficiency level due to the increase, but also on the continuity of transparency, justice, and substantial human control in intelligent ecosystems of processes.

9. FUTURE RESEARCH

Although the Autonomy Gradient Architecture (AGA) offers a deliberated conceptual and valuational framework regarding governance conscious autonomous process systems, a number of directions are available to be explored in the future. To enhance the theoretical generalizability, empirical validation across industries, longitudinal analysis of performance and cross-organizational scalability studies should be carried out. As businesses switch to Industry 5.0 model that are more human-centric and sustainable, autonomy models will need to change to facilitate collaborative intelligence and not complete automation replacement. Also, another issue is that as the integration of organisations between each other increases (via digital platforms, supply chains, and ecosystem alliances) new governance complexities emerge (Sharma, 2024). The suggested future research ought to, therefore, focus on quantitative verification of autonomy metrics, study human-AI co-governance structures, and research autonomy scaling in distributed networks of enterprises. It will be imperative to gain a better insight into socio-technical dynamics, cultural accommodation, harmonization of regulations, and risk-sensitive deployment strategies. The next frontier of the Autonomous Business Process Management research is the development of structured autonomy at the level of ecosystem-wide intelligent coordination.

9.1 Quantitative Validation of AGA Metrics

The subsequent study should pay attention to the empirical validation of the suggested evaluation framework: the Autonomy Index, Governance Alignment Coefficient, Adaptive Cycle Time Ratio, Decision Transparency Score, and Self-Evolution Efficiency Index. The reliability and predictability of these measures can be tested in large-scale quantitative research in finance, healthcare, manufacturing, and logistics (Ravikumar, 2025c). Correlations between the levels of autonomy and the indicators of performance, i.e., throughput, resilience, compliance stability, and optimization of costs, can be investigated with the use of statistical modeling. The longitudinal studies are especially significant to determine whether maturity of autonomy contributes to a lasting competitive advantage or decreasing marginal returns. Experimental research can also compare the implementation of the gradient to unstructured automation strategies to consider the difference in results. Also, statistical benchmarking systems may allow comparing autonomy development across industries. The quantitative validation will make AGA more theoretically sound and offer the practitioners a set of standardized instruments to deploy and evaluate the performance.

9.2 Industry 5.0 and Human–AI Co-Governance

The purpose of industry 5.0 is human-focused innovation, sustainability and collaborative intelligence between people and machines. The next round of research ought to look into how Autonomy Gradient Architecture can be used to facilitate human-AI co-governance models that maintain agency without increasing adaptability capacity. Research can be conducted to examine how good autonomy levels are where human strategic control is complementing machine productivity. The study of behavioral phenomena could include examination of the processes of trust, acceptance of decision, and skill change in high autonomy situations (Ravikumar, 2025b). To accommodate justice, integrity and collaborative governance within smart businesses, ethical models should be extended as well. The adaptive oversight models that need additional design investigation are dynamic intervention triggers and collaborative dashboards. The studies are supposed to examine the role played by the organizational culture, leadership structures and competencies as far as the successful integration of autonomy is concerned (Visagamurthy, 2025). The shift to Industry 5.0 implies that autonomy cannot take human judgment away but support it. Thus, the main focus of inquiry in the future is sustainable and inclusive models of autonomy in the balancing of technological development and socio-ethical accountability.

9.3 Cross-Organizational Autonomy Scaling

Since businesses are moving towards more and more interdependent ecosystems, future studies will need to take on cross-organizational scaling of autonomy. Supply chains, digital platforms, and public-private alliances are autonomous processes that bring about coordination and governance complexities across enterprise-level boundaries. Research must look at interoperability criteria, information exchanging guidelines and distributed governance frameworks that can facilitate autonomy among networked organizations. The autonomous decisions of one organization may also affect the downstream partners and this necessitates the analysis of risk propagation mechanisms. Comparative studies could be done on the centralized and federated autonomy governance systems in the ecosystem settings. The analysis of scalability must be based on regulatory heterogeneity between jurisdictions and sector-specific compliance limitations. (Klun & Trkman, 2018) Also, resilience modeling has seen to be able to measure the implication of distributed autonomy on stability in a system in case of distributed disruption. Knowledge of the autonomy gradients expansion beyond silos will be important in the creation of collaborative and smart ecosystems. Cross-organizational scaling is thus a research strategy that can be taken to develop sustainable autonomous enterprise architectures.

10. MANAGERIAL IMPLICATIONS

The adoption of Autonomy Gradient Architecture (AGA) has considerable managerial consequences on organizations that move away to more adaptive and governance-conscious process ecosystems by adopting automation as a managerial tool. The results of the case indicate that autonomy can only provide quantifiable performance benefits when implemented in a systematic and coordinated way and consistent with risk management models. Managers then have to think of autonomy as not a technological upgrade but a strategic change initiative that incorporates the architecture design, governance embedding, workforce adaptation, and performance measurement. Organized deployment the deployment makes it less fragmented, it curtails uncontrolled expansion of automation and allows scaling to be predictable. The commitment of leadership plays a crucial role in matching the initiatives related to autonomy with organizational aims, regulatory needs, and expectations of the stakeholders. Moreover, the shift to the next level of autonomy will also demand the cultural preparedness, restructuring of skills, and the mechanisms of higher transparency to maintain trust (Uzoamaka et al., 2025). The subsequent subsections present managerial course of action, such as gradient-based deployment planning, systematic maturity evaluation, and governance-focused control measures that would facilitate to make sure that the autonomy progression would be performance-increasing and ethically based.

10.1 Gradient-Based Deployment Planning

The use of gradient-based deployment planning helps managers to use autonomy progressively instead of in large-scale automation efforts. By establishing the levels of structured autonomy, such as assisted decision support to adaptive orchestration, an organization is able to match the degree of deployment with the scales of operational complexities and regulatory sensitivity. Managers would then start with low-risk process where the decision support tools would improve the human performance, and then either move to semi-autonomous execution. Cycle-time reduction, exception latency and accuracy level are performance measures that can be used to make progression decisions (Mutschler & Reichert, 2013). Each stage needs to be followed by risk assessment to have the proper governance mechanisms that are proportionately sized. Such a gradual strategy will eliminate disturbance to the functioning of the organization and enable constant assessment of the performance and compliance results. Planning, which is gradient-based, also makes it easy to achieve stakeholder acceptance as it is more likely to avoid transparency issues and retain the human control over the planning in initial adoption processes. Associating the level of autonomy with the strategic goals and the risk levels, managers are able to turn automation efforts into controlled, quantifiable, and long-term enterprise capabilities.

10.2 Autonomy Maturity Assessment Framework

A maturity assessment framework on autonomy helps organizations to assess their present status within the autonomy gradient and the ways of progressing. The model must include multidimensional measures, such as the autonomy index score, the score of the alignment of governance through the coefficient and the score of adaptive cycle proportions, the transparency indicators. These measures help managers to compare the departments, cross-functional preparedness, and identify a lack of alignment between technological ability and governance preparedness. Maturity assessment also helps in prioritization of investments by establishing the high impact process areas that can be used to scale autonomy. Notably, maturity assessment should have qualitative criteria like workforce flexibility, cultural preparedness and leadership facilitation (Tarasiuk, 2025). Technological maturity may not bring long-term results without organizational alignment. A frequent review helps to guarantee the ongoing enhancement, and avoid stagnation at the stage of partial autonomy. Formalizing the maturity evaluation will allow the managers to shift their ad hoc automation experimentation to formal autonomy governance, which will guarantee the coherent enterprise-wide change in line with long term strategic objectives. Figure 7 that illustrates important quantitative and qualitative measures- autonomy index, governance alignment, adapting cycles, transparency, and qualitative measures- in determining enterprise autonomy maturity.

Figure 7. Autonomy Maturity Assessment Framework

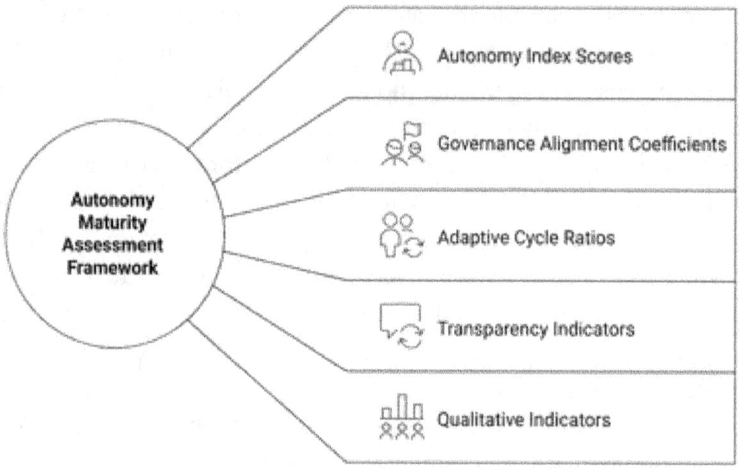

10.3 Governance Checkpoints and Risk Overrides

The most important managerial controls in a more autonomy setting are governance checkpoints and risk override mechanisms. With processes now adaptive and self-optimizing, control systems are to be incorporated in working processes. Governance checkpoints consist of predetermined steps of reviewing where automated decisions are evaluated against policy limits and performance boundaries. Human in the loop validation should be caused by high-risk transactions or unclear situations to maintain accountability. The risk override systems allow quick intervention in the case of anomalies that are beyond acceptable tolerance levels. To maintain transparency, managers are encouraged to put up bias monitoring procedures, explainability dashboard, and traceable audit logs (Ziwoni, 2025). The regulatory compliance requirements should be formalized into policy engines operating in a continuous mode and not in a retroactive manner. There should also be well defined escalation routes in case of unanticipated behaviour of the system.

Organizations may avoid autonomy drift by institutionalizing governance checkpoints. Proper managerial control is the key to making sure that the performance gains do not erode the ethics, integrity in compliance, and trust of the stakeholders in smart enterprise ecosystem. Table 6 Real-world roadmap between the levels of autonomy and managerial activities and corporate controls.

Table 6. Managerial Implementation Roadmap

Phase	Autonomy Level	Managerial Action	Governance Control
Phase 1	Assisted	Deploy analytics tools	Manual audit review
Phase 2	Semi-Autonomous	Automate low-risk tasks	Policy monitoring
Phase 3	Adaptive	Enable dynamic orchestration	Real-time checkpoints
Phase 4	Self-Evolving	Implement RL & digital twins	Risk override systems

11. CONCLUSION

The movement toward business process ecosystems that are self-evolving and no longer traditional automotive can be seen as a fundamental change in how businesses structure, manage, and maintain their performance in operations. It has been argued in this chapter that the concept of autonomy cannot be thought of as dichotomous transition between manual and automated execution, but as an ordered continuum embedded in enterprise architecture. The Autonomy Gradient Architecture (AGA) offers a formalized scheme, which incorporates sensing, intelligence, orchestration, feedback, and governance into a unified and scalable scheme. AGA is able to promote the development of adaptive intelligence through accountability and transparency by applying the levels of autonomy, instilling quantifiable measures of evaluation, and linking progression with governance safeguards. Evidence on the case in the financial services and healthcare industries shows that autonomy, systematically architected, is an achievable source of quantifiable benefits in cycle-time reduction, accuracy, throughput capacity, and resilience. But these advantages can only be maintained in the event that autonomy and governance are performed in the same framework. In the governance, explainability, auditability, human oversight, and risk overrides should be a part of intelligent systems and not external supervisory controls. The results support the finding that there is no way that one can have the autonomy of the true process without the infrastructure governance intelligence. In the case of smart businesses that are working in Industry 5.0 settings, strategic value will be more reliant on the capacity to develop process ecosystems that are always learning, adapting and refining themselves without losing their ethical focus or breaking legal regulations. Autonomy Gradient Architecture is thus a strategic roadmap to the enterprise that would like to progress out of the fragmented automation projects into accountable, robust and self-functioning digital procedure systems to balance machine insight with human strategic administration.

REFERENCES

Aarthi, S., Ravikumar, R. N., & Pardaev, J. (2025). Synergizing multimodal generative AI and blockchain for the future of digital media. In *Advances in Computational Intelligence and Robotics* (pp. 1–34). IGI Global Scientific Publishing. DOI: 10.4018/979-8-3373-1504-1.ch001

Aarthi, S., Ravikumar, R. N., Sobirova, K., & Mamadiyarov, Z. (2025). Empowering a sustainable future through digital learning and the SDGs. In *Harnessing E-Learning to Create a Sustainable Future* (pp. 145–172). IGI Global Scientific Publishing., DOI: 10.4018/979-8-3373-0112-9.ch006

Adesanya, O. S., Akinola, A. S., & Oyeniyi, L. D. (2021). Robotic Process Automation ensuring regulatory compliance within finance by automating complex reporting and auditing. *International Journal of Multidisciplinary Research and Growth Evaluation, 2*(4), 1051–1067. DOI: 10.54660/.IJMRGE.2021.2.4.1051-1067

Ayoubi, M., Shinwari, A. R., & Khan, M. S. (2019). *An event-driven service oriented architecture approach for E-governance systems.* Kardan Journal of Engineering and Technology., DOI: 10.31841/KJET.2021.1

Babalola, O., Iluore, E., Bakare, A., & Udechukwu, L. M. (2022). AI-Enabled Business Process Optimization Engine for risk-Aware Cloud Operations. *Journal of Frontiers in Multidisciplinary Research, 3*(2), 211–216. DOI: 10.54660/.IJFMR.2022.3.2.211-216

Beigi, H., & Franconi, A. (2025). Design for Traceability (DfT). *Proceedings of the 6th Product Lifetimes and the Environment Conference (PLATE2025)*, (6). DOI: 10.54337/plate2025-10322

Bhardwaj, A. (2025). Systematic literature review on explainable AI in finance: Methods, applications, and research gaps. *International Journal For Multidisciplinary Research, 7*(5), 58963. Advance online publication. DOI: 10.36948/ijfmr.2025.v07i05.58963

Debbadi, R. K., & Boateng, O. (2025). Enhancing cognitive automation capabilities with reinforcement learning techniques in robotic process automation using UiPath and automation anywhere. *International Journal of Science and Research Archive, 14*(2), 733–752. DOI: 10.30574/ijsra.2025.14.2.0450

Eshuis, R., Hull, R., & Yi, M. (2019). Reasoning about property preservation in adaptive Case Management. *ACM Transactions on Internet Technology, 19*(1), 1–21. DOI: 10.1145/3177778

Gabriel, V. S.Victor Samuel Gabriel. (2025). AI and distributed manufacturing systems: Strengthening healthcare supply chains for national biosecurity. *World Journal of Advanced Research and Reviews*, *26*(2), 4080–4086. DOI: 10.30574/wjarr.2025.26.2.2033

Gamboa-Cruzado, J., Pineda-Delacruz, V., Salcedo-Mera, H., Alzamora Rivero, C., Coveñas Lalupu, J., & Narro-Andrade, M. (2025). Blockchain and data management security for sustainable digital ecosystems: A systematic literature review. *Sustainability (Basel)*, *18*(1), 185. DOI: 10.3390/su18010185

Gudipudi, S. (2025). Toward autonomous business intelligence: Research trends in automation and cloud integration. *European Journal of Computer Science and Information Technology*, *13*(49), 163–177. DOI: 10.37745/ejcsit.2013/vol13n49163177

Halliday, N. (2023). A conceptual framework for financial network resilience integrating cybersecurity, risk management, and digital infrastructure stability. *International Journal of Advanced Multidisciplinary Research and Studies*, *3*(2), 1253–1263. DOI: 10.62225/2583049X.2023.3.2.4887

Hetman, K. N. E. U. named after V., Sahaidak, M., Maryniuk, I., & Technology, S. U. of E. and. (2025). Conceptual foundations for the formation and development of organizational compliance management systems under digital transformation. In *Integration of Marketing into the Public Administration System: Ukrainian Experience and Prospects* (pp. 69–102). Scientific Center of Innovative Research. DOI: 10.36690/IM-PAS-69-102

Ike, P. N., Ogbuefi, E., Nnabueze, S. B., Olatunde-Thorpe, J., Aifuwa, S. E., Oshoba, T. O., & Akokodaripon, D. (2021). Supplier Relationship Management strategies fostering innovation, collaboration, and resilience in global supply chain ecosystems. *International Journal of Multidisciplinary Evolutionary Research*, *2*(2), 52–62. DOI: 10.54660/IJMER.2021.2.2.52-62

Karras, A., Theodorakopoulos, L., Karras, C., Krimpas, G. A., Giannaros, A., & Bakalis, C.-P. (2025). LLM-driven big data management across digital governance, marketing, and accounting: A spark-orchestrated framework. *Algorithms*, *18*(12), 791. DOI: 10.3390/a18120791

Klun, M., & Trkman, P. (2018). Business process management – at the crossroads. *Business Process Management Journal*, *24*(3), 786–813. DOI: 10.1108/BPMJ-11-2016-0226

Lawal, O. A., & Oduleye, T. E. (2020). Process automation and financial reporting integrity: A conceptual governance model. *International Journal of Multidisciplinary Research and Growth Evaluation, 1*(5), 583–594. DOI: 10.54660/.IJMRGE.2020.1.5.583-594

Lyell, D., Coiera, E., Chen, J., Shah, P., & Magrabi, F. (2021). How machine learning is embedded to support clinician decision making: An analysis of FDA-approved medical devices. *BMJ Health & Care Informatics, 28*(1), e100301. DOI: 10.1136/bmjhci-2020-100301 PMID: 33853863

Munnangi, S. (2018). Seamless Automation: Integrating BPM and RPA with Pega. *Turkish Journal of Computer and Mathematics Education, 9*(3). Advance online publication. DOI: 10.61841/turcomat.v9i3.14971

Mutschler, B., & Reichert, M. (2013). Understanding the costs of business process management technology. In *Studies in Computational Intelligence* (pp. 157–194). Springer Berlin Heidelberg., DOI: 10.1007/978-3-642-28409-0_7

Natta, P. K.BENTONVILLE USA. (2025). Scalable governance frameworks for AI-driven enterprise automation and decision-making. *International Journal of Research Publications in Engineering. Technology and Management, 8*(6). Advance online publication. DOI: 10.15662/IJRPETM.2025.0806022

Nwachukwu, P. S., Chima, O. K., & Okolo, C. H.Priscilla Samuel NwachukwuOnyeka Kelvin ChimaChinelo Harriet Okolo. (2025). The artificial intelligence governance framework for finance: A control-by-design approach to algorithmic decision-making in accounting. *Finance & Accounting Research Journal, 7*(8), 350–379. DOI: 10.51594/farj.v7i8.2016

Okoruwa, P. O., Babatope, O. M., Akokodaripon, D. A., & Akinleye, O. K. (2024). Developing integrated digital platforms for enhancing transparency in procurement and supply chain management. *International Journal of Multidisciplinary Research and Growth Evaluation, 5*(6), 1719–1729. DOI: 10.54660/.IJMRGE.2024.5.6.1719-1729

Oluwafemi, I. O., Clement, T., Adanigbo, O. S., Gbenle, T. P., & Adekunle, B. I. (2021). Evaluating the efficacy of DID chain-enabled blockchain frameworks for real-time provenance verification and anti-counterfeit control in global pharmaceutical supply chains. *International Journal of Multidisciplinary Research and Growth Evaluation, 2*(2), 436–444. DOI: 10.54660/.IJMRGE.2021.2.2.436-444

Ravikumar, R. N., & Aarthi, S. (2025b). AI-driven digital twins and autonomous operations for sustainable logistics and procurement. In *Advances in Computational Intelligence and Robotics* (pp. 65–100). IGI Global Scientific Publishing. DOI: 10.4018/979-8-3373-7006-4.ch003

Ravikumar, R. N., & Aarthi, S. (2025c). Digital Twin applications in urban planning. In *Building Information Modeling and Geographic Information Systems for Next-Generation Spatial Intelligence* (pp. 29–58). IGI Global Scientific Publishing., DOI: 10.4018/979-8-3373-4058-6.ch002

Ravikumar, R. N., & Aarthi, S. (2025d). Real-time IoT, quantum advances, and ethical implications in autonomous AI systems. In *Advances in Computational Intelligence and Robotics* (pp. 185–216). IGI Global Scientific Publishing. DOI: 10.4018/979-8-3373-3146-1.ch007

Ravikumar, R. N., Aarthi, S., Dilshodbek, A., Satheesh Kumar, A., Eshkaraev, S., & Shanmugam, K. B. (2025). Integrating blockchain, smart contracts, and AI for decentralized healthcare in the metaverse. In *Blockchain-Based Healthcare Management in the Metaverse* (pp. 261–294). IGI Global Scientific Publishing., DOI: 10.4018/979-8-3373-1225-5.ch010

Ravikumar, R. N., Aarthi, S., Pardaev, J., & Shukla, P. (2025). Quantum computing for financial modelling. In *Quantum-Driven Financial Intelligence* (pp. 221–252). IGI Global Scientific Publishing., DOI: 10.4018/979-8-3373-2893-5.ch010

Ravikumar, & Aarthi, S. (2025a). Conscious AI Integration in business education. In *Advances in Computational Intelligence and Robotics* (pp. 73–104). IGI Global Scientific Publishing. DOI: 10.4018/979-8-3373-4377-8.ch003

Salameh, A., & Bass, J. M. (2022). An architecture governance approach for Agile development by tailoring the Spotify model. *AI & Society*, *37*(2), 761–780. DOI: 10.1007/s00146-021-01240-x

Sang, L. T., Ravikumar, R. N., Aarthi, S., & Ben, T. L. (2025). Optimizing sales and inventory performance using AI and business intelligence in POS environments. In *Advances in Computational Intelligence and Robotics* (pp. 259–290). IGI Global Scientific Publishing. DOI: 10.4018/979-8-3373-4392-1.ch011

Sharma, D. T. (2024). Embracing artificial intelligence in management: Navigating the futuristic landscape. In *Futuristic Trends in Management Volume 3 Book 13* (pp. 101–116). Iterative International Publisher, Selfypage Developers Pvt Ltd. DOI: 10.58532/V3BFMA13P3CH1

Singh, P. (2024). Using AI to navigate cross-border regulatory complexities in FinTech. *International Journal of Leading Research Publication*, *5*(1), 1596. Advance online publication. DOI: 10.70528/IJLRP.v5.i1.1596

Szelągowski, M., & Lupeikiene, A. (2020). Business process management systems: Evolution and development trends. *Lithuanian Academy of Sciences. Informatica (Vilnius)*, •••, 579–595. DOI: 10.15388/20-INFOR429

Tarasiuk, M. V. (2025). Genesis and evolutionary factors concepts of process-oriented management. *Вісник Київського Національного Лінгвістичного Університету. Серія Історія, Економіка. Філософія*, (30), 166–179. DOI: 10.32589/2412-9321.30.2025.331979

Uzoamaka, N.-O. P., Amujiogu, A. A., & Blessing, E. O. (2025). Governance and ethical oversight of agentic artificial intelligence in business process management: Balancing human-autonomy, organizational accountability and algorithmic transparency. *Journal of Basic and Applied Research International*, *31*(6), 121–133. DOI: 10.56557/jobari/2025/v31i610028

Vidgen, R., & Wang, X. (2006). From business process management to business process ecosystem. *Journal of Information Technology*, *21*(4), 262–271. DOI: 10.1057/palgrave.jit.2000076

Villamil, R. M., Restrepo-Carmona, J. A., Escobar, A., Aponte-Moreno, A., Herrera, J. A., Gutiérrez-Betancur, S. A., & Fletscher, L. (2025). An Enterprise Architecture-driven service integration model for enhancing fiscal oversight in supreme audit institutions. *Applied System Innovation*, *9*(1), 16. DOI: 10.3390/asi9010016

Visagamurthy, G. (2025). Cross-sector digital transformation: Lessons from insurance industry automation and interoperability models. *European Journal of Computer Science and Information Technology*, *13*(41), 131–145. DOI: 10.37745/ejcsit.2013/vol13n41131145

Zetta Software Tlc Shpk, A., Benelli, F., Maciariello, F., New Generations Sensors Italy, 56024, Marketing Area) Italy, S. M. la F. (CE, Marku, R., & Department of Economics and Business Albania, T. B. U. C. (2025). Towards a human-centric Physical Internet: The next digital transition. *Journal of Robotics and Automation Research, 6*(4), 01–09. DOI: 10.33140/JRAR.06.04.10

Ziwoni, C. (2025). Strategic oversight and governance framework for Business Process Management at Old Mutual Namibia. *International Journal of Innovative Science and Research Technology*, *2081–2085*, 2081–2085. Advance online publication. DOI: 10.38124/ijisrt/25jun1428

KEY TERMS AND DEFINITIONS

Autonomous Business Process Management (ABPM): An advanced form of BPM where processes can make decisions, adapt to changes, and optimize themselves with minimal human intervention.

Autonomy Gradient Architecture (AGA): A structured framework that defines different levels of process autonomy and integrates governance, intelligence, and control mechanisms.

Autonomy Gradient: A continuum of autonomy levels ranging from human-assisted decision support to fully adaptive and self-evolving process orchestration.

Autonomy Index: A metric used to measure the degree of decision independence and adaptive capability within a business process.

Governance Alignment Coefficient: An indicator that measures how well autonomous processes comply with regulatory, ethical, and organizational policies.

Adaptive Orchestration: The ability of a system to dynamically reconfigure workflows based on real-time data, performance thresholds, and risk indicators.

Process Intelligence: The use of analytics, machine learning, and event data to monitor, predict, and improve business processes.

Digital Process Twin: A virtual replica of a real-world business process used for simulation, optimization, and predictive analysis.

Reinforcement Learning in BPM: A machine learning approach where process systems improve decisions over time based on feedback and performance outcomes.

Human-in-the-Loop (HITL): A governance mechanism where human oversight is integrated into automated decision-making processes.

Explainability (XAI): The capability of AI systems to provide understandable reasons for their decisions and actions.

Self-Evolving Process System: A business process system that continuously learns from data and adapts its execution strategy over time.

Governance Checkpoints: Predefined control stages where automated decisions are reviewed for compliance and risk management.

Risk Override Mechanism: A control feature that allows human intervention when automated decisions exceed risk thresholds.

Industry 5.0: A human-centric industrial paradigm emphasizing collaboration between humans and intelligent machines.

Chapter 8
Process Sentience: A Governance-Embedded Cognitive-Orchestration Framework for Autonomous Business Process Management

Gafurov Anvar Bazarbayevich
Alfraganus University, Tashkent, Uzbekistan

Aarthi Selvakumar
http://orcid.org/0009-0006-9064-2091
Marwadi University, Rajkot, India

Dilmurod Turaev
Termez University of Economics and Service, Termez, Uzbekistan

Ravikumar Natarajan
http://orcid.org/0009-0009-3705-1681
Marwadi University, Rajkot, India

Karimov Rashid Salayevich
Mamun University, Khiva, Uzbekistan

Mavlanov Normumin
Tashkent State University of Economics, Tashkent, Uzbekistan

ABSTRACT

Process Sentience has been introduced as a higher order of Autonomous Business Process Management whereby workflows become more dynamic, instead of being rule-based automation, and cognitively coordinated and governance-conscious systems. The response to real-time cues and risk response is achieved through dynamic reconfiguration of processes using structured Cognitive-Orchestration Framework of sensing, reasoning, learning, and adaptive action. Evidence of cross-domain proves that there are measurable improvements in resilience, efficiency, accuracy, and compliance. Autonomy gradients guarantee calibrated autonomy aided by transparency, auditability and human monitoring, making BPM a strategic ability in intelligent these days firms.

DOI: 10.4018/979-8-2600-0050-2.ch008

1. INTRODUCTION

The modern world of business is experiencing a shift in the structure of automation of its rules to the autonomy coordinated by cognitive mechanisms. Traditional Business Process Management (BPM) focused on efficiency, standardization and compliance with the use of developed logic and deterministic workflow. Although these strategies enhanced congruency in operations they are constrained in coping with volatility in the environment, intricacy in regulations and exposure to risks that are dynamic. Combination of artificial intelligence, predictive analytics, and process mining has added adaptive characteristics to process designs so that systems are able to predict deviation and assist proactive interventions. This chapter continues the idea of Process Sentience as an organized development of Autonomous Business Process Management (ABPM), in which workflows feel the context and reason through predictive intelligence, learn through feedback to make decisions and take action within governing constraints in the process. The chapter puts autonomy in a novel context by suggesting a Cognitive-Orchestration Framework, which does not view autonomy as the control-free automation, but rather as governance conscious, strategically aligned, and quantifiable enterprise capability, which fosters greater resilience, responsiveness, and sustainable competitive advantage.

1.1 Evolution from Automation to Autonomy

This move toward automation to autonomy has exposed a structural change in enterprise process structures. The first automation efforts were based on pre-programmed logic and deterministic rules to perform repetitive tasks. RPA solutions and workflow engines helped increase the level of operational consistency but still relied on fixed settings (Niehaves et al., 2012). With the maturation of machine learning, predictive analytics and process mining technologies, processes started to integrate components of adaptive decision-making. Predictive monitoring has allowed the detection of deviations early, and the anomaly detector algorithm has brought about dynamic risk assessment. Autonomy is more than predictive capability; however, it deals with self-adjustment systems to enable processes to re-organize pathways in response to contextual cues (Dorrer, 2020). The autonomous system is composed of a coordinated architecture that has sensing layers, reasoning engines, and action modules. The change does not do away with human control, but it reorganizes it and places human beings in the role of the rulers of governance as opposed to being the working machines. The evolution represents transition of programmed execution to situational intelligence that is incorporated within the process ecosystems. As shown in Figure 1, automation, machine learning, predictive analytics, anomaly detection, self-adjustment, and human oversight are designed to develop into completely autonomous business processes.

Figure 1. Automation to Autonomous Processes

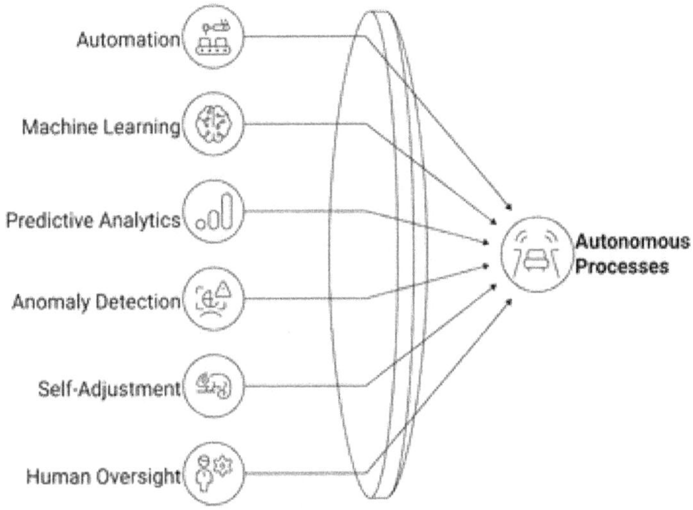

1.2 Limitations of Efficiency-Centric BPM

BPM models focused on efficiency give more emphasis on throughput, cost cutting, and compression in the cycle-time. However, despite the importance of such metrics, they tend to ignore contextual variability, risk sensitivity and strategic alignment. The optimization methodology implies the static condition of the environment and predictability of the working process, which is hardly possible in the modern digital ecosystem. Architectures that are based on efficiency may unintentionally reduce systemic stability through the reduction of redundancy and the restriction of adaptive flexibility. Moreover, past BPM systems focus on checking not on predictive governance (Garcia Medeiro, 2025). Workflow decision rules are prone to fail in new or unknown circumstances and lead to exception escalation and human intervention. Silos optimization is another weakness because individual segments of the processes are optimized without looking at the interdependencies across the ecosystem. This disjointed vision suppresses the ability to be strong and responsive. Efficiency is not a guarantee of sustainability in volatile surroundings where regulatory standards change, there is the threat of cyber attackers, and the markets undergo turbulence. It requires a more comprehensive paradigm that brings together adaptability, context-awareness and responsiveness in governance.

1.3 Emergence of Process Sentience

Process Sentience is a higher form of development of BPM whereby workflows are contextually aware and co-ordinated. Sentient processes combine ways of sensing the operational data, environmental cues and performance deviations in real time. The reasoning engines consume these signals and are able to interpret patterns, predict the future and evaluate the exposure to risk. The learning modules are used to constantly optimize the predictive accuracy and decision thresholds through feedback. Sentient systems do not simply run preplanned instructions as is the case with traditional automation; instead, they make meaning out of the situation, and reconfigure pathways within the bounds of governance. The gradients of autonomy represent the different levels of autonomy, which are decision-support augmentation and conditional self-orchestration (Babalola et al., 2022). Notably, sentience has not been identified with uncontrolled automation. Continuous transparency, explainability and auditability are incorporated as design principles. Through adaptive intelligence and systematic governance, Process Sentience will be able to convert BPM to a strategic endowment that allows them to become more resilient, responsive, and competitive in the long run. Table 1 Comparisons The three automation types: traditional automation, intelligent BPM, and Process Sentience on the architecture, adaptability, and governance aspects.

Table 1. Evolution of BPM from Automation to Process Sentience

Dimension	Rule-Based Automation	Intelligent BPM	Process Sentience
Decision Logic	Predefined rules	Predictive models	Context-aware adaptive reasoning
Flexibility	Low	Moderate	High (dynamic reconfiguration)
Human Role	Execution	Supervision	Strategic oversight
Governance	External compliance	Integrated monitoring	Embedded governance controls
Learning Capability	None	Model-based updates	Continuous adaptive learning
Strategic Alignment	Operational focus	Performance focus	Enterprise strategy alignment

2. BACKGROUND TO THE STUDY

The development of Autonomous Business Process Management (ABPM) is based on the intersection of the theory of BPM, artificial intelligence, and digital governance models. The classical studies of BPM focused on modeling, optimization, and compliance management, giving systematic approaches to the process design and control. The similar advances in the field of artificial intelligence were the introduction of machine learning algorithms that could recognize patterns, predictive-based reasoning, and adaptive learning (Ravikumar, 2025). These areas have been integrated to ensure that processes no longer require a static representation but dynamic systems based on data. At the same time, the systems engineering viewpoints have provided architectural approaches that consider processes as intertwined subunits of digital ecosystems (Bocciarelli & D'Ambrogio, 2014). The governance research also highlights the importance of transparency, accountability and alignment of regulations in more autonomous settings. Combined, these streams of discipline give the theoretical basis of the conceptualization of Process Sentience. Based on this interdisciplinary foundation, the paper presents a Cognitive-Orchestration Framework to establish a universal architectural design of the ecosystem of the intelligent processes based on sensing, reasoning, learning, and governance.

2.1 Foundations of Business Process Management

Business Process Management was a field that developed which aimed at analyzing, modeling and optimization of organizational process. Initial models focused on process mapping, reengineering and performance measurement as ways of improving efficiency in the operations. Six Sigma and Business Process Reengineering methodologies came in with the structured improvement cycles and quality control. BPM adopted lifecycle views of design, implementation, monitoring and optimization later. WF systems computerized implementation logic, and thus provided standardized sequence of tasks and standardized enforcement of compliance. Nevertheless, the original BPM models were mostly based on deterministic logic and business rules that had been in place. These strategies were good when applied in a consistent operation environment, but they were not responsive to changes in the environment (Sang et al., 2025). The direct focus of the theory was on control and predictability not on adaptive intelligence. Nevertheless, BPM developed essential notions, including process visibility, performance measures, and governance gateways, that can currently be used as foundations of more sophisticated autonomous and cognitively coordinated systems.

2.2 AI Integration in Process Architectures

The introduction of artificial intelligence has changed the architecture of processes of rule-based systems to adaptive decision environments. With machine learning algorithms, the predictive process monitoring is through the identification of trends in historical data and the forecasting of possible deviations (Ananthi et al., 2026). The process of natural language processing promotes intelligent document comprehension and automatic communication process. Process mining tools derive information about event logs and uncover latent inefficiency and noncompliance. The technologies can be used when embedded in workflow engines to enable dynamic routing, automated root cause analysis, and prioritization based on risk. Expansive language models also promote the reasoning abilities of processing unstructured data, as well as producing responses within a context. The aspects of AI bring chances of probabilistic decision-making, as opposed to strictly deterministic reasoning. Architectural coordination is however necessary in integrating to avoid fragmented intelligence (Bocharov et al., 2025). The successful use of AI-processes requires coordination layers that will align the predictive results to governance policies and performance goals. AI and BPM overlap, therefore creating the technical foundation upon which Process Sentience can be advanced. There is a diagram in Fig 2 depicting the combination of AI elements including machine learning, NLP, process mining, large language models, and orchestration layers that can be used to convert traditional process architectures into intelligent systems.

Figure 2. AI's Impact on Process Architectures

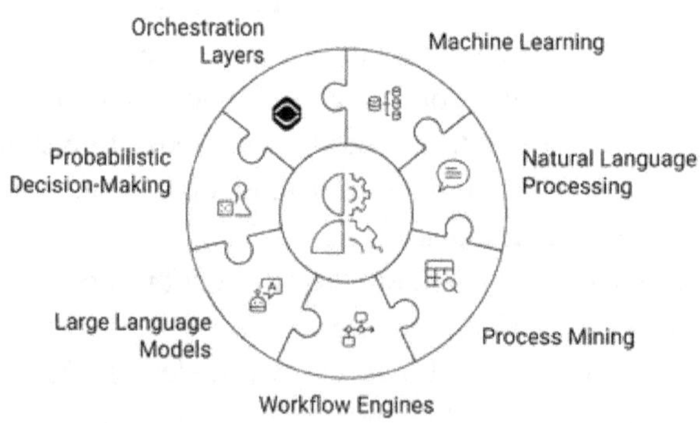

2.3 Digital Governance and Systems Engineering

Digital governance provides the structural protection that is needed in autonomous process ecosystems. These risks of bias, accountability, transparency, and regulatory compliance increase with the degree of increasing autonomy. Its governance systems introduce audit trail, explainability and human overrides facilities in process designs. System engineering introduces the concept of holistic design in which processes are considered to interact, comprise parts that exist in complex adaptive environments. Feedback loops, resiliency modeling and redundancy planning provide the stability of uncertainty. The governance is not defined as a constraining variable but an empowering stratum that causes frequency adjustments on freedom (Sajjanapu, 2025). Organizations can maintain their strategy to the highest point in their workflow lifecycles and still be flexible using the integration of the policy implementation engines and the compliance monitoring into their workflows. The AI ethics also ensure that there is fairness and responsible choices. Cognitively coordinated and sentient process ecosystems can be built on the pillars of governance structures and systems engineering which can deliver sustainable and responsible autonomy.

3. RESEARCH OBJECTIVES

The change between automation-based BPM and cognitively orchestrated and governance-conscious Autonomous Business Process Management (ABPM) requires well-formulated research goals. This chapter attempts to conceptualize Process Sentience as a measurable construct which can be structured as opposed to being a metaphorical extension of automation. The objectives involve theoretical elucidation, architectural planning, integration governance, and empirical assessment procedures that, altogether, still allow passing through to intelligent ecosystems of processes.

- **To conceptualize and formally define Process Sentience constructs**, including sensing capability, contextual reasoning, adaptive learning, and autonomous action within governance constraints.
- **To design a multi-layered Cognitive-Orchestration Framework** integrating data ingestion, predictive analytics, digital process twins, intelligent decision engines, and explainability interfaces.
- **To operationalize autonomy gradients**, defining progressive levels of structured independence from decision-support augmentation to conditional autonomous orchestration.

- **To establish measurable evaluation indices**, including adaptability scores, context-sensitivity metrics, autonomous decision confidence levels, governance compliance ratios, and self-optimization efficiency measures.
- **To embed governance, transparency, and ethical compliance mechanisms** within autonomous workflows, ensuring accountability, auditability, and human override capabilities in sentient process ecosystems.

4. CONCEPTUAL MODEL

The paradigm of Process Sentience in the conceptual model suggested in this chapter is a multi-layered and structured architectural paradigm, not an abstract capability. The model incorporates sensing, reasoning, learning and action mechanisms in a coordination orchestration layer connecting the adaptability of operations and governance limits. The Cognitive-Orchestration model differs with the traditional BPM architecture where the monitoring and execution functions are separated, but intelligence is directly embedded into workflow lifecycles. Every process instance is executed in an adaptive ecosystem that is powered by data ingestion pipes, predictive engine, and contextual interpretation modules as well as policy-aware decision controls (Mendonca et al., 2025). The architecture has a focus on feedback loops, and the processes are able to identify any deviation, interpret any contextual signal, and real-time recalibrate examination channels. Significantly, autonomy is moderated by using the structured control systems that maintain transparency, accountability, and compliance with regulation. The conceptual model thus redefines BPM as a smart system of linked sub systems that can be strategically responsive, resolute and controlled self-evolving within the context of enterprises (Pandit, Ravikumar, Joshi, et al., 2025). Process Sentience in Figure 3 is discussed as a systemic combination of sensing, reasoning, learning and action layers that are governed between the limits of governance to facilitate adaptive and intelligent workflows.

Figure 3. Conceptual Model of Process Sentience

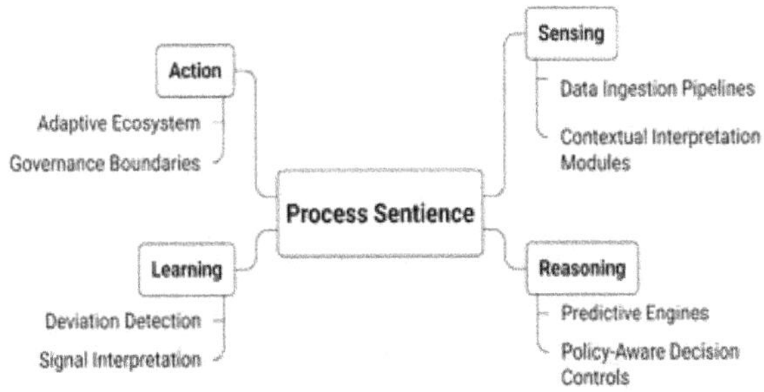

4.1 Sensing–Reasoning–Learning–Action Layers

Sensing-Reasoning-Learning-Action (SRLA) architecture is the cognitive system of the Process Sentience. The sensing layer is the place where it receives both organized and un-organized information concerning the systems being operated, the external world, the interaction between the system and the user and captures the performance. The inputs are then processed in the reasoning layer which deals with predictive analytics, rule engines and machine learning models to determine patterns in the situation and evaluate the risk or opportunity. The learning layer, which refines the predictive accuracy and decision thresholds, is the use of feedback loops, anomaly detection and adaptation of the model used. Controlled learning processes and learner reinforcement will make an individual responsive to the changes in the environment (Mohammed, 2026). Dynamic workflow changes, reroutes activities, policy background interventions with policy background, and dynamic workflow changes are also used in action layer. These layers are not linearly related to one another, however are a chain-reaction to the other, and an architecture that is closed. Integrating perception, cognition, adaptation and execution, the SRLA model is able to alter the workflows within the setting of the stationary automation towards the context-driven and strategically oriented orchestration. Table 2 Provides a brief description of the sensing, reasoning, learning and action layers and the functions of these layers and the enabling technologies.

Table 2. Core Components of the Cognitive-Orchestration Framework

Layer	Function	Enabling Technologies	Outcome
Sensing	Capture real-time data	IoT, APIs, Event logs	Context awareness
Reasoning	Interpret and predict	ML models, Risk engines	Decision intelligence
Learning	Improve over time	Reinforcement learning	Adaptive calibration
Action	Execute workflow changes	Dynamic routing engines	Process reconfiguration
Explainability	Provide transparency	XAI modules	Trust and auditability
Governance	Enforce policy	Compliance engines	Accountability

4.2 Autonomy Gradients and Control Loops

The levels of decision independence in a progressive process ecosystem are determined by the autonomy gradients. There is the use of the AI systems as the bottom-up tools of decision support, which offer suggestions but leave the final decision to human beings. The intermediate levels facilitate conditional automation since they allow invoking adaptive actions whereby, based on predetermined thresholds, which are governed by governance validation. Greater autonomy levels allow self-organized working processes that are not outside the limits of the policies. Such abrupt transitions to such complete independence, can never be allowed by these gradients, but they help in the gradual process of maturity (Uzoamaka et al., 2025). The independence is controlled with the help of the control loops and benchmarking and checking its compliance. Feedback systems recognise the attainment of the strategic objectives or otherwise and change the parameters of decision where appropriate. The loop allows governance checkpoints, which are transparency and auditability. The autonomy gradient model is not close to imposing that various processes need different amounts of autonomy, which depends on the extent of risk exposure and sensitivity to regulation. This independence can be formulated in a manner that it is scaled down to allow the organizations to expand the shapes of cognitions without affecting the accountability and trust. Table 3 Gives the average of progressive autonomy and connected operation of governance protection and human control.

Table 3. Autonomy Gradients and Governance Control Levels

Autonomy Level	Description	Human Involvement	Governance Safeguard
Level 1	Decision Support	Full human control	Manual validation
Level 2	Conditional Automation	Human override	Threshold-based alerts
Level 3	Guided Autonomy	Supervisory oversight	Audit logs

continued on following page

Table 3. Continued

Autonomy Level	Description	Human Involvement	Governance Safeguard
Level 4	Controlled Autonomy	Strategic review	Policy validation
Level 5	Contextual Self-Orchestration	Exception-based	Embedded compliance engine

4.3 Governance-Embedded Process Ecosystem Architecture

Governance-embedded architecture Integrated architecture is process design that includes mechanisms of compliance, explainability and accountability. The model does not take governance as an external audit layer but it encompasses policy executing engines and real-time compliance validation as part of workflow lifecycles. Audit trails document decision paths, model outcomes and activators of interventions in a manner that is traceable. Explainability modules will give interpretations of algorithmic choices in a human understandable format that will guarantee regulatory transparency and trust in the stakeholders. The components of risk assessment are continuously testing the exposure of bias, performance drift, and adherence to ethics. Systems engineering principles can ensure robustness through planning of redundancy and exception management principles. In emergency cases, human override can be employed to permit the human intervention in the case (Smallman & Cook, 2013). The architecture is therefore a moderated strategy of adaptive and structured control. Process ecosystems are strategic in that they create governance on orchestration layers and sensitive to situation. Such integration implies that increased autonomy increases the organizational stability rather than making accountability or adherence to the regulations. There are governance mechanisms that intermittently control autonomous process ecosystems, Figure 4 indicates that policy enforcement, compliance validation, audit trails, explainability, risk assessment, resilience and human override governance mechanisms.

Figure 4. Governance-Embedded Process Ecosystem

5. RESEARCH METHODOLOGY

The methodology as it is research takes an organised conceptual-development and multi-case analytical approach to operationalize Process Sentience in the context of Autonomous Business Process Management (ABPM). Considering the young age of cognitively orchestrated process ecosystems, the given study gives a primary focus to theory-construction buttressed by interdisciplinary synthesis and illustrative empirical underpinning. The theoretical approach combines theoretical modeling, literature alignment, and cross-domain case validation to make sure that the theoretical rigor and the relevance to the context are achieved. An application of the framework-based design logic is used to build the Cognitive-Orchestration model and then evaluable indices of assessment are identified to make the model operationally relevant empirically (He et al., 2023). The latest case examples in the fields of finance, healthcare, manufacturing, and logistics are chosen to prove the architectural viability and contextual flexibility. The methodological focus is on providing a connection between abstract constructs, which include sensing capability, autonomy gradients, and governance responsiveness, with specific process behavior and performance outcome patterns. This strategy will make Process Sentience pronounced as not solely as a theoretical development but also as a practically evaluable and strategically deployable enterprise capability.

5.1 Conceptual Framework Development Approach

The synthesis and the framework of architectural abstraction are used to develop the conceptual framework. The basic insights of Business Process Management, artificial intelligence, systems engineering, and digital governance are checked systematically and aligned to determine the overlapping constructs of adaptability, autonomy, and control. These constructs are planned through conceptual mapping techniques which are applied to structure them into logical architectural layers. The intelligent system architecture and studies of adaptive workflows show that the Sensing-Reasoning-Learning-Action model is the result of comparative analysis. Repetition leads to logical consistency in terms of autonomy gradients and mechanisms of governance control. The framework is done in the form of a modular structure to allow scalability and flexibility of domain (Ruotsalainen & Blobel, 2025). It focuses on a high degree of internal consistency between theoretical constructs and practicable process elements like predictive monitoring engines and explainability modules. The resultant architecture is a conceptual innovation-operational feasibility bridging abstraction.

5.2 Cross-Domain Case Selection Criteria

Case selection is based on purposive and relevancy approach to validate the framework in different operational settings. The domains are identified depending on risk sensitivity variability, regulatory intensity, data richness and complexity of processes. Finance is an example of high compliance and risk-driven workflow; healthcare is an example of context-sensitive and safety-critical processes, manufacturing and logistics exemplify throughput optimization and dynamic coordination problems. Some of the cases chosen have evidence of the use of predictive monitoring, AI-based decision engines, or adaptive orchestration mechanisms (Chineme Scholar Okonkwo et al., 2024). The focus is made on the latest implementations of the modern digital maturity. The cases are not analyzed as statistical samples but as illustrative validation cases of demonstrating structural applicability of Process Sentience constructs. Comparative analysis provides an understanding of the way in which autonomy gradients and governance controls vary in case of the constraint of the context. This cross-domain solution is such that the Cognitive-Orchestration Framework provides conceptual strength and practical flexibility in a wide range of heterogeneous enterprise ecosystems.

5.3 Process Sentience Evaluation Metrics Design

The measures of evaluation are meant to determine abstract sentience constructs in quantifiable measures. Adaptability indices determine how the system can be reconfigured due to contextual deviations. The context-sensitivity scores assess the performance of processes in terms of how well they respond to the environmental and operational signals. The levels of autonomous decision confidence are the measures of predictive reliability and threshold calibration accuracy. Governance compliance ratios are a quantitative measure of compliance with regulatory and policy constraints when carrying out autonomously. The self-optimization efficiency measures compare the performance gains obtained by adapting recalibration to baseline settings. Some of the quantitative measures in metric development include the cycle-time variance reduction and the accuracy of the oddity, as well as the qualitative measures on explainability and transparency (Devi, 2024). The performance system focuses on the multidimensional approach of measuring performance rather than single efficiency indicators. The methodology facilitates practical validity, benchmarking of maturity, and a comparative evaluation of cognitively organized process ecosystems, making the process sentience even more practical through the creation of structured indices. Table 4 Lists includes list of adaptability index, context-sensitivity score, autonomy confidence level, and governance compliance ratio with definition.

Table 4. Process Sentience Evaluation Metrics Framework

Metric	Definition	Measurement Basis	Expected Impact
Adaptability Index	Ability to reconfigure workflows	Deviation correction rate	Increased resilience
Context Sensitivity Score	Accuracy of situational interpretation	Prediction accuracy	Risk reduction
Autonomy Confidence Level	Reliability of automated decisions	Error rate reduction	Trust enhancement
Governance Compliance Ratio	Policy adherence level	Audit conformity %	Regulatory stability
Self-Optimization Efficiency	Improvement over baseline	Cycle time & cost reduction	Performance gain

6. LITERATURE REVIEW

The body of Business Process Management (BPM) literature has continued to shift toward a more intelligent and adaptive process ecosystem than the deterministic workflow optimization. The initial BPM research focused on modeling requirements, lifecycle management and efficiency and compliance-based performance measurement models. As AI kept evolving, studies started incorporating machine learning and analytics into process designs to provide predictiveness and dynamic decision-making (Haripriya, 2025). In modern literature it has been pointed out that the shift of the fixed implementation engines to adaptive systems that can respond to the uncertainty in the environment is being undertaken. Nonetheless, to a great extent, the literature still continues to conceive intelligence as a process of performance improvement as opposed to system-wide cognitive coordination. The nascent field of Autonomous Business Process Management also indicates that a more general paradigm shift in which processes attentively monitor contextual indicators, receive operational feedback, and realign the execution pathways will happen (Semenenko, 2025). This literature review builds upon major advances in intelligent BPM, predictive monitoring and process mining, and AI design that focuses on governance, and aligns these with the goal of the chapter conceptualizing Process Sentience as a structured and quantifiable enhancement of enterprise process architecture.

6.1 Intelligent BPM and Adaptive Workflows

The intelligent BPM research focuses on integrating analytics and learning processes into workflow lifecycles. Adaptive workflows are brought to another level of workflow process beyond fixed rule execution to include contextual stimuli and real-time performance data. Research shows the ability of decision engines to dynamically reroute work, reassign priorities and recommend interventions using probabilistic evaluation. Event-driven architecture has also enhanced responsiveness due to which the workflow can respond to the operational signals instead of adhering to fixed sequences (Kim & Kim, 2025). Digital process twins have also become famous as the virtual analog of results and optimization strategy testing before implementation. Although these have been developed, most smart BPM models still have in place human-based recalibration and established escalation policies. The development of cognitive coordination in full integration, i.e. sensing, reasoning and learning are operating harmoniously within the constraints of governance, has not yet been developed in the literature. This move towards architectural autonomy thus demands a more cohesive approach to architecture.

6.2 Predictive Monitoring and Process Mining

A negative aspect has developed in process intelligence, predictive monitoring. Predictive models are used to determine delays, compliance violations, and performance variations by examining the logs of past events and current data streams.

Process mining techniques play a role in identifying concealed patterns in work flows, bottlenecks, and variance within instances in the process. Conformance checking is a method that improves governance by identifying the nonconformity to preset models. All these methods are able to help in a proactive intervention and not a reactive correction (Ramu & Bansal, 2025). Latest developments combine both anomaly detection and root cause analysis with the automated monitoring pipelines, increasing the accuracy of the diagnosis. Nonetheless, predictive systems typically are used as a form of analytical overlay and not a built-in orchestration layer. The literature shows that there are strong achievements in the area of detection and forecasting but very few studies on adaptive reconfiguration mechanisms (Bobba, 2025). The impossibility of linking predictive insights to automated non-ad-hoc decision execution is a space that needs conceptualizing, which is consistent with the mission of Process Sentience.

6.3 Explainability, Ethics, and AI Governance

With the integration of AI in operational decision-making, the concept of explainability, accountability, and compliance with ethics is discussed in literature. Explainable AI studies aim to convert the complex model outcomes into comprehensible descriptions to the stakeholders. The mechanisms of transparency, audit-trail and bias detection are highlighted to maintain fairness and compliance to regulation. Governance scholarship points to the need to establish policy controls in digital infrastructures instead of considering compliance as an external audit basis. Ethical AI architecture models promote human supervision, risk management, and overriding features especially when dealing with high-stakes financial and health care fields. Most studies overlook governance as a design element although in the recent years, there has been a rise in the interest to address these issues (Nangi et al., 2022). An integrated model that entrenches governance within adaptive orchestration of processes is yet to come. Synchronization of intelligent adaptability and structured accountability is a cornerstone need in facilitation towards sentient process ecosystems.

7. CASE STUDIES

Three case studies in the industry are shown within recent times to demonstrate the process of how the cognitively coordinated workflows can transcend beyond rigid automation to the predictive, adaptive and governance conscious implementation to demonstrate an actual practical viability of Process Sentience. The choices of both of these examples are (i) high regulatory intensity, risk sensitivity (finance), (ii) safety-sensitive, documentation-intensive operations (healthcare) and (iii) throughput-constrained, exception-prone value chains (manufacturing and logistics). A sense of coherence is exhibited in the case studies, all the operational signal and event log sensed in real time; reasoning by prediction analytics and risk scoring; continuous improvement loops create learning; dynamic routing and exception prevention, and policy-limited orchestration provide the action (Liu et al., 2023). To match the model concept to measurable benefits of enterprises, such as a reduction in rework, compression of cycle-time, increase in accuracy, increase in straight through processing and an increase in service reliability are brought out in the case of variability.

7.1 Finance: Predictive Risk-Aware Workflow Orchestration

A massive banking scenario provides a description of predictive risk-conscious orchestration used in individual banking and mortgage experiences. Root causes of root causes of missing applications, abandoned flows, and heavy portions were exposed with the help of process mining. A simulative ability in the form of a digital-twin facilitated test-before-change improvement planning, so that it could proactively track and identify unfinished applications and manual checks focused on the high-risk areas. With the maturity of orchestration, the bank increased account openings by 5% to 10% meaning that it had better conversion and low drop-offs. Personal banking Straight-Through Processing (STP) rose to 43% to 60% and 80% indicating an improved level of automated completion under policy limits. Rework decreased by 45% to 10% and 30% indicating quantifiable decrease in unnecessary corrections and looping exceptions (Paladugu, 2025). Mortgages available to offer decreased by 3-6 days lead time and STP rose to 32% to 40% and 26%, which implies improved flow reliability and fewer handoffs. All these can be seen as an expression of a changed perspective on the rule-driven implementation to predictive prevention when risks are identified earlier on, the intervention is being specific and the workflow routes are being dynamically stabilized without compromising the accountability of governance. Figure 5 displays the percentage improvement in account opening rates, straight-through processing, reduction in rework and efficiency of mortgage following the deployment of predictive risk-aware orchestration.

Figure 5. Finance Case – Predictive Workflow Performance Improvements

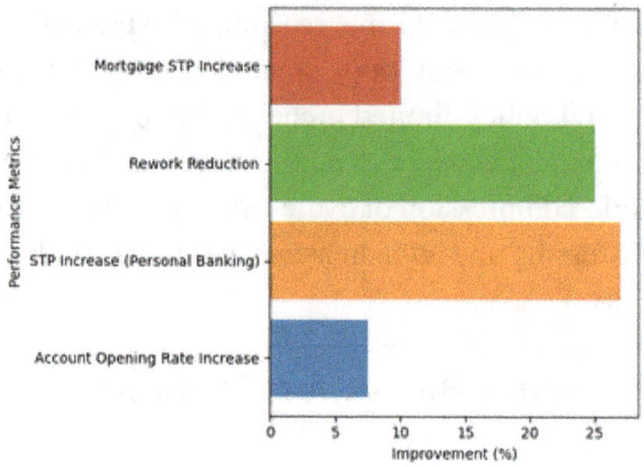

7.2 Healthcare: Cognitive Clinical Process Reconfiguration

An example of recent healthcare revenue-cycle and documentation automation shows that the cognitive orchestration is more effective in enhancing documentation throughput without compromising accuracy or human validation. Using AI-based automation and document insight, Omega Healthcare (handling large-scale healthcare operations) extracted and directed the information in volume correspondence and medical-related documents. Documentation time is said to have decreased 40%, document turnaround was 50% faster, and accuracy of the process was 99.5%, showing that the performance improved not only in speed but in quality as well, under the operational scale. Accounts Receivable (AR), clean claim ratio and claim resolution time were optimized by 30% to 98% and 70% respectively through automation-driven predictive triage and expedited exception handling of workflow lifecycles. The increase in productivity was reported as 100% growth in the advancement of particular operational processes, and also a quantifiable change wherein human-in-the-loop validation was only done when needed (Kerrison et al., 2023). Moreover, the implementation profile shows that 60% to 70% of client processes are to be automated, which implies the use of an autonomy-gradient strategy instead of autonomy-at-once. Process Sentience attributes such as sensing (document and transaction signals), reasoning (extraction/classification and prioritization), learning (process mining-based opportunity identification) and action (automated routing with controlled human override) are found in the case. In Figure 6, percentage changes and increase/decrease in documentation time, claim resolution, productivity, and level of automation, and accuracy are demonstrated post the process of cognitive clinical process reconfiguration.

Figure 6. Healthcare Workflow Performance Improvements

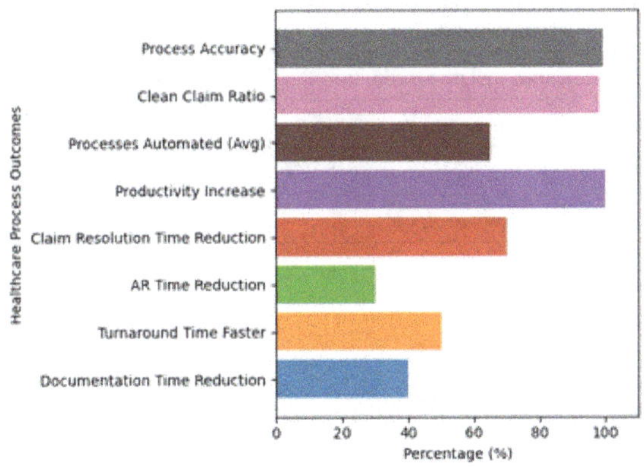

7.3 Manufacturing & Logistics: Autonomous Throughput Optimization

One example of an order-to-cash manufacturing-logistics is autonomous stabilization of throughput through process intelligence and real time exception prevention. In the analysis of DSM, more than 60% of the blocks credit and delivery were identified to be avoidable, and it enabled predicting and giving warnings of the orders that are likely to be delivered late and offered more immediate and controlled action. Measures of such gains as 10% decrease in delivery block and 10% decrease in manual sales-order changes were recorded that indicated a decrease in the churn of exceptions. The first-time-right performance in APAC grew by 18% to 42% (more than 100%), and the quality-at-source. The delivery is now more resilient with greater service reliability and flow resilience in comparison to the 84% rise in the rate of delivery in North America (Besigomwe, 2025). The implication of these results is that predictive surveillance and action loops that are policy-bound can be used to enhance the throughput and produce less disruption without doing away with accountability controls. Table 5 Compares cases of finance, healthcare and manufacturing that have been measured in suggestions of percent improvement of performance and resilience.

Table 5. Cross-Domain Case Study Performance Outcomes

Domain	Key Improvement	Percentage Value	Impact Area
Finance	Straight-Through Processing	60–80%	Efficiency
Finance	Rework Reduction	45% → 10–30%	Quality
Healthcare	Documentation Time Reduction	40%	Speed
Healthcare	Process Accuracy	99.5%	Reliability
Manufacturing	First-Time-Right Improvement	18% → 42%	Throughput
Logistics	OTIF Improvement	84% → 90%	Service Reliability

8. RESULTS AND DISCUSSION

The case studies in finance, healthcare, and manufacturing-logistics all point to the conclusion that process ecosystems that are cognitively coordinated create quantifiable performance, stability, and governance consistent autonomy benefits. The findings are the same: predictive sensing and adaptive routing respectively decrease rework, speed throughput and boost confidence in the decision. Finance workflows had increased straight-through processing and reduced lead times, whereas healthcare implementations were characterized by quicker documentation cycles and more improved accuracy. Cases in manufacturing and logistics had a high level of first-time-right performance and on-time-in-full delivery. In addition to efficiency factors, the results show resilience in operations, operating processes that detect and avert risks before they escalate (Odogwu et al., 2022). The combination of the gradients of autonomy and governance checkpoints made sure that compliance and accountability was not affected by adaptive capabilities. Notably, human control was also tactically integrated especially in the case of extreme situations or those that are highly regulated. The discussion gives emphasis to the fact that Process Sentience is not only generating quantitative benefits but also structural enhancements in responsiveness, stability, and trustworthiness in the various enterprise settings.

8.1 Strategic Efficiency and Stability Enhancement

The case studies posted quantitative results, showing great operational performance and systemic resilience enhancement. Finance Straight-through processing rose to 60-80, rework fell to 45% to as low as 10-30%, and mortgage lead times fell by 3-6 days. Such gains can be said to be early risk identification and predictive intervention in workflow lifecycles. In the healthcare sector, the time spent on documentation was reduced by 40%, turnaround time was also improved by 50%, and process accuracy was 99.5% and this showed better reliability as well as speed. The manufacturing and logistics cases revealed a 10% decrease in delivery blocks, over 100% enhancement in first time right rates in specific areas and upgrading of OTIF performance by 84% to 90%. These results suggest that the realization of resilience was not simply due to automation but automatic correction and reconfiguration. Improved performance was accompanied by a decrease in exception handling as well as a greater process stability under variability (Adekunle et al., 2021). Figure 7 presents the comparison of the key percentage improvements of the finance, healthcare, and manufacturing-logistics cases, which demonstrate how predictive and adaptive orchestration enhances performance and resilience results.

Figure 7. Performance and Resilience Gains Across Case Studies

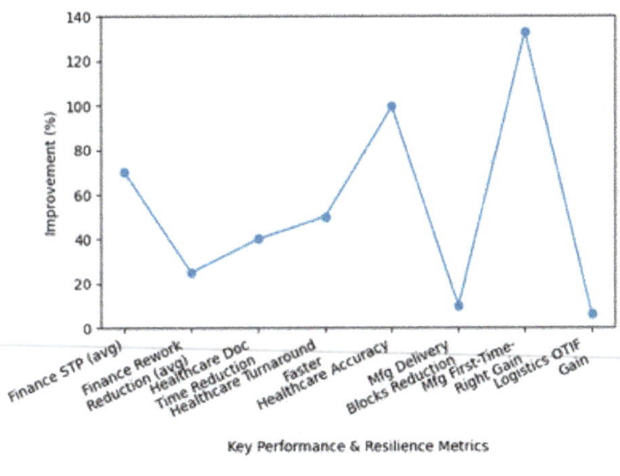

8.2 Autonomous Process Maturity Framework

The case studies portray the incremental levels of autonomy against full automation. Conditional autonomy with governance validation was implemented in finance with intelligent routing and predictive alerts being carried out within policy constrained thresholds. Mid-level autonomy maturity was demonstrated in the implementation of 60-70% of the processes in healthcare automation and human verification of high-stakes decisions. The manufacturing situations were those that demonstrated autonomous exception detection, introspective alerts, however, supervisory controls remained in handling credit and delivery decision making. These trends show that maturation in autonomy is developed in phases, as far as the level of sensing, reasoning, and execution is concerned (Nangi & Obannagari, 2025). These characteristics of maturity are an increase in the predicting accuracy, increase in the rate of straight-through processing and a reduction in the rate of manual processing, and it does not impact the compliance ratios at maturity. The autonomy gradient model upholds systematic autonomy in contrast to unself-regulated automation. Evaluation may thus be a mixture between quantitative performance measures and governance alignment measures and by doing so it is confirmed that Process Sentience does not have any influence on adaptive capability but it does enhance the adherence to accountability and regulatory compliance.

8.3 Hybrid Human–AI Operational Governance

The findings emphasize that Process Sentience does not abolish the human roles and re-organizes them towards strategic direction and exception management. In finance, humans were involved in high-risk applications, and predictive systems were used in the routine approvals. Clinical judgment and integrity of compliance Healthcare professionals certified and reviewed AI-generated documentation and approved complex cases. Supervisory teams were only involved in manufacturing and logistics when critical deviations were detected by intelligent alerts. By reallocating roles, this also decreased cognitive load without taking away decision making responsibility where it was required (Ravikumar & Aarthi, 2025d). Human-AI joint work was, therefore, a complex governance structure with AI improving situational awareness and humans perfecting strategic parameters. The transparency mechanisms, like the audit trail and explainability interface, were used to build trust. The collaboration process proves that engineered independence aids augmentation and not substitution. Organizations realized efficiency gains by integrating human control in adaptive architectures without unethical orientation, and operational legitimacy.

9. FUTURE RESEARCH

The introduction of Process Sentience provides several possibilities of the subsequent theoretical amelioration and experimental confirmation. Although the Cognitive-Orchestration Framework provides a formalized base it is clear that future studies need to further elaborate the technical, governmental, and quantitative aspects of autonomous process ecosystems. The development of digital twin systems, distributed learning systems, and the standardization of evaluation systems will become important in the expansion of sentient capability on enterprise networks. Also, the growing regulatory oversight of the AI-based decision systems requires a more robust combination of adaptive intelligence and privacy-protecting models of governance. Additional research needs to be done on the longitudinal validation of the autonomy gradients, inter-industry cross-validation of the sentience maturity, and experimental validation of the resilience benefits under turbulent conditions (Akinlade et al., 2020). Adaptive stability could also be enhanced with the integration of the emerging technologies like edge intelligence, blockchain-based auditability, and self-healing system architectures. The need to advance interdisciplinary co-operation between scholars in Business Process Management, researchers in artificial intelligence and scholars in governance will be critical in streamlining Process Sentience into an adult and standardized space in the Autonomous Business Process Management.

9.1 Self-Evolving Digital Process Twins

Digital process twins are now being used as simulation environments that reflect the workflow in reality to analyse performance and test scenarios. Future studies ought to be expanded to self-adjusting systems that have the potential to perform autonomous recalibration using live operation feedback. These twins would constantly match the data streams in enterprises, identify performance changes, and suggest structural adjustments before they degrade. By incorporating reinforcement learning mechanisms into digital twins, proactive experimentation in controlled virtual environments could be made possible, which would reduce the risk of implementation. Studies should be conducted to find out how twin architectures may integrate policy simulation and governance constraints (Jayaram, 2025). Multi process ecosystems and cross-organization supply chain scalability also needs to be explored. In the future, self-evolving digital twins might form the predictive heart of sentient firms, where future-adaptive modifications, instead of remedial responses, can be made. The effect of recalibration caused by twins on resilience, throughput stability and long-term strategic alignment need to be determined by empirical studies.

9.2 Federated and Privacy-Aware Process Intelligence

With businesses conducting business in distributed ecosystems, data consolidation will be more limited by regulations of privacy and competition. Federated learning models that can facilitate collaboration in the form of shared intelligence without revealing raw data need to be investigated in the future (Pandit, et al., 2025). Federated process intelligence would enable two or more organizational units or partners to collaboratively refine predictive models without giving up on data sovereignty. Orchestration layers could be integrated with privacy enhancing technologies, such as differential privacy, secure multiparty computation, and homomorphic encryption. Studies need to be conducted to investigate the impact of federated architectures on the accuracy of models, latency, and governance compliance. Also, the autonomy gradients in the federated setting must be scaled with attention to avoid the risk spreading across boundaries. Privacy-conscious intelligence integration is in line with the regulatory schemes and ethical standards of AI (Ravikumar, 2025a). To quantify the effect of federated orchestration on resilience, transparency and trust of distributed business networks, empirical validation is required.

9.3 Standardization of Sentience Metrics

Process Sentience operationalization should be standard measurement models that extend beyond single-performance measures. The second round of the research must be on the development of the adaptability indices, the context-sensitivity score, the autonomy confidence scores and the governance compliance ratios to all the standards which are universally comparable. Terifiatur would enable the evaluation of maturity in the industry and structured certification patterns (Cadet et al., 2024). This necessitates fine tuning of the approaches or procedures in which measurements of multidimensional intelligence and not efficiency gains are gauged. Longitudinal research studies are required to establish how sentience scores can be followed through time and reach correlations to the results of resilience. Benchmark datasets and open evaluation protocols may be able to make reproducibility and comparability possible. Also, sentience indices would have explainability and fairness metrics, which would increase the credibility of governance. Standardized measures would transform Process Sentience into the conceptual innovation into a real strategic capability, permitting organizations to structure their improvement to smart and governance-sensitive process ecosystems.

10. MANAGERIAL IMPLICATIONS

The shift in BPM towards Process Sentience instead of automation has important managerial consequences in the area of governance, organizational design, and strategic positioning. Leaders should understand that autonomous process ecosystems do not consist surrounding solely technological deployments but structural changes on accountability, risk management, and workforce functions (Ullagaddi, 2024). The evidence presented in the case shows that these performance gains are optimized when predictive intelligence is integrated into governance-conscious architectures as opposed to overlaying the existing systems with predictive intelligence. Managers have to thus strike a balance between adaptive autonomy and transparency, compliance, and the ethical protection. Additionally, gradients of autonomy have to be implemented in a calibrated manner, so as not to disrupt the operational process or destroy trust. The maturity of digital capability, organizational preparedness, and the acceptance of AI-based decision system culture are of critical success determinants (Ravikumar & Aarthi, 2025c). In addition to operational efficiency, Process Sentience offers the possibilities of a long-lasting competitive edge by resiliency, responsiveness, and strategic agility. The management issue is to synchronize technology, governance and human interaction into a cohesive system that is able to elevate BPM above being a support service to a strategic resource.

10.1 Governance-First Autonomy Adoption Strategy

Governance-first would ensure that accountability structures are highly defined and autonomy is brought in. Before organizations can amass greater predictive orchestration capabilities, they must first develop an audit trail, explainability interface and compliance gateways. The autonomy grades should be adjusted in tip to the exposure of the process to risk, and the severity of the regulations and sensitivity of decisions (Ravikumar et al., 2026). A human control over the workflows that involve high risks should be more stringent, and less risky and monotonous tasks may be allowed to evolve to conditional autonomy. Governance dashboards can be used to track the performance of autonomy and the rate of deviation and compliance with policy in real-time. Decentralized escalation steps and override policies assist in keeping the managers at check especially in sensitive cases (New Mexico Black Leadership Council United States & Okunlola, 2025). The architectural level of governance assists in the prevention of the uncontrolled automation impression and makes the stakeholders trust. The data presented by the cases of finance and healthcare indicate that the rise in performance is only possible when the compliance ratios are not altered. Governance-first adoption path is also low in reputational risk reduction and intelligent orchestration can be scaled.

10.2 Organizational Readiness and Change Design

Process Sentience can be effectively implemented only when the organization is ready in technological, cultural, and skills aspects. Digital infrastructure should enable real time data ingestion, predictive analytics as well as integration between legacy systems. Workforce readiness entails reskilling workers into supervisory, analytical and exception-management positions as opposed to everyday performance. The design of change must also focus on open communication on the gradients of autonomy and human-AI collaboration patterns (Rathinavel et al., 2025). The uncertainty about possible job displacement is likely to cause resistance; thus, positioning augmentation is necessary (Yanamadala, 2025). Raising experimentation at the levels of pilot deployments and gradual maturity development should be encouraged by the leadership. Risk calibration and ethical compliance can be under the control of cross-functional governance committees. The training programs must improve the data literacy and interpretability of the AI outputs. When there is cultural alignment on the area of evidence-based decision-making, the adaptive workflow becomes easy to accept. Process Sentience will eventually turn out to be a strategic enhancer or an isolated technological project depending on the organizational preparedness.

10.3 Strategic Differentiation through Process Sentience

Process Sentience also provides strategic differentiation through resiliency, responsiveness, and reliability of services in volatile conditions. Businesses that incorporate predictive orchestration into their fundamental processes can react more quickly to market dynamics, regulatory transformations and business interruptions. An increase in the straight-through process rates and a decrease in the rework mean the improvement of customer experience and cost-stability directly. Healthcare and finance examples demonstrate that, anticipatory correction is an effective way of strengthening trust and brand credibility. Sentient process ecosystems facilitate both strategic nimbleness as well as strategic efficiency, allowing scenario simulation and dynamical allocation of resources (Veeravalli, 2025). Organizations are able to detect emergent risks at earlier stages and exploit opportunities more quickly. There are other competitive positioning advantages in governance-built architectures, which prove regulatory compliance and moral responsibility. Structured autonomy is a strategic ability as opposed to an operation characteristic over time. Companies which systematically develop through autonomy scales may create smart corporations with enduring benefit in data-abundant and swiftly altering surroundings. Table 6 Relates strategic priorities, organizational readiness factors and governance mechanisms to autonomy maturity levels.

Table 6. Managerial Implication Matrix for Governance-Aware Autonomy

Strategic Priority	Required Capability	Governance Mechanism	Expected Benefit
Risk Reduction	Predictive monitoring	Real-time alerts	Early intervention
Compliance Stability	Policy engines	Audit trails	Regulatory trust
Operational Agility	Dynamic routing	Threshold control	Faster response
Workforce Transition	Reskilling programs	Oversight protocols	Human–AI balance
Competitive Differentiation	Cognitive orchestration	Embedded governance	Strategic advantage

11. CONCLUSION

The shift of the rule-based automation into cognitively directed Process Sentience is an underlying change in how Autonomous Business Process Management (ABPM) has evolved. The chapter furthers the theoretical research by theorizing Process Sentience as a structured capacity based on the sensing, reasoning, learning and adaptive action within governance limits. The Cognitive-Orchestration Framework, in contrast to the conventional BPM paradigm, which focuses on streamlining efficiency processes, incorporates predictive intelligence, autonomy gradients, digital process twins, and control loops embedded in governance into a single architectural paradigm. The empirical examples of finance, healthcare, and manufacturing show that quantifiable benefits in straight-through processing, reduction in rework, decrease in turnaround time, and operational stability are obtained when autonomy is organized rather than unrestricted. The paradigm shifts BPM out of a science of procedural control into a science of strategic domain that deals with adaptive coordination and situational intelligence. The model incorporates transparency, explainability and human override, which makes accountability not weaken in increasing autonomy. Process Sentience thus allows businesses to move in the direction of smart systems that can have predictive responsiveness and risk-responsive decision-making organizing. Given the volatility that faces organizations, regulatory complexity, and information overload, structured autonomy has become a competitive necessity. The trend toward awareness enterprise systems does not only indicate technological growth but an overhaul of the system of designing, governing and aligning the processes to the long-term organizational goals.

REFERENCES

Adekunle, B. I., Chukwuma-Eke, E. C., Balogun, E. D., & Ogunsola, K. O. (2021). Machine learning for automation: Developing data-driven solutions for process optimization and accuracy improvement. *International Journal of Multidisciplinary Research and Growth Evaluation, 3*(1), 800–808. DOI: 10.54660/.IJMRGE.2021.2.1.800-808

Akinlade, O. F., Filani, O. M., & Nwachukwu, P. S. (2020). Big data approaches enhancing procurement responsiveness through optimized cycle time reduction. *International Journal of Multidisciplinary Futuristic Development, 1*(2), 29–39. DOI: 10.54660/IJMFD.2020.1.2.29-39

Ananthi, M., Lakshmana Kumar, R., Muthu, B., & Punitha, P. (2026). Adaptive marine intelligence and sensing architecture for autonomous underwater ecosystem monitoring using AI and IoT integration. *Intelligent Data Analysis, 30*(1), 162–181. DOI: 10.1177/1088467X251339271

Babalola, O., Iluore, E., Bakare, A., & Udechukwu, L. M. (2022). AI-Enabled Business Process Optimization Engine for risk-Aware Cloud Operations. *Journal of Frontiers in Multidisciplinary Research, 3*(2), 211–216. DOI: 10.54660/.IJFMR.2022.3.2.211-216

Besigomwe, K. (2025). Self-healing digital twins for manufacturing process resilience. *Cognizance Journal of Multidisciplinary Studies, 5*(3), 23–38. DOI: 10.47760/cognizance.2025.v05i03.003

Bobba, A. R. (2025). Cloud-native AI across industries: Healthcare, finance, and manufacturing. *European Modern Studies Journal, 9*(5), 619–627. DOI: 10.59573/emsj.9(5).2025.56

Bocciarelli, P., & D'Ambrogio, A. (2014). A model-driven method for enacting the design-time QoS analysis of business processes. *Software & Systems Modeling, 13*(2), 573–598. DOI: 10.1007/s10270-013-0345-5

Bocharov, N., Torres, P., & Matos, J. (2025). A practical approach to industrial digitalization through data acquisition and systems integration for predictive maintenance. *EAI Endorsed Transactions on Digital Transformation of Industrial Processes, 1*(2). Advance online publication. DOI: 10.4108/dtip.9700

Cadet, E., Etim, E. D., Essien, I. A., Erigha, E. D., Babatunde, L. A., Ajayi, J. O., & Obuse, E. (2024). Large Language Models for cybersecurity policy compliance and risk mitigation. *International Journal of Scientific Research in Humanities and Social Sciences, 1*(2), 612–643. DOI: 10.32628/IJSRSSH242560

Chineme Scholar Okonkwo, C. S. O., Jibril Agbabiaka, J. A., Winner Mayo, W. M., & Obinna ThankGod Okeke, O. T. O. (2024). Supply Chain Automation Framework Using Service Management Platforms. *Shodhshauryam International Scientific Refereed Research*, (SHISRRJ2472157), 157. DOI: 10.32628/SHISRRJ2472157

Devi, A. K. R. S.Arun Kumar Ramachandran Sumangala Devi. (2024). The future of healthcare data governance AI-enabled ETL testing frameworks on data warehousing testing automation using ML, AI and NLP. *International Journal For Multidisciplinary Research*, *6*(5), 28559. Advance online publication. DOI: 10.36948/ijfmr.2024.v06i05.28559

Dorrer, M. G. (2020). The digital twin of the business process model. *Journal of Physics: Conference Series*, *1679*(3), 032096. DOI: 10.1088/1742-6596/1679/3/032096

Garcia Medeiro, J. (2025). Business process governance in the age of digital transformation. *Journal of Interdisciplinary Debates*, *6*(04), 46–57. DOI: 10.51249/jid.v6i04.2688

Haripriya. (2025). Transforming governance: The role of automation in tax, compliance, and public sector efficiency. In *Rethink 2025: Navigating Technology Trends Through Multidisciplinary Insights* (pp. 94–99). Royal Book Publishing. DOI: 10.26524/300.13

He, X., Zheng, X., & Ding, H. (2023). Existing barriers faced by and future design recommendations for direct-to-consumer health care artificial intelligence apps: Scoping review. *Journal of Medical Internet Research*, *25*, e50342. DOI: 10.2196/50342 PMID: 38109173

Jayaram, Y. (2025). AI-powered ECM automation with agentic AI for adaptive, policy-driven content processing pipelines. *International Journal of Artificial Intelligence, Data Science, and Machine Learning*, *6*, 124–134. DOI: 10.63282/3050-9262.IJAIDSML-V6I3P118

Kerrison, S., Jusak, J., & Huang, T. (2023). Blockchain-enabled IoT for rural healthcare: Hybrid-channel communication with digital twinning. *Electronics (Basel)*, *12*(9), 2128. DOI: 10.3390/electronics12092128

Kim, E. J., & Kim, J. (2025). Privacy-by-design framework for large language model chatbots in urology. *International Neurourology Journal*, *29*(Suppl 2), S65–S72. DOI: 10.5213/inj.2550274.137 PMID: 41355258

Liu, Z., Hansen, D. W., & Chen, Z. (2023). Leveraging digital twins to support industrial symbiosis networks: A case study in the Norwegian wood supply chain collaboration. *Sustainability (Basel)*, *15*(3), 2647. DOI: 10.3390/su15032647

Mendonca, R. S., da Silva, M., Ayres, F. A. C.Jr, Bessa, I. V., Medeiros, R. L. P., & Lucena, V. F.Jr. (2025). Development of a novel retrofit framework considering Industry 4.0 concepts: A case study of a modular production system. *Processes (Basel, Switzerland)*, *13*(1), 136. DOI: 10.3390/pr13010136

Mohammed, B. (2026). The sentient infrastructure paradigm: A review of novel sensors, digital twins, and AI for civil structural health monitoring. *Sensor Review*, •••, 1–19. DOI: 10.1108/SR-10-2025-0822

Nangi, P. R., & Obannagari, C. K. R. N. (2025). AI-driven security automation for continuous compliance monitoring in regulated cloud environments. *International Journal of Emerging Trends in Computer Science and Information Technology*, *6*, 95–105. DOI: 10.63282/3050-9246.IJETCSIT-V6I2P112

Nangi, P. R., Obannagari, C. K. R. N., & Settipi, S. (2022). Self-auditing deep learning pipelines for automated compliance validation with explainability, traceability, and regulatory assurance. *International Journal of Artificial Intelligence, Data Science, and Machine Learning*, *3*, 133–142. DOI: 10.63282/3050-9262.IJAIDSML-V3I1P114

Niehaves, B., Plattfaut, R., & Becker, J. (2012). Business process governance: A comparative study of Germany and Japan. *Business Process Management Journal*, *18*(2), 347–371. DOI: 10.1108/14637151211225234

Odogwu, R., Ogeawuchi, J. C., Abayomi, A. A., Agboola, O. A., & Owoade, S. (2022). Optimizing productivity in asynchronous remote project teams through. *International Journal of Multidisciplinary Research and Growth Evaluation*, *3*(4), 628–634. DOI: 10.54660/.IJMRGE.2022.3.4.628-634

Okunlola, F.New Mexico Black Leadership Council United States. (2025). A cyber-physical framework for data assurance and emergency response readiness in critical energy infrastructure. *Journal of Engineering and Applied Sciences (Asian Research Publishing Network)*, *4*(1), 1–10. DOI: 10.70560/6ag3wg21

Paladugu, N. (2025). Intelligent data governance frameworks for multi-cloud financial environments: An AI-driven approach to compliance automation. *European Modern Studies Journal*, *9*(4), 1304–1315. DOI: 10.59573/emsj.9(4).2025.121

Pandit, S., Ravikumar, R. N., Joshi, N., Pandit, S., Kishore, S., & Aarthi, S. (2025). Retrieval-Augmented Generation (RAG) for enterprise intelligence. In *Leveraging LLMs for Business Innovation* (pp. 69–106). IGI Global Scientific Publishing. DOI: 10.4018/979-8-3373-3912-2.ch003

Pandit, S., Ravikumar, R. N., Pandit, S., Joshi, N., Kishore, S., & Aarthi, S. (2025). Explainable Federated Learning for secure and transparent wireless edge networks. In *Advances in Computational Intelligence and Robotics* (pp. 127–168). IGI Global Scientific Publishing. DOI: 10.4018/979-8-3373-3576-6.ch004

Ramu, D. S., & Bansal, P. (2025). A study on AI's Transformative Impact on Strategic Decision-Making. *International Journal of Science & Technoledge*, *16*(2), 6335. Advance online publication. DOI: 10.71097/IJSAT.v16.i2.6335

Rathinavel, P., Ravikumar, R. N., Aarthi, S., Habib, M., Salaeva, N., & Pardaev, J. (2025). Augmenting human capabilities. In *Advances in Computational Intelligence and Robotics* (pp. 1–30). IGI Global Scientific Publishing. DOI: 10.4018/979-8-3373-2372-5.ch001

Ravikumar, R. N., & Aarthi, S. (2025a). Federated Learning for cross-enterprise supply chain collaboration. In *Advances in Computational Intelligence and Robotics* (pp. 47–84). IGI Global Scientific Publishing. DOI: 10.4018/979-8-3373-2833-1.ch002

Ravikumar, R. N., & Aarthi, S. (2025b). Fine-tuning large Language Models for domain-specific business applications. In *Leveraging LLMs for Business Innovation* (pp. 107–136). IGI Global Scientific Publishing., DOI: 10.4018/979-8-3373-3912-2.ch004

Ravikumar, R. N., & Aarthi, S. (2025c). Human-AI synergy for organizational resilience. In *Advances in Computational Intelligence and Robotics* (pp. 287–322). IGI Global Scientific Publishing. DOI: 10.4018/979-8-3373-7416-1.ch011

Ravikumar, R. N., & Aarthi, S. (2025d). Towards collaborative intelligence. In *Advances in Computational Intelligence and Robotics* (pp. 1–32). IGI Global Scientific Publishing. DOI: 10.4018/979-8-3373-5747-8.ch001

Ravikumar, R. N., Aarthi, S., & Mamadiyarov, Z. (2026). Building the foundations of augmented organizations through AI empowerment. In *Advances in Computational Intelligence and Robotics* (pp. 101–130). IGI Global Scientific Publishing. DOI: 10.4018/979-8-3373-8607-2.ch006

Ruotsalainen, P., & Blobel, B. (2025). A system model and requirements for transformation to human-centric digital health. *Journal of Medical Internet Research*, *27*, e68661. DOI: 10.2196/68661 PMID: 40294402

Sajjanapu, M. (2025). Cloud-native government: Digital infrastructure transforming public service delivery. *Journal of Computer Science and Technology Studies*, *7*(7), 82–90. DOI: 10.32996/jcsts.2025.7.7.6

Sang, L. T., Ravikumar, R. N., Aarthi, S., & Ben, T. L. (2025). Optimizing sales and inventory performance using AI and business intelligence in POS environments. In *Advances in Computational Intelligence and Robotics* (pp. 259–290). IGI Global Scientific Publishing. DOI: 10.4018/979-8-3373-4392-1.ch011

Semenenko, Y. (2025). Modeling digital twins of business processes as a tool for managing company efficiency. *Scientific Notes of Ostroh Academy National University Series Economics*, *1*(38(66)), 99–106. DOI: 10.25264/2311-5149-2025-38(66)-99-106

Smallman, H. S., & Cook, M. B. (2013). Proactive supervisory decision support from trend-based monitoring of autonomous and automated systems: A tale of two domains. In *Virtual, Augmented and Mixed Reality. Systems and Applications* (pp. 320–329). Springer Berlin Heidelberg. DOI: 10.1007/978-3-642-39420-1_34

Ullagaddi, P. (2024). Digital transformation in the pharmaceutical industry: Ensuring data integrity and regulatory compliance. *International Journal of Business and Management*. Advance online publication. DOI: 10.5539/ijbm.v19n5p16

Uzoamaka, N.-O. P., Amujiogu, A. A., & Blessing, E. O. (2025). Governance and ethical oversight of agentic artificial intelligence in business process management: Balancing human-autonomy, organizational accountability and algorithmic transparency. *Journal of Basic and Applied Research International*, *31*(6), 121–133. DOI: 10.56557/jobari/2025/v31i610028

Veeravalli, P. K.Pradeep Kiran Veeravalli. (2025). The synergistic integration of artificial intelligence and cloud orchestration in process automation. *Journal of Computer Science and Technology Studies*, *7*(2), 316–322. DOI: 10.32996/jcsts.2025.7.2.32

Yanamadala, P. (2025). Strategic pathways for AI modernization: A framework for enterprise evolution. *Journal of Computer Science and Technology Studies*, *7*(10), 680–688. DOI: 10.32996/jcsts.2025.7.10.68

KEY TERMS AND DEFINITIONS

Action Layer: The execution component that dynamically adjusts workflows based on intelligent insights.

Adaptability Index: A measurement of how effectively a process can adjust to changing conditions.

Autonomous Business Process Management (ABPM): An advanced form of BPM where workflows operate with structured independence using AI and predictive intelligence.

Autonomy Gradients: Different levels of process independence, ranging from human-assisted decision support to conditional autonomous execution.

Cognitive Orchestration: The coordination of sensing, reasoning, learning, and action layers to enable adaptive and intelligent workflow behavior.

Digital Process Twin: A virtual model of a real process used for simulation, monitoring, and optimization.

Explainability (XAI): The ability of AI systems to provide clear and understandable reasoning behind decisions.

Governance-Embedded Architecture: A design approach that integrates compliance, audit trails, and policy controls directly into process systems.

Human–AI Collaboration: A working model where AI supports decision-making while humans retain strategic oversight and control.

Learning Layer: The mechanism that improves decision accuracy over time through feedback and model updates.

Predictive Process Monitoring: The use of analytics to forecast delays, deviations, or risks in ongoing workflows.

Process Mining: A technique that analyzes event logs to discover, monitor, and improve business processes.

Process Sentience: The ability of business processes to sense data, interpret situations, learn from experience, and adapt actions within governance limits.

Reasoning Layer: The analytical engine that interprets data, predicts outcomes, and assesses risks.

Sensing Layer: The component that collects real-time operational and environmental data from multiple sources.

Chapter 9
Governance, Risk Management, and Ethical Considerations in ABPM Data Management

Gautam Shivaraj
http://orcid.org/0009-0000-5184-8448
Adarsh Institute of Management and Information Technology, Chamarajpet, India

ABSTRACT

The emergence of Autonomous Business Process Management (ABPM) has significantly transformed the way organizations manage their operations and data. This paper explores the integration of governance, risk management, and ethical considerations in ABPM, emphasizing the challenges and solutions associated with data management. Governance frameworks play a crucial role in ensuring that ABPM systems operate within legal, regulatory, and organizational boundaries, while robust risk management strategies are necessary to mitigate potential threats associated with automation and data privacy concerns. Furthermore, ethical considerations, particularly in data handling and decision-making algorithms, must be at the forefront of ABPM implementation to ensure transparency, fairness, and accountability. This study examines best practices in managing data in autonomous environments, proposing an integrated approach that aligns technological advancements with ethical and governance standards to foster trust and sustainability in business processes.

DOI: 10.4018/979-8-2600-0050-2.ch009

1. INTRODUCTION

1.1 Background and Context

In contemporary digital ecosystems, data is not merely a byproduct of operations, it is a strategic asset that drives decision-making, automation, performance optimization, and innovation. Within this landscape, ABPM data management, whether referring to *Autonomous Business Process Management* in intelligent enterprise systems represents a particularly sensitive and high-impact domain of data governance.

ABPM systems continuously generate time-series data, contextual metadata, user inputs, and system logs. These datasets are often:

- High in volume and velocity
- Integrated across multiple platforms
- Used for automated or semi-automated decision-making
- Subject to regulatory oversight
- Sensitive in nature (health, behavioral, operational, or personal data)

Because of these characteristics, ABPM data management requires structured governance mechanisms, robust risk controls, and carefully embedded ethical safeguards (Raman et al., 2025).

1.2 Why Governance in ABPM Data Matters

ABPM systems do not operate in isolation. They sit within broader digital infrastructures involving:

- Edge devices and sensors
- Cloud platforms and analytics engines
- APIs and third-party integrations
- Dashboards, alerting systems, and decision-support tools

Poor governance in such environments can result in:

- Data inaccuracies affecting critical decisions
- Privacy breaches and reputational damage
- Regulatory penalties
- Automation errors with real-world consequences
- Erosion of stakeholder trust

Effective governance ensures that ABPM data remains:

- **Accurate** (data integrity and validation)
- **Secure** (confidentiality and cybersecurity)
- **Available** (resilience and continuity)
- **Compliant** (alignment with laws and standards)
- **Ethically responsible** (aligned with societal values)

Governance therefore becomes a foundational pillar, not a bureaucratic add-on.

1.3 Unique Characteristics of ABPM Data

ABPM data management differs from traditional data environments in several ways:

1. Continuous and Real-Time Nature

ABPM systems often generate streaming data. Real-time processing introduces challenges related to:

- Latency
- Synchronization
- Edge-to-cloud security
- Immediate decision impact

2. Sensitivity and Personalization

ABPM datasets frequently include personally identifiable information (PII), health metrics, behavioral indicators, or performance metrics tied to individuals or departments.

3. Decision-Influencing Outputs

In many deployments, ABPM data feeds automated systems that:

- Trigger alerts
- Prioritize workflows
- Recommend interventions
- Allocate resources

This elevates governance concerns because errors or biases can directly influence outcomes.

4. Multi-Stakeholder Ecosystem

ABPM environments typically involve:

- End users (patients, employees, customers)
- Technical teams
- Vendors
- Regulators
- Management and policy-makers

Each stakeholder group carries distinct expectations and obligations.

1.4 The Convergence of Governance, Risk, and Ethics

Table 1. Traditionally, governance, risk management, and ethics were treated as separate organizational functions. However, in ABPM data management, these elements converge:

Governance	Risk Management	Ethics
Defines structure, policies, accountability	Identifies and mitigates threats	Ensures moral and societal alignment
Establishes data ownership	Assesses impact and likelihood	Protects human dignity and fairness
Controls access and usage	Monitors vulnerabilities	Prevents misuse and harm

A mature ABPM data strategy integrates all three into a single operational framework (Abourida et al., 2026).

For example:

- A governance policy may require encryption.
- Risk management may assess breach probability.
- Ethical oversight ensures sensitive data is not collected unnecessarily.

This integrated approach reduces silos and strengthens accountability.

Conceptual Governance–Risk–Ethics (GRE) Integration Model

The Governance–Risk–Ethics (GRE) Integration Model provides a unified framework for responsible ABPM data management. Rather than treating governance, risk management, and ethics as isolated organizational functions, the model integrates them into a continuous oversight cycle supporting operational resilience, regulatory compliance, and ethical accountability.

Conceptual Diagram — Governance–Risk–Ethics (GRE) Integration Model

The Governance–Risk–Ethics (GRE) Integration Model illustrates how governance, risk management, and ethics operate as mutually reinforcing functions within a continuous oversight cycle. Unlike sequential or siloed approaches, the GRE model embeds a feedback mechanism that enables adaptive improvement across all three pillars. The diagram below provides a visual representation of this integrated framework:

Figure 1. Governance–Risk–Ethics (GRE) Integration Model for ABPM Data Management

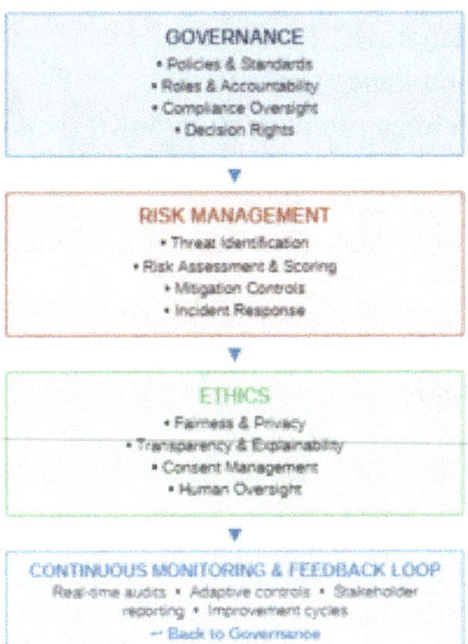

Governance establishes the institutional structure — policies, standards, decision rights, and accountability mechanisms — that anchors all data management activity.

Risk Management operationalizes that structure by identifying threats, assessing likelihood and impact, and deploying mitigation controls. Ethics ensures that system behavior remains aligned with human values, protecting fairness, privacy, and transparency throughout. The feedback loop completes the cycle: continuous monitoring captures performance signals and governance gaps, returning corrective intelligence to the governance layer for adaptive refinement. This cyclical, integrated structure prevents the governance–risk–ethics triad from operating as isolated silos and promotes organizational resilience.

The model demonstrates that governance establishes institutional structure and accountability, risk management operationalizes threat mitigation and resilience, while ethics ensures that system behavior remains aligned with human values and societal expectations. Continuous monitoring creates a feedback loop that enables adaptive improvement and long-term sustainability.

1.5 Regulatory and Compliance Drivers

ABPM data environments are often subject to national and international regulations. Depending on the domain and geography, these may include:

- Data protection and privacy regulations
- Health information regulations
- Cybersecurity standards
- Industry-specific compliance frameworks
- AI governance guidelines (if automation or ML is involved)

Failure to comply can lead to:

- Financial penalties
- Legal liability
- Operational suspension
- Loss of public trust

Governance must therefore embed regulatory mapping into everyday operational processes rather than treating compliance as an afterthought (Li et al., 2026).

1.6 Ethical Imperatives in the Age of Automation

As ABPM systems increasingly incorporate analytics and automation, ethical challenges intensify. These include:

- Bias in automated decisions
- Over-reliance on algorithmic outputs
- Inadequate transparency
- Lack of explainability
- Consent ambiguity
- Data over-collection

Ethical governance in ABPM requires proactive measures such as:

- Human oversight for high-impact decisions
- Transparency in data use
- Periodic fairness audits
- Clear consent frameworks
- Minimization of unnecessary data collection

Ethics becomes particularly critical when ABPM systems influence health interventions, employment decisions, workflow prioritization, or financial outcomes.

1.7 Objectives of This Chapter

This chapter aims to:

1. Define structured governance mechanisms for ABPM data management
2. Identify and categorize risks specific to ABPM ecosystems
3. Provide actionable risk mitigation frameworks
4. Integrate ethical principles into operational data practices
5. Present implementation guidance for organizations at different maturity levels

Rather than offering purely theoretical discussion, the chapter emphasizes:

- Practical governance models
- Risk register structures
- Policy frameworks
- Ethical-by-design approaches
- Implementation roadmaps

1.8 Scope and Structure

The remainder of this chapter is organized as follows:

Section 2: ABPM data lifecycle and ecosystem overview
Section 3: Governance structures, roles, and policy frameworks
Section 4: Risk identification, assessment, and mitigation strategies
Section 5: Ethical considerations and responsible data stewardship
Section 6: Implementation roadmap and maturity progression
Section 7: Conclusion and future outlook

ABPM data management sits at the intersection of technology, human welfare, and organizational accountability. As digital systems become more interconnected and autonomous, the consequences of mismanaged data increase proportionally. Governance ensures order and structure. Risk management safeguards resilience and security. Ethical considerations protect human dignity and fairness. Together, these elements form the foundation for sustainable, trustworthy, and responsible ABPM data ecosystems (Cunningham et al., 2025)

2. ABPM DATA LANDSCAPE AND LIFECYCLE

Understanding the ABPM data landscape requires more than a technical inventory of systems and storage locations. It demands a holistic appreciation of how data is generated, transmitted, interpreted, transformed, and ultimately retired within a living ecosystem of stakeholders, technologies, and regulatory environments. Whether ABPM refers to Ambulatory Blood Pressure Monitoring in clinical contexts or Autonomous/Advanced Business Process Management in enterprise automation environments, the data lifecycle is dynamic, interconnected, and deeply consequential (Krúpová & Koman, 2026).

At its foundation, ABPM data originates in environments where measurement or event capture occurs continuously or at high frequency. In healthcare deployments, devices collect physiological signals at scheduled intervals across extended periods. In enterprise settings, workflow engines capture process events, performance metrics, and user interactions in near real time. In both scenarios, data generation is not episodic but persistent. This persistence introduces both value and vulnerability. Continuous streams provide richer insights and improved predictive capacity, yet they also increase exposure to error accumulation, system drift, and security intrusion (Karim & Asaduzzaman, 2026).

The architecture of ABPM ecosystems typically spans multiple layers. At the edge, data is generated through devices, applications, or workflow triggers. This layer may operate under constrained computational resources or unstable network conditions. Data at this stage may be temporarily stored, filtered, or formatted before transmission. Governance at the edge is particularly significant because errors introduced at the point of capture often propagate invisibly downstream. Timestamp misalignment, calibration issues, or corrupted firmware updates can create systemic distortions that later appear as analytic anomalies. For this reason, validation mechanisms and device authentication controls must be embedded as early as possible in the lifecycle (Dhaigude & Kamath, 2025).

Once data leaves the point of origin, it traverses transmission channels that may include wireless networks, secured APIs, cloud ingestion endpoints, or private enterprise connections. Transmission represents one of the most risk-sensitive phases in the lifecycle because it exposes data to interception, latency distortions, and integrity compromise. Encryption in transit is essential, but encryption alone is insufficient without certificate management, authentication protocols, and network segmentation. The lifecycle perspective emphasizes that data security must travel with the data itself, rather than being confined to storage endpoints (Fontoura et al., 2025).

After transmission, data enters processing and storage environments. These environments are rarely uniform. Structured relational databases may coexist with unstructured data lakes. Operational systems optimized for real-time responsiveness may run parallel to analytical warehouses designed for historical analysis. In some cases, cold storage archives retain data for regulatory or longitudinal purposes. The heterogeneity of storage environments complicates governance because each storage tier carries different access rules, performance characteristics, and retention expectations. Without clear classification frameworks and lifecycle documentation, data may inadvertently migrate between tiers in ways that obscure ownership or accountability (Manning et al., 2023).

Processing and transformation represent another critical lifecycle stage. Raw ABPM data often has limited standalone value. It becomes meaningful only after aggregation, contextualization, normalization, and sometimes model-driven interpretation. Derived metrics, risk scores, alerts, or predictions are generated through algorithms or rule engines. These transformations introduce a layer of abstraction between original data and actionable output. Governance challenges intensify here because transformation logic may evolve over time. If documentation is inadequate or version control practices are inconsistent, organizations may lose the ability to explain how a specific output was produced. In regulated or ethically sensitive environments, such opacity can undermine trust and create compliance exposure (Pasas-Farmer & Jain, 2025).

Data lineage therefore becomes a central element of lifecycle integrity. Lineage mapping traces the journey of data from its origin through each transformation step to its ultimate use. It enables traceability, supports audit processes, and accelerates incident investigation. In ABPM contexts where automated decisions may trigger interventions, the ability to reconstruct the decision pathway is essential for both accountability and learning. Lineage documentation is not simply a technical artifact; it is a governance instrument that links technical processes to organizational responsibility (X. Wang et al., 2024).

As data progresses toward application and decision layers, its impact becomes tangible. Dashboards display trends, alerts prompt action, and automated systems may adjust workflows or recommend interventions. At this stage, ABPM data ceases to be passive information and becomes an operational driver. Errors or biases embedded earlier in the lifecycle can now produce material consequences. A misinterpreted physiological reading may trigger unnecessary escalation. A flawed workflow metric may reprioritize resources incorrectly. The lifecycle perspective emphasizes that quality assurance must not be confined to ingestion but must extend into output validation and decision auditing (Rouzky et al., 2025).

Access and sharing form another dimension of the ABPM lifecycle. Data is rarely consumed exclusively by its creators. Analysts, administrators, supervisors, external vendors, regulators, and partners may all interact with subsets of the dataset. Each interaction represents both an opportunity and a risk. Overly permissive access controls can lead to unauthorized disclosure or misuse, while overly restrictive controls may impair operational efficiency. Periodic access reviews and clearly documented role-based permissions help maintain balance. Furthermore, when data crosses organizational boundaries, contractual safeguards and due diligence mechanisms must accompany the transfer. Accountability remains with the originating organization even when infrastructure is outsourced (Kadłubek, 2021).

Retention and archival practices extend the lifecycle beyond immediate operational needs. Regulatory frameworks often require that certain data be preserved for defined periods. Longitudinal analysis may also depend on historical datasets. However, indefinite retention introduces privacy risks and storage costs. Proportionality must guide retention decisions. Archival systems should ensure immutability and integrity while preventing unnecessary exposure. Equally important is the final stage of deletion and destruction. Secure erasure processes, cryptographic deletion methods, and documented destruction records protect against residual risk. Failure to delete data responsibly can be as damaging as failing to secure it during active use (Lee & Lee, 2025).

Throughout the lifecycle, data quality management operates as a continuous discipline rather than a one-time checkpoint. Accuracy, completeness, consistency, timeliness, and validity must be monitored dynamically. Automated quality checks at ingestion can identify anomalies early, but downstream validations are equally important. Data drift, schema evolution, and integration changes may degrade quality over time. Mature ABPM ecosystems implement feedback loops in which quality incidents inform process improvement and system refinement (Huang et al., 2025).

The ABPM data landscape is further complicated by vendor integration and cross-border considerations. Cloud providers, device manufacturers, analytics platforms, and third-party service providers may each process or store segments of the data. This distributed architecture requires coordinated governance frameworks that extend beyond internal policies. Vendor assessments, security certifications, and contractual data processing agreements become integral to lifecycle management. In multinational deployments, geographic data residency and cross-border transfer restrictions introduce additional layers of complexity. Mapping data flows geographically ensures that regulatory obligations are met without disrupting operational continuity (N. Wang, 2023).

As organizations evolve, their approach to lifecycle management typically matures. Early-stage deployments may rely on ad hoc documentation and reactive issue resolution. Over time, structured lifecycle models, automated monitoring, and integrated governance tools emerge. In advanced stages, ethical oversight mechanisms, bias monitoring frameworks, and real-time anomaly detection systems become embedded within the lifecycle itself. At this point, governance is no longer an external control imposed on the data environment; it becomes an intrinsic property of the system's design (N. Wang, 2023).

The ABPM data lifecycle spans nine interconnected stages, each carrying distinct governance obligations and risk exposures. The table below provides a structured visual overview of these stages, the primary risks associated with each, and a qualitative risk severity rating to guide prioritization of controls.

Figure 2. ABPM Data Lifecycle — Stages, Risk Points, and Risk Severity Ratings

Lifecycle Stage	Description	Key Risk Points	Risk Level
1. Data Generation	Devices, sensors, workflow triggers capture raw ABPM data continuously	Calibration errors, timestamp misalignment, firmware vulnerabilities	⚠ HIGH
2. Transmission	Data traverses wireless networks, APIs, cloud ingestion endpoints	Interception, latency distortion, integrity compromise in transit	⚠ HIGH
3. Ingestion & Validation	Data enters processing pipelines, validation rules applied	Schema mismatches, missing fields, corrupted records pass through	⚠ MEDIUM
4. Storage	Structured DBs, data lakes, cold archives retain datasets	Unauthorized access, data leakage, improper tier migration	⚠ HIGH
5. Processing & Transformation	Aggregation, normalization, ML model scoring, alerts generation	Bias propagation, opaque logic, version drift, loss of lineage	⚠ HIGH
6. Access & Sharing	Analysts, vendors, regulators consume data subsets	Over-permissive access, unauthorized disclosure, third-party risk	⚠ MEDIUM
7. Application & Decision	Dashboards, automated alerts, recommendations drive actions	Biased outputs, misinterpretation, over-reliance on automation	⚠ HIGH
8. Retention & Archival	Regulatory or longitudinal preservation in immutable stores	Privacy risk from excessive retention, storage cost escalation	⚠ MEDIUM
9. Deletion & Destruction	Secure erasure, cryptographic deletion, destruction records	Residual data exposure, inadequate destruction documentation	⚠ MEDIUM

Risk levels reflect the combination of likelihood and potential impact. High-severity stages (data generation, transmission, storage, processing, and decision output) warrant the strongest preventive and detective controls. Medium-severity stages require structured oversight but carry lower inherent exposure given appropriate policy alignment. Organizations should use this lifecycle map as a foundation for risk register development and control gap analysis.

Ultimately, the ABPM data lifecycle reflects a continuum of responsibility. Data does not merely exist; it travels, transforms, influences, and eventually expires. Each stage introduces unique vulnerabilities and obligations. By conceptualizing ABPM data management as a lifecycle rather than a collection of isolated technical tasks, organizations gain the clarity needed to align operational efficiency with regulatory compliance and ethical integrity. The lifecycle model thus provides the structural backbone for the governance, risk management, and ethical frameworks developed in subsequent chapters. (Dumitrache et al., 2023).

3: GOVERNANCE FRAMEWORKS FOR ABPM DATA MANAGEMENT

Effective governance is the structural foundation upon which ABPM data management operates. While technology enables data collection, processing, and automation, governance ensures that these capabilities are aligned with organizational objectives, regulatory requirements, and ethical standards. In ABPM ecosystems—whether deployed in healthcare settings or enterprise automation environments—governance must address complexity, scale, accountability, and trust simultaneously. It is not merely about compliance; it is about designing a coherent system of authority, responsibility, oversight, and control that sustains data integrity throughout the

Table 2. Governance Roles and Responsibilities in ABPM Data Management

Governance Role	Primary Responsibilities	Key Accountability
Executive Leadership	Define governance vision, approve policies, allocate resources	Strategic oversight
Data Owner	Approve data usage, ensure lawful processing, define access requirements	Data accountability
Data Steward	Maintain data quality, metadata, and lifecycle documentation	Operational integrity
Security Officer	Implement cybersecurity controls and incident response	Information security
Compliance Officer	Monitor adherence to legal and regulatory obligations	Regulatory compliance
Ethics Committee	Review fairness, transparency, and ethical implications	Ethical governance
IT Operations Team	Maintain infrastructure availability and system reliability	Technical continuity

continued on following page

Table 2. Continued

Governance Role	Primary Responsibilities	Key Accountability
Risk Management Team	Conduct risk assessments and monitor mitigation controls	Risk resilience
Internal Auditor	Evaluate governance effectiveness and policy enforcement	Independent assurance
Third-Party Vendors	Process or store data according to contractual obligations	External compliance

lifecycle (Turienzo et al., 2023)

At its core, governance defines who has the authority to make decisions about data and under what conditions those decisions are made. In many organizations, the absence of clearly articulated governance structures leads to ambiguity. Data may be collected without defined ownership, accessed without accountability, or modified without traceability. Such fragmentation increases operational risk and erodes trust in analytical outputs. A formal governance framework mitigates these vulnerabilities by establishing clear decision rights and institutionalizing oversight mechanisms (Ullah et al., 2025).

A mature governance structure typically operates across multiple levels. At the strategic level, executive leadership sets the vision for responsible data stewardship. This includes defining the organization's risk appetite, ethical commitments, and compliance priorities. Without executive sponsorship, governance efforts often lack authority and sustainability. Strategic direction ensures that data governance is treated as a core organizational function rather than a peripheral technical initiative (Rožman et al., 2025).

At the operational level, governance translates strategic vision into enforceable policies and structured processes. This requires clearly defined roles such as data owners, data stewards, security officers, and compliance managers. The data owner is typically accountable for the overall quality, accessibility, and lawful use of a specific dataset or domain. The data steward manages day-to-day oversight, including maintaining definitions, resolving quality issues, and coordinating lifecycle updates. Security and privacy officers ensure that confidentiality, integrity, and regulatory alignment are embedded within operational processes. These roles must not exist in isolation; they require coordination mechanisms to resolve conflicts, manage escalations, and adapt policies as systems evolve (Hassen et al., 2024)

Central to governance is the establishment of comprehensive policy frameworks. Policies articulate expectations and constraints regarding data classification, access control, retention, sharing, and auditing. Classification policies determine how data is categorized based on sensitivity and risk. In ABPM environments where sensitive physiological data or performance metrics may be processed, classification becomes critical to determining appropriate safeguards. Access control policies operationalize the principle of least privilege, ensuring that individuals access only the data necessary for their roles. Retention policies define how long data should be preserved, balancing regulatory obligations against privacy considerations and storage efficiency (Chen et al., 2025).

Governance also requires formalized documentation systems that promote transparency and consistency. Data dictionaries and metadata catalogs ensure that terms, metrics, and calculations are defined uniformly across departments. Without standardized definitions, analytical outputs may vary depending on interpretation, undermining trust in decision-making processes. Documentation not only enhances clarity but also supports auditability. Regulators and internal auditors increasingly expect organizations to demonstrate how data flows through systems and how controls are applied at each stage (Dziembek & Turek, 2025).

Another critical element of governance is change management. ABPM systems are rarely static. Devices are updated, workflows are redesigned, algorithms are retrained, and integrations are added. Each change introduces potential risks. Governance frameworks must therefore incorporate structured change control procedures that require documentation, review, and approval before modifications are deployed. Version control mechanisms preserve historical configurations, enabling rollback in the event of unintended consequences. This discipline reduces operational disruption and strengthens accountability (Acitelli et al., 2025).

Governance frameworks must also address cross-functional coordination. ABPM data often traverses departmental boundaries, linking IT teams, analytics units, operational managers, compliance officers, and external partners. Without coordinated governance, siloed decision-making can result in inconsistent standards and duplicated effort. Establishing governance councils or cross-functional committees provides a structured forum for resolving disputes, prioritizing initiatives, and harmonizing policies. These forums also facilitate the integration of ethical considerations into technical decision-making, ensuring that system design reflects broader organizational values (Kumar & Kotler, 2024).

Technology plays a supportive but not substitutive role in governance. Automated monitoring tools, identity management systems, data lineage trackers, and access review platforms enhance enforcement capabilities. However, governance is fundamentally about accountability and oversight, which cannot be fully automated. Human judgment remains essential, particularly when interpreting ambiguous regulations or evaluating ethical trade-offs. The integration of technological controls with human oversight creates a balanced governance ecosystem.

Accountability mechanisms are another defining feature of robust governance frameworks. Audit trails, logging systems, and periodic reviews provide visibility into who accessed data, what changes were made, and when those changes occurred.

Transparency in access and modification activities deters misuse and supports forensic analysis in the event of incidents. Governance frameworks should define clear escalation pathways for violations, ensuring that policy breaches are investigated promptly and consistently (Antczak et al., 2025).

In addition to internal controls, governance must extend to external relationships. Vendors, cloud providers, device manufacturers, and integration partners frequently process or store ABPM data. Governance frameworks should therefore incorporate third-party risk management processes, including due diligence assessments, contractual safeguards, and ongoing monitoring. Delegating infrastructure responsibilities does not transfer accountability. Organizations remain responsible for ensuring that external partners adhere to equivalent standards of security and ethical conduct.

Ethical oversight increasingly forms an integral component of governance frameworks. As ABPM systems incorporate automation and advanced analytics, decisions influenced by data may affect individuals directly. Governance structures should therefore include mechanisms for reviewing high-impact use cases, assessing fairness, and evaluating unintended consequences. Ethical review committees or advisory boards can provide independent scrutiny, reinforcing transparency and public trust. Embedding ethical checkpoints within governance processes ensures that innovation does not outpace responsibility.

Maturity in governance is achieved gradually. Early-stage organizations may focus primarily on compliance documentation. Over time, governance evolves into a proactive discipline that anticipates risks and integrates ethical reflection into system design. In advanced stages, governance frameworks become adaptive, incorporating continuous monitoring, performance metrics, and feedback loops that drive improvement. Governance then shifts from reactive enforcement to strategic enablement, facilitating innovation while safeguarding integrity.

Ultimately, governance frameworks for ABPM data management are not static rulebooks but living systems. They must evolve alongside technological advancements, regulatory changes, and organizational growth. Effective governance balances structure with flexibility, control with empowerment, and compliance with ethical stewardship. By institutionalizing accountability, clarifying responsibilities, and embedding oversight across the data lifecycle, governance frameworks ensure that ABPM systems operate reliably, transparently, and responsibly in increasingly complex digital environments (Alla, 2026).

Table 3. Governance Roles and Responsibilities in ABPM Data Management

Governance Role	Primary Responsibilities	Key Accountability	Organizational Unit
Executive Leadership	Define governance vision, approve policies, allocate resources	Strategic oversight	Board / C-Suite
Data Owner	Approve data usage, ensure lawful processing, define access requirements	Data accountability	Department Head
Data Steward	Maintain data quality, metadata, and lifecycle documentation	Operational integrity	Data Management
Security Officer	Implement cybersecurity controls and incident response procedures	Information security	IT / CISO
Compliance Officer	Monitor adherence to legal and regulatory obligations	Regulatory compliance	Legal / Compliance
Ethics Committee	Review fairness, transparency, and ethical implications of data use	Ethical governance	Cross-functional
IT Operations Team	Maintain infrastructure availability and system reliability	Technical continuity	IT Operations
Risk Management Team	Conduct risk assessments and monitor mitigation controls	Risk resilience	Risk / Audit
Internal Auditor	Evaluate governance effectiveness and policy enforcement independently	Independent assurance	Internal Audit
Third-Party Vendors	Process or store data according to contractual and regulatory obligations	External compliance	External Partners

Role clarity reduces accountability gaps and prevents the fragmented stewardship that often results in data quality failures, compliance breaches, or ethical blind spots. Organizations should formally document these roles in governance charters, reviewed and reaffirmed annually or following significant system or regulatory changes. Where roles overlap — such as between Data Steward and IT Operations — escalation pathways and authority boundaries should be explicitly defined.

4: RISK MANAGEMENT FRAMEWORKS IN ABPM DATA MANAGEMENT

Risk management within ABPM data environments is not a peripheral compliance exercise but a central operational necessity. Because ABPM systems often operate continuously, integrate multiple technological layers, and influence real-time decisions, the risks associated with data mismanagement are amplified. These risks extend beyond technical malfunction and include legal exposure, ethical harm, operational disruption, reputational damage, and financial loss. A comprehensive risk management framework provides a structured methodology for identifying, assessing, mitigating, monitoring, and responding to these vulnerabilities in a disciplined and sustainable manner.

At its foundation, risk management begins with recognition that risk is inherent in all data-driven systems. ABPM ecosystems generate and process high volumes of dynamic data. This constant movement introduces exposure points at every stage of the lifecycle—from data capture and transmission to transformation and sharing. Rather than attempting to eliminate risk entirely, effective frameworks seek to understand risk in context and manage it proportionately. The objective is resilience rather than absolute prevention.

The first stage of a structured risk management framework is risk identification. In ABPM contexts, risks can arise from technical failures, human error, malicious activity, environmental disruptions, regulatory shifts, or ethical blind spots. Identification requires a multidisciplinary approach. Technical teams may detect infrastructure vulnerabilities, compliance teams may highlight regulatory gaps, and operational users may identify workflow weaknesses. Workshops, system audits, historical incident reviews, and threat modeling exercises contribute to building a comprehensive inventory of risks. Importantly, identification must consider both direct and indirect impacts. A minor synchronization error at the data ingestion stage may escalate into major decision inaccuracies if left unaddressed (Correa-Peralta et al., 2025).

Once risks are identified, systematic assessment follows. Assessment evaluates both the likelihood of occurrence and the potential impact. In ABPM environments, impact should not be measured solely in financial terms. A security breach may have reputational consequences; a biased algorithm may cause ethical harm; a system outage may disrupt critical services. A balanced assessment framework therefore incorporates operational, legal, financial, ethical, and reputational dimensions. Quantitative scoring systems can help prioritize risks, but qualitative judgment remains essential, particularly when evaluating high-stakes outcomes (Pasas-Farmer & Jain, 2025).

Mitigation represents the next phase of the framework. Mitigation strategies may involve preventive controls, detective mechanisms, corrective processes, or risk transfer arrangements. Preventive controls include encryption protocols, authentication systems, validation rules, and change management procedures. Detective mechanisms such as logging, anomaly detection, and automated monitoring enable rapid identification of emerging threats. Corrective processes, including rollback capabilities and disaster recovery plans, reduce the duration and severity of incidents. In some cases, risk may be transferred through insurance or contractual agreements, particularly when third-party vendors are involved. The selection of mitigation strategies should align with the organization's defined risk appetite and operational capacity (Rouzky et al., 2025).

A distinctive feature of ABPM systems is the interdependence of risks. Technical vulnerabilities may create compliance exposure, and operational weaknesses may amplify ethical concerns. For example, insufficient access controls may not only increase the probability of data breaches but also undermine fairness if unauthorized modifications alter decision-making outputs. Risk management frameworks must therefore avoid siloed thinking. Integrated risk mapping, which visualizes relationships between vulnerabilities, enhances understanding of systemic exposure and prevents fragmented mitigation efforts.

Continuous monitoring is essential in dynamic ABPM ecosystems. Risk profiles evolve as systems are updated, new integrations are added, or regulations change. Monitoring mechanisms may include automated alerts, periodic access reviews, vulnerability scans, penetration testing, and key risk indicators. These indicators provide early warning signals that control effectiveness may be degrading. Monitoring also supports adaptive governance, allowing organizations to recalibrate controls in response to shifting threat landscapes. Without ongoing oversight, mitigation efforts may become outdated, leaving previously controlled risks unguarded.

Incident response planning forms another core component of the framework. Even with robust preventive measures, incidents are inevitable in complex digital systems. A well-designed response plan outlines clear escalation pathways, communication protocols, containment strategies, and recovery procedures. Timely and transparent communication during incidents can significantly reduce reputational harm. Post-incident analysis is equally important. Lessons learned should be documented and integrated into revised controls, ensuring that risk management evolves from experience rather than remaining static (Lee & Lee, 2025).

Third-party risk management deserves particular emphasis in ABPM environments. External vendors often manage storage, analytics, device firmware, or integration services. Each external relationship extends the risk perimeter beyond internal boundaries. Risk frameworks should therefore incorporate vendor due diligence assessments, contractual obligations regarding data protection, and periodic compliance reviews. Monitoring third-party performance ensures that external dependencies do not become hidden vulnerabilities (Huang et al., 2025).

Cybersecurity risk constitutes one of the most visible categories within ABPM data management. Threat actors may target sensitive data, exploit configuration weaknesses, or deploy ransomware attacks. Cyber risk management includes layered defenses such as network segmentation, endpoint protection, encryption, multifactor authentication, and continuous vulnerability assessment. However, cybersecurity must be complemented by human-centric controls. Phishing attacks, credential misuse, and social engineering tactics exploit behavioral weaknesses. Training programs and awareness initiatives therefore serve as integral components of risk mitigation (Huang et al., 2025).

Operational risk also merits careful attention. System downtime, integration failures, or software defects can disrupt data flows and compromise decision-making reliability. Business continuity planning ensures that ABPM systems maintain essential functions during disruptions. Redundant infrastructure, failover mechanisms, and periodic disaster recovery drills enhance resilience. The goal is not merely to restore functionality but to maintain confidence in system reliability under adverse conditions.

Regulatory and compliance risks further complicate ABPM environments. Data protection laws, sector-specific standards, and cross-border transfer regulations impose obligations that vary across jurisdictions. Failure to adhere to these requirements may result in penalties or legal sanctions. Risk management frameworks should include regulatory mapping exercises to identify applicable obligations and track evolving legal developments. Integrating compliance checks into routine operational processes reduces the likelihood of inadvertent violations (N. Wang, 2023).

Ethical risks increasingly intersect with traditional risk categories. As ABPM systems incorporate automation and advanced analytics, decisions influenced by data may affect individuals directly. Bias in predictive models, lack of transparency in automated outputs, or overreliance on algorithmic recommendations can produce unintended harm. Ethical risk assessment should therefore be embedded within the broader framework. Evaluating fairness, proportionality, and transparency strengthens responsible system design and fosters trust among stakeholders (Dumitrache et al., 2023).

Risk culture ultimately determines the effectiveness of any framework. Policies and technical controls are insufficient if organizational attitudes toward risk are complacent. Leadership must model a proactive approach to risk awareness, encouraging reporting of vulnerabilities and near-misses without punitive repercussions. When employees feel empowered to highlight concerns, risk identification becomes more comprehensive and timelier. Maturity in ABPM risk management evolves through stages. Initial efforts may focus on compliance-driven checklists and reactive incident handling. Over time, organizations develop structured risk registers, quantitative assessment models, and integrated monitoring systems. In advanced stages, predictive risk analytics and scenario simulations enhance preparedness. Continuous improvement cycles ensure that risk management remains aligned with technological innovation and organizational growth (Turienzo et al., 2023).

In conclusion, risk management frameworks in ABPM data environments serve as dynamic systems of anticipation, mitigation, and adaptation. They acknowledge that risk is an inherent feature of complex digital ecosystems and seek to manage it through structured, integrated processes. By combining technical safeguards, organizational accountability, ethical evaluation, and continuous monitoring, risk management strengthens the resilience and trustworthiness of ABPM systems. In doing so, it supports not only operational continuity but also the broader objectives of governance and ethical stewardship that define responsible data management (Rožman et al., 2025).

5: ETHICAL CONSIDERATIONS IN ABPM DATA MANAGEMENT

Ethical considerations in ABPM data management extend beyond compliance obligations and technical safeguards; they address the fundamental question of how data-driven systems affect individuals, communities, and institutional trust. In both healthcare-oriented ABPM deployments and enterprise automation environments, data does not exist in isolation from human experience. It represents physiological conditions, behavioral patterns, performance indicators, and sometimes deeply personal attributes. As such, the management of ABPM data carries moral responsibilities that require deliberate reflection and structured oversight. Ethical governance ensures that technological advancement does not outpace accountability and that efficiency does not overshadow human dignity (Abourida et al., 2026).

One of the primary ethical principles guiding ABPM data management is respect for autonomy. Individuals whose data is collected must retain meaningful agency over how that data is used. This principle requires transparency regarding collection practices, clarity about intended purposes, and accessible mechanisms for consent and withdrawal where applicable. In healthcare contexts, autonomy is closely tied to informed consent, ensuring that individuals understand how monitoring systems function and how their information may influence care decisions. In enterprise environments, autonomy relates to fair disclosure about how workflow data or performance metrics may affect evaluation or resource allocation. Ethical stewardship demands that consent is not reduced to a procedural formality but is treated as an ongoing relationship between data subjects and institutions (Ma et al., 2025).

Closely linked to autonomy is the principle of purpose limitation. ABPM systems often generate vast quantities of granular data, and the temptation to repurpose that data for secondary uses can be significant. Ethical data management requires restraint. Data collected for a specific operational or clinical objective should not be repurposed without clear justification, transparency, and, where necessary, renewed authorization. Purpose limitation protects individuals from unforeseen uses of their data that may conflict with their expectations or interests. It also reinforces trust by demonstrating that data custodians respect boundaries (Dhaigude & Kamath, 2025).

Privacy forms another central pillar of ethical ABPM governance. Privacy is not merely the absence of unauthorized access but the preservation of contextual integrity. Individuals share information within specific contexts—clinical, professional, or operational—and expect that information to remain confined to those contexts.

Ethical data management requires minimizing collection to what is genuinely necessary, anonymizing or pseudonymizing data when possible, and implementing safeguards that prevent re-identification. Even when data is technically secured, excessive retention or unnecessary granularity may erode privacy expectations. Ethical reflection therefore complements technical controls by asking whether the scope of data collection itself is justified (Mossa et al., 2025).

Fairness and equity are increasingly critical in ABPM environments that incorporate analytics and automation. When data feeds predictive models or automated decision systems, biases embedded within datasets or algorithms may produce unequal outcomes. Ethical governance requires proactive examination of representativeness and potential disparities. In healthcare, biased training data may lead to inaccurate assessments for underrepresented populations. In enterprise systems, workflow analytics may disadvantage certain teams or roles if contextual factors are not adequately considered. Ethical review processes should include regular fairness assessments, documentation of known limitations, and corrective actions when inequities are identified. Fairness is not achieved through neutrality alone; it requires intentional design and continuous evaluation (Junjia et al., 2025).

Transparency and explainability further strengthen ethical integrity. Individuals affected by data-driven decisions should be able to understand how those decisions were reached. In ABPM systems, particularly those that generate automated alerts or recommendations, opaque processes can undermine confidence and accountability. Explainability does not necessarily require revealing proprietary algorithms but does require providing meaningful descriptions of decision logic, input factors, and confidence levels. Transparent communication fosters trust and enables informed engagement with system outputs. It also enhances organizational learning by making assumptions and limitations visible. (Caramihai et al., 2020)

Table 4. Ethics-by-Design Checkpoints for ABPM Systems

Ethical Dimension	Operational Checkpoint	Measurement Indicator	Review Frequency
Bias and Fairness	Conduct algorithmic bias audits	Bias variance reports across demographic groups	Quarterly
Transparency	Maintain explainability logs for automated decisions	Percentage of decisions with explainable outputs	Monthly
Consent Management	Track consent collection and withdrawal status	Consent compliance rate	Continuous
Data Minimization	Review unnecessary data collection practices	Reduction in non-essential fields collected	Semi-Annual

continued on following page

Table 4. Continued

Ethical Dimension	Operational Checkpoint	Measurement Indicator	Review Frequency
Human Oversight	Require manual review for high-impact decisions	Percentage of escalated decisions reviewed by humans	Monthly
Accountability	Maintain ethical incident reporting mechanisms	Number of unresolved ethical incidents	Quarterly
Privacy Protection	Monitor anonymization effectiveness	Re-identification risk score	Quarterly
Stakeholder Inclusion	Conduct stakeholder ethical review sessions	Number of stakeholder consultations conducted	Annual

Accountability constitutes the structural backbone of ethical governance. Ethical principles must be operationalized through clear assignment of responsibility. When data misuse occurs or automated systems produce harmful outcomes, accountability ensures that corrective action is taken and lessons are integrated into future practice. Ethical oversight committees, review boards, or designated officers can provide structured evaluation of high-impact use cases. These mechanisms create forums for deliberation, particularly when new technologies introduce uncertain consequences. Accountability transforms ethical reflection from abstract ideals into actionable governance (Antczak et al., 2025).

The principle of non-maleficence, or the obligation to avoid harm, is particularly salient in ABPM contexts. Continuous monitoring systems may influence decisions that carry tangible consequences. Erroneous alerts may cause unnecessary interventions, while missed anomalies may delay critical responses. Ethical stewardship requires rigorous validation, ongoing monitoring of system performance, and human oversight where stakes are high. Overreliance on automation without sufficient human judgment can amplify harm. A balanced approach recognizes the strengths of data-driven systems while preserving human responsibility for interpretation and intervention.

Proportionality provides another ethical lens for evaluating ABPM practices. The intrusiveness of monitoring and the granularity of data collection should be proportionate to the benefits derived. Excessive surveillance, even if technically permissible, may undermine trust and erode morale in enterprise contexts or create discomfort in healthcare settings. Ethical proportionality encourages organizations to ask whether the intensity of monitoring is justified by the intended outcomes and whether less intrusive alternatives could achieve similar objectives (Alla, 2026).

Justice and inclusivity also play essential roles in ethical ABPM governance. Stakeholder engagement during system design and deployment can reveal diverse perspectives and potential unintended consequences. Involving representatives from affected groups ensures that ethical considerations reflect lived experiences rather than abstract assumptions. Inclusive design strengthens legitimacy and reduces the risk of systemic bias (Alla, 2026).

Ethical data management is further intertwined with organizational culture. Policies and oversight mechanisms are effective only when supported by a culture that values integrity and transparency. Training programs, open communication channels, and leadership commitment to ethical principles foster an environment in which concerns can be raised without fear of retaliation. Encouraging dialogue about ethical dilemmas reinforces collective responsibility and continuous improvement.

Measurable Ethics-by-Design Checkpoints

Ethics-by-Design is a proactive governance philosophy that embeds ethical evaluation directly into the operational and technical lifecycle of ABPM systems, rather than treating ethical review as a retrospective compliance exercise. The checkpoints below translate abstract ethical principles — fairness, transparency, consent, accountability — into measurable, time-bound operational commitments. Each checkpoint is assigned a responsible party, a quantifiable success metric, and a review cadence to ensure consistent enforcement.

Table 5. (Enhanced). Ethics-by-Design Checkpoints for ABPM Systems

Ethics-by-Design Checkpoint	Operational Action	Measurable Indicator / Target	Review Frequency	Responsible Party	Method
Bias & Fairness Audit	Run algorithmic bias detection tests across demographic segments	Bias variance report (< 5% disparity threshold across groups)	Quarterly	Ethics Committee	Automated + Manual
Consent Tracking	Log consent collection, version, timestamp, and withdrawal status per data subject	Consent compliance rate ≥ 99%; zero unresolved withdrawals	Continuous	Compliance Officer	Automated
Explainability Logs	Capture decision rationale for all automated outputs flagged as high-impact	% of high-impact decisions with documented explanation ≥ 95%	Monthly	Data Steward	Automated
Data Minimization Review	Audit data fields collected against stated collection purpose	Reduction in non-essential fields; proportionality score improvement	Semi-Annual	Data Owner	Manual

continued on following page

Table 5. Continued

Ethics-by-Design Checkpoint	Operational Action	Measurable Indicator / Target	Review Frequency	Responsible Party	Method
Human Oversight Escalation	Require manual review for decisions above risk threshold (clinical or HR)	% of escalated decisions reviewed by humans within SLA	Monthly	Operations Lead	Hybrid
Privacy Impact Assessment	Conduct PIA before new data collection or system changes	PIA completion rate = 100% for new use cases	Per Change	Compliance Officer	Manual
Re-identification Risk Test	Test anonymized datasets for re-identification vulnerability	Re-identification risk score < 0.01 (k-anonymity ≥ 5)	Quarterly	Security Officer	Automated
Ethical Incident Reporting	Track and resolve all reported ethical concerns within defined SLAs	Zero unresolved ethical incidents beyond 30-day resolution window	Quarterly	Ethics Committee	Manual
Stakeholder Consultation	Conduct structured ethical review sessions with affected stakeholder groups	Min. 2 consultations per year; documented outcomes and actions	Annual	Ethics Committee	Manual
AI Transparency Disclosure	Publish or document AI system capabilities, limitations, and data use	Transparency statement current and accessible to all stakeholders	Annual	Legal / Comms	Manual

These checkpoints should be operationalized through dedicated tooling — consent management platforms, bias monitoring dashboards, explainability logging frameworks — and reviewed in ethics committee meetings on the schedule specified. Where checkpoints fall below defined thresholds, corrective action plans must be documented and tracked to closure. The integration of measurable ethics checkpoints into governance reporting elevates ethical stewardship from organizational value statement to accountable operational practice.

As ABPM technologies evolve, ethical considerations will expand in scope and complexity. Artificial intelligence, predictive analytics, and cross-system integrations introduce new dimensions of uncertainty. Ethical frameworks must therefore remain adaptive, revisited regularly to account for technological and societal change.

Ethical governance is not a static achievement but an ongoing process of reflection, dialogue, and refinement (Dziembek & Turek, 2025).

In conclusion, ethical considerations in ABPM data management provide the moral compass guiding governance and risk frameworks. They ensure that data practices honor autonomy, protect privacy, promote fairness, maintain transparency, and uphold accountability. By embedding ethical principles within operational processes, organizations cultivate trust, strengthen legitimacy, and ensure that ABPM systems serve human interests rather than undermine them. Ethical stewardship thus completes the triad of governance, risk management, and responsible data practice that defines sustainable ABPM ecosystems (Abourida et al., 2026).

6: CONCLUSION AND FUTURE DIRECTIONS

The preceding chapters have examined governance structures, risk management frameworks, and ethical considerations within ABPM data management. Together, these elements form an integrated foundation for responsible stewardship of complex, data-driven ecosystems. As ABPM systems continue to evolve—whether in healthcare monitoring environments or enterprise automation infrastructures—the importance of structured oversight will only intensify. The convergence of continuous data generation, real-time analytics, cross-system integration, and automated decision-making creates both unprecedented opportunity and heightened responsibility (Abourida et al., 2026).

At its core, ABPM data management is not merely a technical challenge but an institutional commitment. Governance establishes clarity of ownership, policy alignment, and accountability. Risk management introduces disciplined anticipation and mitigation of vulnerabilities. Ethical oversight ensures that operational efficiency remains aligned with human dignity, fairness, and societal trust. When these three pillars operate cohesively, they transform ABPM systems from isolated technological tools into sustainable, trustworthy infrastructures. One of the most significant insights emerging from this integrated framework is the interdependence of governance, risk, and ethics. Governance without risk awareness becomes rigid bureaucracy. Risk management without ethical reflection may focus narrowly on financial or operational exposure while overlooking human impact. Ethical principles without governance structures lack enforceability. Sustainable ABPM ecosystems require these dimensions to reinforce one another through structured processes and cultural commitment. Looking forward, several trends will shape the future of ABPM data management. The increasing incorporation of artificial intelligence and predictive analytics will demand more sophisticated oversight mechanisms. As models become more complex, explainability and fairness will become central governance concerns. Organizations will need to adopt standardized documentation practices, model evaluation protocols, and bias monitoring systems to maintain accountability in automated decision-making environments.

Another emerging direction involves real-time governance integration. Traditional governance models often rely on periodic reviews and retrospective audits. However, the velocity of ABPM data flows necessitates dynamic oversight capabilities. Automated compliance monitoring, real-time anomaly detection, and adaptive access controls will increasingly become embedded features of governance frameworks. Such integration allows organizations to move from reactive correction to proactive prevention (Cunningham et al., 2025).

Data localization and cross-border regulatory fragmentation will also influence ABPM strategies. As jurisdictions enact diverse data protection regulations, organizations operating globally must harmonize compliance efforts without sacrificing operational efficiency. This will require sophisticated data mapping, jurisdiction-aware storage strategies, and continuous regulatory monitoring. Governance frameworks must remain agile to adapt to evolving legal landscapes. Cybersecurity resilience will remain a foundational priority. The expansion of interconnected devices and cloud platforms increases the attack surface for malicious actors. Future ABPM systems will likely rely on advanced threat detection, zero-trust architectures, and continuous authentication models. However, technological defenses must be complemented by strong organizational culture and training to mitigate human-centric vulnerabilities (Karim & Asaduzzaman, 2026).

Ethical governance will expand beyond fairness and transparency to encompass broader societal considerations. Questions of surveillance intensity, data commodification, and algorithmic autonomy will become more prominent. Institutions may establish independent ethics committees or advisory boards to evaluate emerging use cases and assess long-term societal implications. Embedding ethical deliberation into strategic planning will help ensure that innovation remains aligned with public trust. Organizational maturity will also influence the trajectory of ABPM data management. Early-stage systems may prioritize compliance stabilization and risk containment. As maturity increases, organizations will likely shift toward optimization, innovation enablement, and value extraction while maintaining strong safeguards. Continuous improvement cycles, performance metrics for governance effectiveness, and integrated feedback mechanisms will define advanced maturity levels. Ultimately, the sustainability of ABPM ecosystems depends on trust. Stakeholders whether patients, employees, regulators, or partners—must believe that their data is handled responsibly, securely, and ethically. Trust is built through transparency, consistent enforcement of policies, visible accountability, and demonstrated responsiveness to emerging challenges. Governance, risk management, and ethical frameworks serve as instruments for cultivating and maintaining this trust (Fontoura et al., 2025).

In conclusion, ABPM data management represents a dynamic intersection of technology, law, and human values. As digital infrastructures grow more autonomous and interconnected, the consequences of mismanagement grow proportionately. By institutionalizing structured governance, embedding comprehensive risk management processes, and integrating ethical principles into everyday operations, organizations can ensure that ABPM systems contribute positively to innovation, operational excellence, and societal well-being. The future of ABPM data management will not be defined solely by technological advancement, but by the strength and integrity of the frameworks that guide its use (Kadłubek, 2021).

REFERENCES

Abourida, M., Short, M., Klymenko, O. V., Khamis, N. M., Mahammedi, C., Al-Mhdawi, M. K. S., & Sakr, A. H. (2026). The evolving landscape of AI-driven risk management in the biogas production: A systematic and bibliometric review. *Waste Management Bulletin*, *4*(1), 100271. Advance online publication. DOI: 10.1016/j.wmb.2025.100271

Acitelli, G., Alman, A., Maggi, F. M., & Marrella, A. (2025). Achieving framed autonomy in AI-augmented business process management systems through automated planning. *Information Systems*, *133*, 102573. Advance online publication. DOI: 10.1016/j.is.2025.102573

Alla, P. B. (2026). Augmenting Intelligent Process Automation through Generative AI for Human-in-the-Loop Decision Systems. *Digital Engineering*, *8*, 100071. Advance online publication. DOI: 10.1016/j.dte.2025.100071

Antczak, J., Kabus, J., Miciuła, I., Lenort, R., Brzozowska, A., Wolniak, R., & Kana, R. (2025). Artificial Intelligence in business management - SWOT analysis. *Procedia Computer Science*, *270*, 4410–4420. DOI: 10.1016/j.procs.2025.09.566

Ben Hassen, M., Zahaf, S., & Gargouri, F. (2024). Business View Specification of Enterprise Information Systems Based on Core Ontologies. *Procedia Computer Science*, *237*, 380–388. DOI: 10.1016/j.procs.2024.05.118

Caramihai, S. I., Dumitrache, I., Moisescu, M. A., & Sacala, I. S. (2020). Bio-inspired autonomous enterprise systems. *IFAC-PapersOnLine*, *53*(2), 10879–10884. DOI: 10.1016/j.ifacol.2020.12.2820

Chen, R., Zhang, J., & Wang, H. (2025). Autonomous fleet management system in smart ports: Practical design and analytical considerations. *Multimodal Transportation*, *4*(3), 100211. Advance online publication. DOI: 10.1016/j.multra.2025.100211

Correa-Peralta, M., Vinueza-Martínez, J., & Castillo-Heredia, L. (2025). Evolution, topics and relevant research methodologies in business intelligence and data analysis in the academic management of higher education institutions. A literature review. In *Results in Engineering* (Vol. 25). Elsevier B.V., DOI: 10.1016/j.rineng.2024.103782

Cunningham, J., Davies, N., Devaney, S., Harding, M., Holm, S., Neumann, V., & Ainsworth, J. (2025). The development of a blockchain-based business model simulator with an application to civic medical data management. *Digital Business*, *5*(1), 100112. Advance online publication. DOI: 10.1016/j.digbus.2025.100112

Dhaigude, A. S., & Kamath, G. B. (2025). Mapping responsible artificial intelligence in business and management: Trends, influence, and emerging research directions. In *Journal of Open Innovation: Technology, Market, and Complexity* (Vol. 11, Number 4). Elsevier B.V. DOI: 10.1016/j.joitmc.2025.100640

Dumitrache, I., Predescu, L. S., Caramihai, S. I., & Moisescu, M. A. (2023). Collaborative Decisions in Knowledge Management for Intelligent Cyber-Enterprises. *Procedia Computer Science*, *221*, 955–962. DOI: 10.1016/j.procs.2023.08.074

Dziembek, D., & Turek, T. (2025). A Model for Integrating Artificial Intelligence with ERP Systems - Towards Autonomous Business Management Systems. *Procedia Computer Science*, *270*, 6260–6269. DOI: 10.1016/j.procs.2025.10.096

Fontoura, L., Luiz de Mattos Nascimento, D., Neto, J. V., & Gusmão Caiado, R. G. (2025). Energy Gen-AI technology framework: A perspective of energy efficiency and business ethics in operation management. *Technology in Society*, *81*, 102847. Advance online publication. DOI: 10.1016/j.techsoc.2025.102847

Huang, F., Luo, X., Xiao, M., & Dong, W. (2025). Dual performance of business model innovation in emerging market enterprises: A configurational approach. *Journal of Innovation & Knowledge*, *10*(4), 100755. Advance online publication. DOI: 10.1016/j.jik.2025.100755

Junjia, Y., Jiawen, L., Alias, A. H., Haron, N. A., & Abu Bakar, N. (2025). Reinforcement learning in risk management for pharmaceutical construction projects: Frontiers, challenges, and improvement strategies. In *Sustainable Futures* (Vol. 10). Elsevier Ltd., DOI: 10.1016/j.sftr.2025.101534

Kadłubek, M. (2021). Expectations for the use of Intelligent Transport Systems applications in the management of freight transport enterprises. *Procedia Computer Science*, *192*, 2318–2329. DOI: 10.1016/j.procs.2021.08.222

Karim, D. N., & Asaduzzaman, M. (2026). Linking talent management practices to autonomous work motivation and proactive customer service behavior: Insights from hospital nurses of Bangladesh. *Social Sciences & Humanities Open*, *13*, 102444. Advance online publication. DOI: 10.1016/j.ssaho.2026.102444

Krúpová, S., & Koman, G. (2026). The Role of IoT and Big Data in Smart City Transport Management: Lessons Learned from Case Studies in Slovakia and the EU. *Transportation Research Procedia*, *93*, 948–959. DOI: 10.1016/j.trpro.2025.12.027

Kumar, V., & Kotler, P. (2024). A market management approach to transformative business operations. *Marketing Strategy Journal*, *1*, 100005. DOI: 10.1016/j.msj.2025.100005

Lee, J. W., & Lee, K. (2025). Building a consensus: Harmonizing AI ethical guidelines and legal frameworks in Korea for enhanced governance. *Government Information Quarterly*, *42*(3), 102060. Advance online publication. DOI: 10.1016/j.giq.2025.102060

Li, M., Zhu, Y., Chen, F., & Anaba, O. A. (2026). Research on the multiple promotion paths of green innovation performance in Chinese high-end equipment manufacturing enterprises: A configuration study based on TOE framework. *European Research on Management and Business Economics*, *32*(1), 100300. DOI: 10.1016/j.iedeen.2026.100300

Ma, T., Wang, P., Yang, D., & Wang, Z. (2025). Towards inclusive conservation governance in forest landscape co-management: Insights from the Yarlung Tsangpo river basin. *Trees, Forests and People*, *19*, 100762. Advance online publication. DOI: 10.1016/j.tfp.2024.100762

Manning, L., Brewer, S., Craigon, P. J., Frey, J., Gutierrez, A., Jacobs, N., Kanza, S., Munday, S., Sacks, J., & Pearson, S. (2023). Reflexive governance architectures: Considering the ethical implications of autonomous technology adoption in food supply chains. In *Trends in Food Science and Technology* (Vol. 133, pp. 114–126). Elsevier Ltd., DOI: 10.1016/j.tifs.2023.01.015

Mossa, Y., Smith, P., & Bland, K. (2025). Reconceptualizing Enterprise Resource Planning (ERP) Systems from a Software Architecture Perspective Using a Framework Based on ERP System Characteristics. *Procedia Computer Science*, *256*, 174–189. DOI: 10.1016/j.procs.2025.02.110

Pasas-Farmer, S., & Jain, R. (2025). From discovery to delivery: Governance of AI in the pharmaceutical industry. *Green Analytical Chemistry*, *13*, 100268. Advance online publication. DOI: 10.1016/j.greeac.2025.100268

Raman, R., Mandal, S., Gunasekaran, A., Papadopoulos, T., & Nedungadi, P. (2025). Transforming business management practices through metaverse technologies: A Machine Learning approach. In *International Journal of Information Management Data Insights* (Vol. 5, Number 1). Elsevier B.V. DOI: 10.1016/j.jjimei.2025.100335

Rouzky, R., Sarrafzadeh, A., Sowells-Boone, E., Pasandi, H. B., & Goins, G. (2025). Data governance in smart farming: A systematic review of challenges and gaps. In *Information Processing in Agriculture*. China Agricultural University., DOI: 10.1016/j.inpa.2025.11.006

Rožman, N., Corn, M., Kozjek, D., Vrabič, R., & Podržaj, P. (2025). Autonomous production unit: An architecture for blockchain-based shared manufacturing. *Robotics and Computer-integrated Manufacturing*, *96*, 103035. Advance online publication. DOI: 10.1016/j.rcim.2025.103035

Turienzo, J., Cabanelas, P., & Lampón, J. F. (2023). Business models in times of disruption: The connected and autonomous vehicles (uncertain) domino effect. *Journal of Business Research*, *156*, 113481. Advance online publication. DOI: 10.1016/j.jbusres.2022.113481

Ullah, I., Arishi, A., Singh, S. K., Alharbi, F., Ibrahim, A. H., Islam, M., Daradkeh, Y. I., & Choi, C. (2025). Autonomous network management for 6G communication: A comprehensive survey. *Digital Communications and Networks*, *11*(6), 1917–1940. DOI: 10.1016/j.dcan.2025.07.001

Wang, N. (2023). Design of an Intelligent Processing System for Business Data Analysis Based on Improved Clustering Algorithm. *Procedia Computer Science*, *228*, 1215–1224. DOI: 10.1016/j.procs.2023.11.105

Wang, X., Shi, X., Chen, J., Guo, X., & Li, D. (2024). Exploring optimal pathways for enterprise procurement management systems based on fast neural modeling and semantic segmentation. *Heliyon*, *10*(7), e26474. Advance online publication. DOI: 10.1016/j.heliyon.2024.e26474 PMID: 38689967 [illegible] (2020). [illegible] production [illegible] architecture for [illegible] based [illegible] manufacturing. [illegible] Manufacturing, [illegible], 102085. Advance online publication. DOI: 10.1016/j.rcim.2020.102085

[illegible] (2021). Business models [illegible] of [illegible] autonomous vehicles [illegible] conceptual [illegible]. [illegible] of Business Research, [illegible]. Advance online publication. DOI: 10.1016/j.jbusres.2021.[illegible]

[illegible] (2022). [illegible] of [illegible] manufacturing [illegible]. [illegible] DOI: [illegible]

[illegible]

[illegible]

Chapter 10
Sustainable and Green Autonomous BPM:
Designing Process Intelligence for Environmental Performance

Othman Boulitama
http://orcid.org/0000-0003-3541-1582
Hassan II University of Casablanca, Morocco

Amina Ouatiq
http://orcid.org/0000-0003-3470-6178
Hassan I University, Morocco

Driss Rahli
http://orcid.org/0000-0002-3306-4460
Hassan II University of Casablanca, Morocco

Karim Sabri
Hassan II University of Casablanca, Morocco

ABSTRACT

This chapter develops the concept of Sustainable and Green Autonomous BPM as a higher-order process capability through which environmental objectives are embedded into the sensing, analysis, adaptation, execution, and governance of business processes. It argues that environmental performance should not be treated as a peripheral reporting outcome but as an internal dimension of process intelligence. Building on BPM, Green BPM, self-adaptive process systems, and process-level environmental measurement, the chapter explains how environmentally intelligent process systems can detect sustainability-relevant conditions, interpret their operational significance, and adjust execution within defined governance boundaries. It

DOI: 10.4018/979-8-2600-0050-2.ch010

also highlights the role of process mining, environmental metrics, enabling analytics, human oversight, and staged implementation. The chapter concludes that the future of BPM will increasingly depend on its ability to integrate ecological intelligence into adaptive and accountable process management.

INTRODUCTION

Business Process Management (BPM) has traditionally focused on the design, execution, control, and continuous improvement of organizational processes, with a dominant emphasis on efficiency, standardization, quality, and cost reduction. This orientation has made BPM a central instrument of operational excellence and digital transformation. However, such a performance logic is no longer sufficient in a context where organizational processes also constitute major sources of environmental burden through energy consumption, material use, transport intensity, waste generation, and emissions embedded in everyday operations (Akinbowale et al., 2026). Once environmental degradation is understood as being enacted through recurrent process execution rather than only through final products or aggregate corporate outcomes, the process itself becomes a legitimate and necessary unit of sustainability analysis and intervention (Troshani & Rowbottom, 2024). This recognition has stimulated the emergence of Green BPM, which extends conventional BPM by incorporating environmental sustainability into the process lifecycle and by examining how processes can be modeled, monitored, improved, and governed in ways that reduce ecological harm (Couckuyt & Looy, 2019). However, prior research has demonstrated the relevance of GBPM, the field remains stronger at explaining why processes matter for sustainability than at clarifying how environmentally intelligent process systems should function under conditions of variability, uncertainty, and runtime change (Brocke et al., 2013).

This limitation becomes particularly visible when sustainability objectives must be pursued in environments shaped by fluctuating demand, resource volatility, supply disruptions, regulatory pressures, and changing operating conditions. In such contexts, environmental process management cannot rely solely on static process models, periodic redesign, or ex post reporting routines. It requires process systems capable of detecting relevant signals, interpreting their environmental significance, selecting appropriate responses, and adjusting execution logic while operations are still unfolding (Yang & Xue, 2026). Recent work on self-adaptive business processes confirms the growing importance of this issue but also shows that the field remains fragmented across mechanisms, solutions, and evaluation criteria, which hinders the development of cohesive strategies for integrating sustainability into business process management. A parallel limitation concerns the measurement architecture needed to support environmentally responsive BPM (Dauerer, 2025). Sustainability cannot shape process behavior unless it is translated into indicators that are sufficiently granular, timely, and operationally meaningful. Advances in process mining are important in this respect because they show how event data can be connected to sustainability metrics to reveal environmental impacts at the level of actual process execution (Areta Hiziroglu & Dogan, 2025). Even so, visibility alone does not amount to autonomy. A process may be observable and analytically rich without being capable of adapting itself in environmentally preferable ways.

The central gap, therefore, is not the absence of sustainability-oriented BPM research, self-adaptive process research, or process-level environmental measurement taken separately. Existing studies have generated important insights within each of these domains independently; however, the literature still lacks sufficiently integrated conceptual explanations clarifying how environmental intelligence, adaptive process behavior, analytical measurement infrastructures, and governance mechanisms can jointly operate within a coherent BPM architecture. The deeper problem lies in the lack of an integrative conceptual perspective explaining how these streams can be combined into a coherent model of environmentally aware and intelligently adaptive process management. Existing GBPM research provides important foundations for embedding ecological concerns into process thinking, self-adaptation research clarifies mechanisms of runtime adjustment, and process mining strengthens empirical observability. However, the literature still offers limited guidance on how environmental objectives can be embedded within the intelligence structure of BPM itself so that sensing, analysis, decision logic, execution, and learning become explicitly aligned with ecological performance. This chapter addresses that gap by advancing the concept of sustainable and green autonomous BPM. It argues that environmental performance should not be treated as a peripheral reporting outcome assessed after process execution but as a constitutive dimension of process intelligence. In this view, a process system becomes environmentally intelligent when it can detect sustainability-relevant conditions, interpret their operational implications, adapt execution paths within defined governance boundaries, and accumulate knowledge from prior interventions in order to improve future responses. Autonomy is thus understood in a bounded and accountable sense, not as the elimination of managerial oversight, but as the design of process systems capable of context-sensitive, explainable, and goal-directed adjustments while remaining subject to strategic, ethical, and organizational control.

The originality of this chapter lies in three interrelated contributions. First, it conceptualizes Sustainable and Green Autonomous BPM as a distinct higher-order process capability situated at the intersection of BPM, Green BPM, self-adaptive process systems, and environmental performance management. Second, it develops an integrated conceptual framework that specifies the constitutive dimensions, architectural logic, measurement requirements, analytical mechanisms, governance conditions, and implementation pathway necessary for environmentally intelligent and adaptively governed process management. Third, it advances a forward-looking research agenda by identifying the principal theoretical, methodological, and empirical directions through which this emerging domain can be refined and validated. In doing so, the chapter repositions sustainability from an external evaluative constraint to an internal logic that shapes how process systems sense, interpret, decide, execute, and learn.

The remainder of the chapter is organized as follows. Section 2 establishes the theoretical foundations of Sustainable and Green Autonomous BPM. Section 3 conceptualizes the construct through its constitutive dimensions and architectural logic. Section 5 examines environmental performance measurement together with the enabling analytical mechanisms that support adaptive sustainability management. Section 5 addresses governance, risks, and human oversight. Section 6 presents an implementation roadmap for organizations. Finally, Section 7 integrates the research agenda, future directions, and conclusion in order to consolidate the chapter's implications for theory and practice.

THEORETICAL FOUNDATIONS OF SUSTAINABLE AND GREEN AUTONOMOUS BPM

The theoretical foundations of sustainable and green autonomous BPM are best established through an integrated reading of three complementary streams of scholarship: Business Process Management, Green BPM, and self-adaptive process systems. Such an approach is necessary because none of these streams, taken independently, provides a sufficient basis for explaining how process systems can be designed to pursue environmental objectives through intelligent and adaptive forms of execution (Viteri-Sánchez & Novillo-Villegas, 2025). Conventional BPM offers the essential, process-centered view. It treats organizational performance as the outcome of structured, coordinated, and analyzable activity systems whose configuration shapes efficiency, quality, compliance, and responsiveness (Aalst, 2013). Within this tradition, the process is not a mere operational sequence but a managerial and analytical unit through which organizational value creation can be understood, governed, and improved. This premise is fundamental for the present chapter because it establishes that environmental performance, once linked to resource use and recurring patterns of execution, must also be examined at the process level rather than only at the level of products, projects, or aggregated organizational reporting (Petropoulos et al., 2026).

Building on this foundation, Green BPM broadens the evaluative horizon of process management by incorporating ecological consequences into process analysis, design, and control. Recent systematic review studies further indicate that Green BPM research has progressively evolved from a narrow focus on environmentally efficient workflows toward broader concerns involving sustainability governance, process intelligence, digital technologies, organizational capability development, and environmentally informed process innovation across multiple sectors. Seidel et al., (2012) made a decisive contribution in arguing that environmentally sustainable organizations require process perspectives capable of accounting for ecological impacts embedded in routine organizational action. In a similar vein, (Couckuyt & Van Looy, 2020) demonstrated that Green BPM has evolved into a distinct research domain, with concerns spanning process modeling, improvement, deployment, monitoring, and managerial alignment. The theoretical significance of this stream lies in its reformulation of process performance itself. A Green BPM perspective asserts that process excellence extends beyond cost, speed, quality, and compliance. It must also encompass the environmental effects generated through energy consumption, material intensity, emissions, waste, and process variation. The central implication is that environmental sustainability should be regarded as an intrinsic dimension of process value, not as an external criterion applied after execution has taken place (Vrchota et al., 2020).

A second theoretical foundation derives from the literature on self-adaptive process systems. Traditional BPM approaches have generally been built on the assumption that process logic, exception handling, and decision paths can be sufficiently specified during design and periodically revised through formal improvement cycles. This assumption becomes increasingly fragile in contexts characterized by volatility, uncertainty, supply disruptions, shifting regulations, and changing environmental conditions (Szelagowski & Berniak-Woźny, 2020). Under such circumstances, processes must do more than comply with predefined models. They must be capable of detecting contextual changes, interpreting their implications, and modifying aspects of execution while operations remain in progress. The self-adaptation literature provides the conceptual resources for this shift. Oukharijane et al. (2025) show that self-adaptive business process research has developed into a substantial field while also remaining fragmented across adaptation triggers, mechanisms, levels, and assessment criteria. This diagnosis is theoretically important because it reveals that adaptation is not a secondary technical feature but a fundamental reorientation in the understanding of process systems. Autonomy, in this sense, should be conceptualized with precision and restraint. It does not denote unrestricted machine agency or the disappearance of managerial judgment. Rather, it refers to the bounded capacity of a process system to sense relevant variation, evaluate alternatives, and adjust behavior within predefined strategic, ethical, and governance parameters.

A third pillar concerns environmental observability and the measurement architecture required to support sustainability-oriented adaptation. No process system can become environmentally intelligent unless ecological effects are translated into signals that are sufficiently granular, timely, and operationally meaningful to influence process behavior (Abushab & Hamdan, 2025). This requirement explains the growing relevance of process mining and event-based analytics in sustainability research. Recent studies have shown that environmental indicators can be connected to process execution data in ways that make visible the ecological implications of actual process flows rather than only aggregate corporate outcomes. In particular, the framework proposed by Areta Hiziroglu & Dogan (2025) is significant because it integrates process mining with Global Reporting Initiative metrics to strengthen the granularity and operational relevance of sustainability assessment. The theoretical importance of this contribution lies in its relocation of environmental measurement from a retrospective reporting role to an informational layer embedded in process execution itself. However, observability alone is not equivalent to adaptive intelligence. A process may be measurable, traceable, and analytically transparent while remaining unable to reconfigure itself in response to environmentally significant deviations (Miedema et al., 2026). For this reason, environmental measurement must be linked to process-level decision logic, executional flexibility, and learning mechanisms if it is to support genuinely autonomous and sustainability-oriented process management.

Taken together, these three streams provide the basis for theorizing Sustainable and Green Autonomous BPM as an organizational process capability. It is best understood as a higher-order capability because it combines process visibility, environmental interpretive capacity, adaptive response mechanisms, and governance-bounded learning into a coherent mode of process management. BPM contributes the structural logic of coordinated organizational action. Green BPM contributes the normative and evaluative lens through which ecological performance becomes a legitimate process objective. Self-adaptive process research contributes the runtime logic that enables processes to respond intelligently to changing conditions. Their integration supports a more advanced theoretical position in which sustainability is not treated as an external reporting concern or a peripheral optimization target, but as an embedded criterion that shapes how process systems sense, analyze, decide, execute, and learn. This interpretation also aligns with broader capability-based perspectives in organizational theory, particularly dynamic capability logic, according to which organizations develop higher-order capacities to sense environmental change, seize adaptive opportunities, and reconfigure operational processes under evolving external conditions. Within this perspective, Sustainable and Green Autonomous BPM may be understood as a sustainability-oriented adaptive capability through which ecological intelligence becomes embedded within organizational process reconfiguration and decision-making. On this basis, Sustainable and Green Autonomous BPM can be conceptualized as the foundation of environmental process intelligence, thereby preparing the ground for a more precise definition of the construct and its analytical boundaries in the following section.

CONCEPTUALIZING SUSTAINABLE AND GREEN AUTONOMOUS BPM: CONSTITUTIVE DIMENSIONS AND ARCHITECTURAL LOGIC

Sustainable and Green Autonomous BPM may be defined as a higher-order process management capability through which organizations embed environmental objectives into the sensing, interpretation, adaptation, execution, and governance of business processes. This conceptualization departs both from narrow understandings of automation and from approaches that treat sustainability as an ex post evaluative layer (Viteri-Sánchez & Novillo-Villegas, 2025). In the BPM tradition, processes are the structured vehicles through which organizational value is created and controlled, while Green BPM extends this logic by recognizing that process performance must also be assessed in relation to ecological effects generated during execution (Brocke et al., 2013; Couckuyt & Looy, 2019). In parallel, recent work on self-adaptation in business processes demonstrates that adaptive process solutions must be distinguished according to their triggers, mechanisms, layers, and assessment criteria, which makes conceptual precision indispensable when theorizing autonomy in process environments (Oukharijane et al., 2025). On this basis, autonomy should not be interpreted as unrestricted machine discretion or the displacement of managerial judgment. It should be understood as the bounded capacity of a process system to recognize environmentally relevant conditions, evaluate their operational significance, and modify process behavior within explicit strategic, ethical, and governance constraints.

From this perspective, Sustainable and Green Autonomous BPM is best specified through a set of constitutive dimensions rather than a descriptive list of technical features. The first dimension is environmental process visibility (Couckuyt & Looy, 2019). A process cannot be managed in environmentally intelligent ways unless its ecological effects are rendered visible at the level of activities, variants, resource allocations, and execution patterns. For example, in manufacturing environments, process mining systems connected to energy monitoring infrastructures may identify production variants associated with excessive electricity consumption or avoidable material waste, thereby enabling environmentally informed process redesign and operational adjustment. This requirement follows directly from Green BPM scholarship, which has consistently argued that ecological impact must be integrated into process analysis rather than isolated in aggregate sustainability statements (Seidel et al., 2012; Couckuyt & Van Looy, 2020). The second dimension is context-sensitive sensing. Autonomous sustainability-oriented process management presupposes the capacity to capture signals arising from operational events, resource conditions, process deviations, and environmental thresholds. Without such sensing, adaptation remains disconnected from the actual circumstances in which environmental performance deteriorates or improves. The third dimension is sustainability-aware interpretive and decision logic. Environmental data become theoretically meaningful only when they influence process choices, which implies that adaptation decisions cannot be assessed exclusively through conventional criteria such as speed, cost, and throughput (Geremicca et al., 2026). They must also be evaluated in relation to emissions intensity, material use, waste generation, energy demand, and comparable process-level ecological consequences. Recent process mining research reinforces this point by showing that sustainability metrics can be connected to operational process data in order to make environmental performance analytically actionable within ongoing process execution (Areta Hiziroglu & Dogan, 2025).

Three additional dimensions complete the construct. The fourth is bounded autonomy, which gives the concept its analytical specificity. In scholarly terms, a process is not autonomous simply because it executes predefined rules automatically. It becomes autonomous when it can modify selected aspects of its behavior in response to changing conditions while remaining interpretable, auditable, and normatively governed (Sapkota et al., 2026). This interpretation is consistent with the self-adaptation literature, where the challenge lies not merely in technical reactivity but in the controlled design of adaptive behavior under explicit constraints (Oukharijane et al., 2025). The fifth dimension is executional reconfigurability. A process system capable of environmentally informed adaptation must be able to alter task sequencing, resource deployment, routing choices, prioritization rules, or escalation logic when ecological objectives require adjustment. The sixth dimension is cumulative learning (Noetzold et al., 2026). Autonomous Green BPM would remain incomplete without a knowledge-based mechanism through which prior interventions, their conditions, and their environmental consequences inform future action. In this sense, sustainability-oriented autonomy is not only reactive but developmental, because the process system improves its capacity for environmentally responsible adjustment through accumulated organizational knowledge (Cheng et al., 2023). Taken together, these six dimensions clarify that Sustainable and Green Autonomous BPM is neither reducible to Green BPM nor equivalent to generic intelligent automation. This distinction is theoretically important because much of the existing literature still examines sustainability-oriented BPM, intelligent automation, and adaptive process systems as partially separate domains, thereby limiting the development of integrative explanations capable of connecting environmental intelligence, process adaptation, governance, and organizational learning within a unified process management framework. It constitutes a distinct process capability in which environmental performance is embedded within the very logic of process observation, evaluation, adaptation, and learning.

These constitutive dimensions can be translated into an architectural logic that explains how the capability operates at the process system level. At its foundation lies a sensing and data acquisition layer that captures event logs, process traces, contextual resource data, and environmental indicators generated during process execution (Liu et al., 2026). This informational layer is followed by an analytical and interpretive layer, where process deviations, sustainability-relevant anomalies, and ecological performance patterns are identified and evaluated. A planning and decision layer then determines whether adaptation is required and selects an appropriate response in light of environmental priorities and governance parameters. An execution layer subsequently implements the selected intervention through process rerouting, resource reallocation, task modification, prioritization changes, or escalation to human actors where necessary. Across these layers, a knowledge layer stores process models, adaptation rules, environmental thresholds, historical outcomes, and governance constraints, thereby ensuring continuity, traceability, and learning. This architecture is theoretically coherent with the adaptive logic emphasized in the self-adaptation literature and operationally reinforced by recent work linking sustainability metrics to process mining infrastructures (Areta Hiziroglu & Dogan, 2025; Oukharijane et al., 2025).

The academic value of this integrated formulation lies in its ability to consolidate three discussions that have often remained separate in prior scholarship. Green BPM has established the importance of ecological process performance; systematic review evidence has clarified the domain's principal capability areas; self-adaptation research has specified the runtime logic of controlled process change; and recent sustainability-oriented process mining work has improved the measurement basis for environmental process intelligence (Seidel et al., 2012; Couckuyt & Van Looy, 2020; Oukharijane et al., 2025; Areta Hiziroglu & Dogan, 2025). What this section adds is a more explicit conceptual synthesis: sustainability is not treated as a peripheral reporting requirement, and autonomy is not treated as a purely technical property of software-enabled execution. Rather, Sustainable and Green Autonomous BPM is theorized here as a higher-order organizational capability through which environmental criteria become internal to the way process systems sense, analyze, decide, execute, and learn. In that sense, the construct provides not only a conceptual clarification but also an intellectual bridge between environmental process management and adaptive process intelligence, thereby offering a stronger foundation for the measurement and governance discussions developed in the subsequent parts of the chapter.

MEASURING ENVIRONMENTAL PERFORMANCE AND ENABLING ANALYTICAL MECHANISMS IN SUSTAINABLE AND GREEN AUTONOMOUS BPM

The capacity of Sustainable and Green Autonomous BPM to generate meaningful environmental outcomes depends on a prior analytical condition: ecological effects must be rendered measurable at the level of process execution. This requirement is more demanding than conventional sustainability reporting because aggregate disclosure rarely reveals how environmental burdens are produced, amplified, or reduced within concrete process instances (Areta Hiziroglu & Dogan, 2025). Green BPM has long argued that environmental sustainability must be examined through the logic of processes themselves, since resource use, waste generation, emissions, and operational inefficiencies are embedded in recurrent patterns of organizational action rather than only in final outputs or corporate summaries (Couckuyt & Van Looy, 2020; Seidel et al., 2012). From this standpoint, process level measurement is not an auxiliary reporting exercise. It is the epistemic foundation of environmental process intelligence, because a process system cannot adapt in environmentally responsible ways unless it can first identify where ecological degradation emerges, how it propagates across activities, and under which execution conditions it becomes more severe.

For that reason, environmental performance in autonomous BPM should be conceptualized as a multidimensional measurement architecture rather than as a single indicator or dashboard. At a minimum, this architecture should capture direct process impacts such as energy consumption, material use, water use, emissions, and waste associated with specific activities, cases, or variants. It should also include eco-efficiency indicators that relate ecological burden to process output, such as energy per transaction, carbon intensity per completed case, or material intensity per service delivered (Czvetkó et al., 2025). A further layer concerns environmentally consequential process behavior, including rework loops, waiting time, avoidable transport, redundant approvals, idle resource consumption, and deviation patterns that increase ecological cost without generating commensurate operational value. This analytical differentiation is essential because it shifts environmental assessment from broad sustainability aspirations to observable process phenomena that can support diagnosis, comparison, and intervention. Recent work on process mining and sustainability measurement reinforces this point by showing that operational data can be linked to sustainability metrics to improve granularity and move beyond static or purely ex post forms of evaluation (Denkena et al., 2022; Zhang et al., 2023).

Within the conceptual logic of Sustainable and Green Autonomous BPM, measurement acquires theoretical significance only when it is coupled with analysis and adaptation. Environmental metrics become academically relevant not merely because they describe ecological impact but because they can function as triggers, constraints, and evaluation criteria within a broader adaptive process cycle (Ghobakhloo et al., 2022). A process system becomes environmentally intelligent when measured deviations are interpreted in relation to process goals, alternative responses are assessed in both operational and ecological terms, and interventions are selected accordingly. This stage is where analytical mechanisms become indispensable (Zimek & Baumgartner, 2024). Descriptive analytics support the visibility of environmental hotspots across activities and variants. In logistics operations, for instance, descriptive and predictive analytics may reveal transport routes, warehouse activities, or delivery sequences associated with disproportionately high carbon emissions, thereby supporting environmentally optimized routing and resource allocation decisions. Diagnostic analytics identify the process conditions and causal patterns associated with ecological inefficiency. Predictive analytics estimates the likely environmental consequences of ongoing execution under changing operational conditions. Prescriptive analytics support the selection of environmentally preferable courses of action among competing process alternatives. In conceptual terms, these layers transform measurement from passive observation into decision-relevant intelligence. Without such transformation, environmental indicators remain informative but analytically inert (Islam et al., 2021; Relich, 2023). This limitation remains visible in many contemporary sustainability initiatives, where organizations generate extensive environmental reporting data but still struggle to translate such information into real-time operational adaptation and process-level decision-making.

This interpretation also clarifies why process mining occupies a central place in the architecture of Sustainable and Green Autonomous BPM. Its value does not lie simply in visualizing process flows but in reconstructing actual execution behavior from event data and connecting that behavior to environmental consequences. Through this capability, organizations can identify ecologically costly variants, compare process pathways, detect waste-intensive loops, and monitor the environmental implications of noncompliant or suboptimal executions. The framework proposed by Areta Hiziroglu & Dogan (2025) is especially relevant because it explicitly links Global Reporting Initiative metrics with process mining, thereby addressing the long-standing problem of insufficient granularity in sustainability performance assessment. In theoretical terms, such work is important because it shifts environmental measurement closer to the runtime logic of process management and opens the possibility of using environmental information not only for retrospective evaluation but also for adaptive control.

The enabling analytical mechanisms of Sustainable and Green Autonomous BPM should therefore be interpreted according to their role in the adaptive process cycle rather than as an inventory of digital tools. Rule-based engines translate environmental thresholds, strategic priorities, and governance constraints into executable decision logic (Valmohammadi & Mortaz Hejri, 2023). Machine learning techniques strengthen anomaly detection, forecasting, and pattern recognition when environmental and process data are sufficiently rich (Malki et al., 2022). Sensor-based infrastructures and Internet of Things environments are especially valuable where ecological performance depends on near-real-time signals related to energy use, emissions, equipment load, temperature, movement, or transport conditions (Ma et al., 2026). Simulation and digital twin environments add a further level of analytical depth by allowing organizations to test alternative process configurations and estimate their likely environmental implications before large-scale implementation (Yeung et al., 2025). The theoretical relevance of these mechanisms lies in their capacity to convert environmental observability into adaptive capability. In other words, they matter not because they are technologically advanced but because they mediate the transition from environmental information to environmentally informed process behavior (Rame et al., 2024).

This discussion leads to the conclusion that theorizing autonomous Green BPM should not separate measurement and analytics. Measurement provides the evidentiary substrate through which ecological effects become visible, comparable, and governable. Analytical mechanisms transform that substrate into interpretation, anticipation, and choice. Enabling technologies operationalize these functions within ongoing process execution. The academic value of this integrated perspective is that it avoids two common reductions: treating sustainability as a reporting layer detached from process management and treating digital intelligence as a technical enhancement disconnected from environmental purpose. Sustainable and Green Autonomous BPM requires a more demanding configuration in which process-level environmental metrics, analytical reasoning, and adaptive mechanisms jointly support ecologically responsible execution. On that basis, environmental measurement is no longer a peripheral control device, and enabling analytics are no longer neutral technical add-ons; instead, they are essential components that drive decision-making and improve sustainability outcomes in business process management. Together, they form the operational core through which environmental process intelligence becomes actionable.

GOVERNANCE, RISKS, AND HUMAN OVERSIGHT

Governance constitutes a central condition of Sustainable and Green Autonomous BPM because no process system can be considered environmentally intelligent merely by virtue of its analytical sophistication or adaptive capacity (Viteri-Sánchez & Novillo-Villegas, 2025). Once processes are allowed to sense conditions, interpret signals, and modify elements of execution, the question is no longer only whether adaptation is technically feasible but whether it remains strategically aligned, normatively bounded, and organizationally accountable (Hao et al., 2025a). This issue is particularly important in the context of Green BPM, where environmental improvement cannot be reduced to isolated optimization routines but must be integrated into broader managerial structures of strategy, monitoring, and control. Recent work on Green BPM in manufacturing SMEs identifies governance, alongside strategy, modeling, monitoring, and optimization, as one of the core factors shaping business sustainability through processes, which confirms that environmental process performance depends on formal managerial arrangements as much as on operational techniques (Couckuyt & Van Looy, 2020; Sohns et al., 2023)

From this perspective, governance in Sustainable and Green Autonomous BPM should be understood as the set of structures, rules, and decision rights that determine how environmental objectives are translated into process behavior, how adaptive interventions are authorized, and how responsibilities are distributed across human and digital actors (Tumpa & Naeni, 2025). This interpretation is consistent with recent work on responsible AI governance, which conceptualizes governance through structural, procedural, and relational practices rather than through compliance checks alone. Structural practices define roles, authority, and oversight bodies. Procedural practices set up routines for reviews, documentation, escalation, and audits. Relational practices organize communication, trust, and coordination across stakeholders involved in design, deployment, and use. Applied to autonomous BPM, the framework means that process autonomy must remain embedded in an explicit governance architecture capable of specifying what environmental goals matter, which trade-offs are acceptable, when intervention is permitted, and who remains accountable for the consequences of adaptive execution (OECD, 2019; Papagiannidis et al., 2025).

The risk profile of Sustainable and Green Autonomous BPM is correspondingly multidimensional. A first group of risks concerns environmental misalignment. A process system may optimize local indicators such as energy intensity or transport reduction while undermining broader process performance or shifting ecological burdens elsewhere in the value chain (Couckuyt & Looy, 2019). For example, a supply chain process may reduce transportation emissions through shipment consolidation while simultaneously increasing inventory holding times, storage energy consumption, or delivery delays, thereby illustrating the complexity of sustainability-related process trade-offs. A second group concerns analytical risks, including poor data quality, weak metric design, inappropriate thresholds, and misleading causal inference, all of which may distort adaptive decisions. A third group concerns organizational and ethical risks, such as opacity in decision logic, unclear accountability, excessive dependence on automated recommendations, or the normalization of process changes that have not been adequately validated (Moreno-Sánchez et al., 2026). In adaptive digital environments, such risks cannot be treated as exceptional side effects. They are constitutive governance concerns because the capacity to intervene at runtime also creates the possibility of intervening wrongly, disproportionately, or without sufficient justification. Recent AI governance initiatives and the NIST AI Risk Management Framework converge on the necessity for governance mechanisms that can mitigate organizational, societal, legal, and technical risks throughout the system's lifecycle, rather than solely in response to visible harm (Liao et al., 2025; National Institute of Standards and Technology, 2023, 2024).

Human oversight therefore becomes a constitutive requirement rather than a residual safeguard. In Sustainable and Green Autonomous BPM, oversight should not be reduced to the simplistic image of a human approving or rejecting machine outputs at the end of the process. Empirical evidence from sustainability-oriented digital transformation initiatives further suggests that organizations achieving more effective environmental process adaptation are typically those combining analytical automation with strong managerial oversight, cross-functional coordination, and continuous process validation mechanisms. More rigorous recent work on AI human collaboration shows that the issue extends beyond a narrow human-in-the-loop model toward a broader process of sensemaking in which humans interpret, challenge, contextualize, and redirect system behavior when needed. In process environments, balancing ecological objectives against service continuity, cost pressures, regulatory obligations, and operational resilience is particularly relevant. In these situations, oversight must include the ability to look at adaptation triggers, verify the accuracy of environmental indicators, question the reasons behind recommendations, stop inappropriate actions, and change the rules of governance when the organization's priorities change (Hao et al., 2025b; Papagiannidis et al., 2025).

Emerging regulatory and policy frameworks also reflect this understanding. The European Union describes the AI Act as a risk-based legal framework for AI, and Article 14 requires human oversight for high-risk AI systems to prevent or minimize risks to health, safety, or fundamental rights. The updated OECD AI Principles in 2024 continue to support trustworthy AI based on human rights, democratic values, transparency, robustness, and accountability. At the same time, the OECD's work on governing with AI stresses the need for guardrails that fit the situation instead of one-size-fits-all control mechanisms, which can help ensure that AI applications are tailored to specific contexts and risks associated with their use. Although Sustainable and Green Autonomous BPM is not reducible to AI regulation, these frameworks are relevant because they reinforce a core principle of this chapter: autonomous process intelligence must remain bounded by explainability, contestability, proportionality, and accountable human supervision (European Commission, 2025; OECD, 2024b, 2024a)

For this reason, human oversight in autonomous Green BPM should be designed as a layered organizational function. At the strategic level, managers define environmental objectives, acceptable trade-offs, and escalation thresholds. At the tactical level, process owners and sustainability leaders review rules, indicators, and model performance. At the operational level, designated actors retain the authority to intervene in ongoing execution when process adaptation creates environmental, ethical, or organizational concerns (Lawter & Garnjost, 2025). Such a layered model prevents the false alternative between full automation and permanent manual control. It supports a more robust interpretation of autonomy as governed delegation, where process systems are empowered to act within specified boundaries while humans remain responsible for boundary design, exception handling, and institutional learning (Brandao, 2025). This approach aligns with the broader governance literature, which increasingly treats oversight as a dynamic organizational capability rather than a one-time approval function (Papagiannidis et al., 2025).

A further implication concerns auditability and traceability. If process systems are expected to adapt in environmentally significant ways, organizations must be able to reconstruct why a given intervention occurred, which data informed it, what alternatives were considered, and what ecological and operational consequences followed. Without such traceability, governance cannot move beyond abstract principles. Process-level logs, rule repositories, model documentation, override histories, and post-intervention assessments are essential tools for organizational accountability. In the context of sustainability, this framework also has a reporting dimension, as recent work on ESG reporting through process mining indicates that process data can strengthen transparency by linking sustainability disclosures to underlying operational evidence. Consequently, auditability in Sustainable and Green Autonomous BPM should be regarded as both a governance tool and a link between internal process control and external accountability for sustainability (Areta Hiziroglu & Dogan, 2025; Lewańska & Stróżyna, 2025).

The governance challenge of Sustainable and Green Autonomous BPM is to determine the right balance between being able to adapt and having institutional control. Excessive rigidity undermines responsiveness and reduces the value of process intelligence. Unbounded autonomy, by contrast, weakens accountability and increases the probability of environmentally or organizationally harmful decisions. The most appropriate position is therefore one of governed adaptability, in which environmental criteria are embedded in process intelligence, risks are assessed across the system lifecycle, and human oversight remains active at the points where interpretation, contestation, and responsibility are indispensable. On this basis, governance, risk management, and human oversight are not peripheral additions to autonomous Green BPM. They are the organizational conditions that make environmentally responsible process autonomy possible in the first place (European Commission, 2025; National Institute of Standards and Technology, 2024; OECD, 2024b, 2024a).

IMPLEMENTATION ROADMAP FOR ORGANIZATIONS

The implementation of Sustainable and Green Autonomous BPM should be approached as a staged organizational transformation rather than as an isolated technology project. Existing research on Green BPM shows that environmentally oriented process improvement depends on a combination of strategic, governance, modeling, monitoring, and optimization capabilities, while recent work on sustainable BPM adoption proposes a progression from basic process documentation toward more mature configurations supported by advanced technologies. This indicates that organizations are unlikely to achieve environmentally intelligent process autonomy through direct large-scale deployment without first establishing the managerial, informational, and process foundations on which such autonomy depends (Sohns et al., 2023; Viteri-Sánchez & Novillo-Villegas, 2025).

Figure 1 presents the implementation roadmap for Sustainable and Green Autonomous BPM as a staged and cumulative organizational transformation process. Rather than treating implementation as a stand alone technology deployment, the figure clarifies that environmentally intelligent process autonomy emerges through successive phases involving readiness assessment, process diagnosis, metric integration, governed experimentation, and institutionalization. In this way, the roadmap provides a structured representation of how organizations can progressively develop the managerial, analytical, and governance capabilities required to embed environmental objectives into adaptive process management.

Figure 1. Implementation roadmap for Sustainable and Green Autonomous BPM.

Stage	Activities
1. Organizational Readiness and Scoping	• Define environmental objectives • Assess BPM maturity • Identify governance actors • Clarify process ownership • Evaluate data availability
2. Process Selection and Baseline Diagnosis	• Select high impact processes • Map current process variants • Identify environmental hotspots • Establish baseline ecological performance • Assess redesign feasibility
3. Metric Design, Data Integration, and Pilot Configuration	• Define process-level environmental indicators • Link metrics to event and operational data • Set thresholds and escalation rules • Configure pilot scope • Assign monitoring responsibilities
4. Governed Experimentation and Iterative Redesign	• Test adaptive interventions • Evaluate environmental and operational effects • Refine routing, resource allocation, and decision rules • Document unintended consequences • Adjust governance mechanisms
5. Scaling, Institutionalization, and Continuous Capability Development	• Extend to additional processes and units • Embed governance and auditability routines • Strengthen process analytics capabilities • Formalize learning mechanisms • Institutionalize environmental process intelligence

A first stage is organizational readiness and scoping. At this stage, the organization clarifies why Sustainable and Green Autonomous BPM is being pursued, which environmental objectives are strategically relevant, and which governance actors will be responsible for steering the initiative. This phase also requires an assessment of current BPM maturity, data availability, sustainability priorities, and process ownership. The importance of this preparatory stage is supported by evidence showing that Green BPM progress is shaped by organizational pressures, barriers, and maturity conditions rather than by technical ambition alone. A readiness assessment therefore helps prevent a common implementation error, namely the premature selection of digital solutions before the organization has determined its process priorities, environmental goals, and decision rights.

A second stage is process selection and baseline diagnosis. Target processes selectively, as they offer varying environmental leverage. Organizations should therefore identify a limited number of high-relevance processes based on ecological burden, operational criticality, data accessibility, and redesign feasibility. Once these processes are selected, a baseline diagnosis should establish how environmental burdens are generated across current process variants, activities, and resource configurations. Recent frameworks in sustainability-oriented BPM stress the need for process-level analysis before redesigning. SOPA, for example, was developed because practical approaches often lacked a clear and applicable mechanism for sustainability-driven process analysis and redesign, and it addresses this limitation by combining life cycle assessment and activity-based costing within the BPM lifecycle. In a similar vein, the PMG3 guidelines for managing greenhouse gas emissions provide detailed steps for defining business and data requirements, analyzing inefficiencies, and formulating sustainability improvement recommendations through process mining (Costache et al., 2025; Klessascheck et al., 2025)

A third stage is metric design, data integration, and pilot configuration. Once a target process has been diagnosed, the organization must specify which environmental indicators will guide monitoring and adaptation, how those indicators will be linked to operational data, and which thresholds or rules will trigger intervention. This stage is decisive because environmental autonomy is impossible when ecological criteria remain detached from runtime process information. Research on process mining and sustainability confirms that organizations need explicit methods for connecting environmental metrics to operational process data if they are to improve granularity, reporting quality, and process-level control. In practical terms, this phase means enriching event logs with environmental variables, defining decision rules, assigning escalation responsibilities, and configuring a limited pilot in which adaptive interventions can be tested under controlled conditions before broader deployment (Areta Hiziroglu & Dogan, 2025; Costache et al., 2025).

A fourth stage is controlled experimentation and iterative redesign. In this phase, the organization moves from analytical visibility to controlled intervention by testing environmentally informed changes in routing logic, resource allocation, sequencing, exception handling, or human escalation. The objective is not to maximize automation immediately, but to examine whether adaptive changes produce measurable environmental improvements without generating unacceptable operational or governance risks. This is consistent with design-oriented BPM research, which treats sustainability improvement as an iterative cycle of analysis, redesign, and evaluation rather than as a one-time implementation exercise. It is also consistent with sustainable BPM adoption research, which associates more advanced stages of BPM with the gradual integration of digital and analytical capabilities rather than abrupt transformation. Pilot experimentation is therefore best understood as a learning stage in which the organization refines its indicators, validates its rules, documents unintended consequences, and adjusts governance mechanisms before scaling the approach (Klessascheck et al., 2025; Viteri-Sánchez & Novillo-Villegas, 2025). In practice, a manufacturing firm may initially pilot environmentally adaptive process routing within a single production unit before extending the approach across multiple facilities, thereby allowing the organization to validate environmental gains, operational feasibility, and governance controls under limited-risk conditions.

A fifth stage is scaling, institutionalization, and continuous capability development. Once pilot evidence demonstrates that environmentally informed process adaptation is both effective and governable, the organization can extend the approach to additional processes, units, or value chain interfaces. At this point, Sustainable and Green Autonomous BPM should no longer be treated as a local initiative but as an emerging organizational capability supported by governance routines, process ownership, auditability, and continuous learning. Research on Green BPM maturity shows that long-term progress depends on embedding sustainability into process strategy, governance, monitoring, and optimization. For that reason, scaling should be accompanied by capability building in process analytics, environmental measurement, cross-functional coordination, and managerial oversight. The practical implication is that implementation success depends less on the isolated deployment of advanced tools than on the institution of a repeatable organizational logic through which environmental objectives become part of how processes are designed, monitored, adapted, and improved over time (Sohns et al., 2023; Viteri-Sánchez & Novillo-Villegas, 2025).

Taken together, these stages form a roadmap that is cumulative rather than linear in a narrow sense. Organizations may revisit earlier stages as data quality improves, governance arrangements mature, or sustainability priorities change. The value of the roadmap lies in offering a disciplined implementation sequence that begins with readiness and process diagnosis, proceeds through metric integration and governed experimentation, and culminates in institutionalized environmental process intelligence. Such a progression is more consistent with the evidence available in contemporary BPM research than approaches that assume sustainable and autonomous process management can be achieved through technology adoption alone. Sustainable and Green Autonomous BPM gains operational credibility when implementation is regarded as a capability-building trajectory, wherein strategy, process analysis, environmental metrics, adaptive mechanisms, and governance develop concurrently (Costache et al., 2025; Klessascheck et al., 2025; Sohns et al., 2023; Viteri-Sánchez & Novillo-Villegas, 2025).

Research Agenda, Future Directions, and Conclusion

The framework developed in this chapter opens a substantive research agenda that calls for deeper theoretical refinement, stronger methodological development, and broader empirical examination. A first priority concerns the further specification and validation of Sustainable and Green Autonomous BPM as a higher-order process capability. Although the chapter has conceptualized this capability through dimensions such as environmental process visibility, context-sensitive sensing, sustainability-aware decision logic, bounded autonomy, executional reconfigurability, and cumulative learning, future research must determine more rigorously whether these dimensions operate as distinct yet interrelated components, as hierarchical subcapabilities, or as contingent configurations whose relevance varies across process contexts. Such work is essential for preventing the construct from remaining overly expansive and for strengthening its explanatory precision within the BPM and sustainability literature. A second line of inquiry concerns measurement and method. The chapter has argued that environmentally intelligent process adaptation depends on process-level metrics capable of linking ecological effects to concrete execution patterns. This claim invites systematic work on indicator development, metric validation, and methodological integration, especially through the combination of process mining, life cycle assessment, activity-based costing, and sustainability reporting frameworks. Research in this area should clarify which indicators are most suitable for different process types, under what conditions environmental metrics become sufficiently reliable to guide adaptive intervention, and how process-level measurement can support both operational control and strategic sustainability assessment. A third priority concerns adaptive mechanisms and decision architectures. Future studies should investigate how environmentally informed adaptation actually unfolds in practice, what triggers activate intervention, how trade-offs between ecological and operational objectives are managed, and what forms of analytical and technological coupling make adaptation both effective and governable. In parallel, a fourth line of inquiry should address governance, responsibility, and human oversight with greater analytical depth. More work is needed on the allocation of decision rights, the design of escalation mechanisms, the conditions under which human judgment should override automated adaptation, and the auditability requirements that arise when environmental criteria become embedded in runtime process management. Finally, future research should examine contextual and sectoral variation more systematically, since the feasibility, visibility, and organizational value of Sustainable and Green Autonomous BPM are unlikely to be identical across manufacturing, logistics, healthcare, professional services, or public administration. Comparative and longitudinal studies would be particularly valuable for identifying distinct implementation trajectories, maturity patterns, and boundary conditions across settings. In particular, future empirical research could investigate how environmentally adaptive BPM systems operate in manufacturing, logistics, healthcare, and energy-intensive industries through longitudinal case studies, process mining analyses, and real-time sustainability performance evaluation.

Taken together, these directions suggest that the future development of Sustainable and Green Autonomous BPM will depend on cumulative evidence capable of explaining not only whether autonomous BPM can support environmental sustainability but also under which organizational conditions, through which mechanisms, and with what consequences for process design, governance, and ecological performance. Against this background, the present chapter has advanced a central argument: environmental sustainability should no longer be treated as a peripheral concern appended to process management after execution has taken place but as an internal criterion that shapes how processes are monitored, interpreted, adapted, and governed. On this basis, the chapter has theorized Sustainable and Green Autonomous BPM as a higher-order process capability through which organizations can detect environmentally relevant conditions, interpret their process implications, adjust execution paths within explicit governance boundaries, and learn from prior interventions over time. It has also clarified the theoretical foundations of this capability, specified its constitutive dimensions, connected environmental measurement to enabling analytical mechanisms, and proposed an implementation logic structured around readiness, diagnosis, experimentation, and institutionalization. The broader implication is that the next generation of BPM will need to be evaluated not only by its capacity to improve efficiency, compliance, or coordination, but also by its ability to incorporate ecological intelligence into process behavior in ways that remain accountable, interpretable, and strategically aligned. In this respect, Sustainable and Green Autonomous BPM offers a conceptual bridge between environmental responsibility and adaptive process intelligence, thereby repositioning BPM as a strategic architecture for environmentally informed organizational action rather than as a purely operational instrument of control.

REFERENCE

Abushab, A. M., & Hamdan, A. M. (2025). A Model for Integrating Sustainability Metrics into Architectural Project Lifecycle Management. *Engineering And Technology Journal, 10*(8), 6083–6105. DOI: 10.47191/etj/v10i08.07

Akinbowale, O. E., Zerihun, M. F., & Mashigo, P. (2026). Business process management framework: Systematic review of the trends, potentials and future. *Cogent Business & Management, 13*(1), 2627025. DOI: 10.1080/23311975.2026.2627025

Areta Hiziroglu, O., & Dogan, O. (2025). A Framework for Sustainability Performance Measurement Through Process Mining: Integration of GRI Metrics in Operational Processes. *Systems 2025, Vol. 13, Page 547, 13*(7), 547. DOI: 10.3390/systems13070547

Brandao, P. R. (2025). The Impact of Artificial Intelligence on Modern Society. *AI 2025, Vol. 6, Page 190, 6*(8), 190. DOI: 10.3390/ai6080190

Cheng, C., Ahmad, S. F., Irshad, M., Alsanie, G., Khan, Y., Ahmad, A. Y. A. B., & Aleemi, A. R. (2023). Impact of Green Process Innovation and Productivity on Sustainability: The Moderating Role of Environmental Awareness. *Sustainability 2023, Vol. 15, Page 12945, 15*(17), 12945. DOI: 10.3390/su151712945

Costache, I., Turetken, O., Aysolmaz, B., & Winter, K. (2025). Process Mining Guidelines for Greenhouse Gas Emission Management in Production Processes. *Lecture Notes in Business Information Processing, 533*, 711–724. DOI: 10.1007/978-3-031-82225-4_52

Couckuyt, D., & Looy, A. Van. (2019). Green BPM as a Business-Oriented Discipline: A Systematic Mapping Study and Research Agenda. *Sustainability 2019, Vol. 11, Page 4200, 11*(15), 4200. DOI: 10.3390/su11154200

Couckuyt, D., & Van Looy, A. (2020). A systematic review of Green Business Process Management. *Business Process Management Journal, 26*(2), 421–446. DOI: 10.1108/BPMJ-03-2019-0106

Czvetkó, T., Sebestyén, V., & Abonyi, J. (2025). Key factors of industry 5.0-based organizational sustainability. *Technology in Society, 83*, 102966. DOI: 10.1016/j.techsoc.2025.102966

Dauerer, A. (2025). A systematic literature review of performance measurement systems and the integration of ESG factors. *Environmental and Sustainability Indicators, 27*, 100746. DOI: 10.1016/j.indic.2025.100746

Denkena, B., Wichmann, M., Kettelmann, S., Matthies, J., & Reuter, L. (2022). Ecological Planning of Manufacturing Process Chains. *Sustainability 2022, Vol. 14, Page 2681, 14*(5), 2681. DOI: 10.3390/su14052681

European Commission. (2025, February 2). *AI Act | Shaping Europe's digital future*. European Commission. https://digital-strategy.ec.europa.eu/en/policies/regulatory-framework-ai

Geremicca, F., Fascetti, A., Brigham, J. C., & Bilec, M. M. (2026). Digital twins for sustainable buildings: From framework to strategy guidelines and application. *Energy and Building*, *359*, 117332. DOI: 10.1016/j.enbuild.2026.117332

Ghobakhloo, M., Iranmanesh, M., Mubarak, M. F., Mubarik, M., Rejeb, A., & Nilashi, M. (2022). Identifying industry 5.0 contributions to sustainable development: A strategy roadmap for delivering sustainability values. *Sustainable Production and Consumption*, *33*, 716–737. DOI: 10.1016/j.spc.2022.08.003

Hao, X., Demir, E., & Eyers, D. (2025a). Beyond human-in-the-loop: Sensemaking between artificial intelligence and human intelligence collaboration. *Sustainable Futures : An Applied Journal of Technology, Environment and Society*, *10*, 101152. DOI: 10.1016/j.sftr.2025.101152

Hao, X., Demir, E., & Eyers, D. (2025b). Beyond human-in-the-loop: Sensemaking between artificial intelligence and human intelligence collaboration. *Sustainable Futures : An Applied Journal of Technology, Environment and Society*, *10*, 101152. DOI: 10.1016/j.sftr.2025.101152

Islam, R., Hossain, M. N., & Tusar, M. Z. H. (2021). PREDICTIVE ANALYTICS IN SUPPLY CHAIN MANAGEMENT A REVIEW OF BUSINESS ANALYST-LED OPTIMIZATION TOOLS. *Review of Applied Science and Technology*, *6*(1), 34–73. DOI: 10.63125/5aypx555

Klessascheck, F., Weber, I., & Pufahl, L. (2025). SOPA: A framework for sustainability-oriented process analysis and re-design in business process management. *Information Systems and e-Business Management*, *23*(2), 489–537. DOI: 10.1007/s10257-024-00695-x

Lawter, L., & Garnjost, P. (2025). Green Human Resource Management and Organizational Performance: A Systematic Review †. *Sustainability (Basel)*, *17*(7), 3132. DOI: 10.3390/su17073132

Lewańska, E., & Stróżyna, M. (2025). Enhancing ESG Reporting Through Process Mining: A Taxonomy of ESG Events. *Procedia Computer Science*, *270*, 3450–3458. DOI: 10.1016/j.procs.2025.09.470

Liao, S. M., Haykel, I., Cheung, K., & Matalon, T. (2025). Navigating the complexities of AI and digital governance: The 5W1H framework. *Journal of Responsible Technology*, *23*, 100127. DOI: 10.1016/j.jrt.2025.100127

Liu, X., Vasantha, G., Goh, K., Lin, W., & Marzano, A. (2026). Data-driven smart product design, smart service design, and smart product-service system design – A comprehensive review. *Advanced Engineering Informatics*, *73*, 104534. DOI: 10.1016/j.aei.2026.104534

Ma, H., Wang, S., Gu, J., Xie, Q., Wang, W., Dong, S., Sun, C., Li, J., Ma, C., Dong, J., & Liu, H. (2026). IoT-based smart environmental control and air emission management in animal farming: A systematic review. *Information Processing in Agriculture*. Advance online publication. DOI: 10.1016/j.inpa.2026.02.007

Malki, A., Atlam, E. S., & Gad, I. (2022). Machine learning approach of detecting anomalies and forecasting time-series of IoT devices. *Alexandria Engineering Journal*, *61*(11), 8973–8986. DOI: 10.1016/j.aej.2022.02.038

Miedema, E., Waschull, S., & Emmanouilidis, C. (2026). Towards trustworthy artificial intelligence for decision-making: A lifecycle perspective on knowledge- and data-driven artificial intelligence systems. *Computers in Industry*, *174*, 104409. DOI: 10.1016/j.compind.2025.104409

Moreno-Sánchez, P. A., Del Ser, J., van Gils, M., & Hernesniemi, J. (2026). A design framework for operationalizing trustworthy artificial intelligence in healthcare: Requirements, tradeoffs and challenges for its clinical adoption. *Information Fusion*, *127*, 103812. DOI: 10.1016/j.inffus.2025.103812

National Institute of Standards and Technology. (2023). *Artificial Intelligence Risk Management Framework (AI RMF 1.0)*. DOI: 10.6028/NIST.AI.100-1

National Institute of Standards and Technology. (2024). *Artificial Intelligence Risk Management Framework: Generative Artificial Intelligence Profile*. DOI: 10.6028/NIST.AI.600-1

Noetzold, D., Leithardt, V. R. Q., Santana, J. F. de P., & Barbosa, J. L. V. (2026). Oraculum: A model for self-adaptive system optimization in smart environments. *Expert Systems with Applications*, *315*, 131705. DOI: 10.1016/j.eswa.2026.131705

OECD. (2019). Artificial Intelligence in Society. In *Artificial Intelligence in Society*. OECD Publishing., DOI: 10.1787/eedfee77-en

OECD. (2024a, June 11). *Artificial Intelligence in Society*. OECD Publishing. *AI & Society*. Advance online publication. DOI: 10.1787/eedfee77-en

OECD. (2024b, June 11). *Recommendation of the Council on Artificial Intelligence*. OECD Publishing. *AI & Society*. Advance online publication. DOI: 10.1787/eedfee77-en

Oukharijane, J., Chaâbane, M. A., Ben Said, I., Andonoff, E., & Bouaziz, R. (2025). An assessment taxonomy for self-adaptation business process solutions. *Data & Knowledge Engineering, 155*, 102374. DOI: 10.1016/j.datak.2024.102374

Papagiannidis, E., Mikalef, P., & Conboy, K. (2025). Responsible artificial intelligence governance: A review and research framework. *The Journal of Strategic Information Systems, 34*(2), 101885. DOI: 10.1016/j.jsis.2024.101885

Petropoulos, F., Akkermans, H., Aksin, O. Z., Ali, I., Babai, M. Z., Barbosa-Povoa, A., Battaïa, O., Besiou, M., Boysen, N., Brammer, S., Brandon-Jones, A., Briskorn, D., Browning, T. R., Buijs, P., Centobelli, P., Chiarini, A., Cousins, P., Cudney, E. A., Davies, A., & Zheng, T. (2026). Operations & supply chain management: Principles and practice. *International Journal of Production Research, 64*(1), 330–513. DOI: 10.1080/00207543.2025.2555531

Rame, R., Purwanto, P., & Sudarno, S. (2024). Industry 5.0 and sustainability: An overview of emerging trends and challenges for a green future. *Innovation and Green Development, 3*(4), 100173. DOI: 10.1016/j.igd.2024.100173

Relich, M. (2023). Predictive and Prescriptive Analytics in Identifying Opportunities for Improving Sustainable Manufacturing. *Sustainability 2023, Vol. 15, Page 7667, 15*(9), 7667. DOI: 10.3390/su15097667

Sapkota, R., Roumeliotis, K. I., & Karkee, M. (2026). AI Agents vs. Agentic AI: A Conceptual taxonomy, applications and challenges. *Information Fusion, 126*, 103599. DOI: 10.1016/j.inffus.2025.103599

Seidel, S., Recker, J., & Vom Brocke, J. (2012). Green Business Process Management. *Green Business Process Management: Towards the Sustainable Enterprise, 9783642274886*, 3–13. DOI: 10.1007/978-3-642-27488-6_1

Sohns, T. M., Aysolmaz, B., Figge, L., & Joshi, A. (2023). Green business process management for business sustainability: A case study of manufacturing small and medium-sized enterprises (SMEs) from Germany. *Journal of Cleaner Production, 401*, 136667. DOI: 10.1016/j.jclepro.2023.136667

Szelagowski, M., & Berniak-Woźny, J. (2020). The adaptation of business process management maturity models to the context of the knowledge economy. *Business Process Management Journal, 26*(1), 212–238. DOI: 10.1108/BPMJ-11-2018-0328

Troshani, I., & Rowbottom, N. (2024). Corporate sustainability reporting and information infrastructure. *Accounting, Auditing & Accountability Journal*, *37*(4), 1209–1237. DOI: 10.1108/AAAJ-01-2023-6244

Tumpa, R. J., & Naeni, L. (2025). Improving decision-making and stakeholder engagement at project governance using digital technology for sustainable infrastructure projects. *Smart and Sustainable Built Environment*, *14*(4), 1292–1329. DOI: 10.1108/SASBE-10-2024-0451

Valmohammadi, C., & Mortaz Hejri, F. (2023). Designing a conceptual green process model in software development: A mixed method approach. *International Journal of Information Management Data Insights*, *3*(2), 100204. DOI: 10.1016/j.jjimei.2023.100204

van der Aalst, W. M. P. (2013). Business Process Management: A Comprehensive Survey. *International Scholarly Research Notices*, *2013*(1), 507984. DOI: 10.1155/2013/507984

Viteri-Sánchez, C., & Novillo-Villegas, S. (2025). A Framework for a Sustainable Adoption of Business Process Management. *Sustainability 2025, Vol. 17, Page 9827*, *17*(21), 9827. DOI: 10.3390/su17219827

Vom Brocke, J., Seidel, S., & Recker, J. (2013). Green business process management: Towards the sustainable enterprise. *Green Business Process Management: Towards the Sustainable Enterprise*, *9783642274886*, 1–251. DOI: 10.1007/978-3-642-27488-6/COVER

Vrchota, J., Pech, M., Rolínek, L., & Bednář, J. (2020). Sustainability Outcomes of Green Processes in Relation to Industry 4.0 in Manufacturing: Systematic Review. *Sustainability 2020, Vol. 12, Page 5968*, *12*(15), 5968. DOI: 10.3390/su12155968

Yang, Y., & Xue, G. (2026). How does environmental regulation reshape global value chain resilience? *Cleaner Environmental Systems*, *20*, 100395. DOI: 10.1016/j.cesys.2026.100395

Yeung, T., Martinez, J. G., Schlenger, J., Borrmann, A., & Sacks, R. (2025). Integrating digital twin and agent-based simulation to support adaptive production system design in building projects. *Automation in Construction*, *180*, 106550. DOI: 10.1016/j.autcon.2025.106550

Zhang, D., Frei, R., Wills, G., Gerding, E., Bayer, S., & Senyo, P. K. (2023). Strategies and practices to reduce the ecological impact of product returns: An environmental sustainability framework for multichannel retail. *Business Strategy and the Environment*, *32*(7), 4636–4661. DOI: 10.1002/bse.3385

Zimek, M., & Baumgartner, R. J. (2024). Systemic sustainability assessment: Analyzing environmental and social impacts of actions on sustainable development. *Cleaner Production Letters*, *7*, 100064. DOI: 10.1016/j.clpl.2024.100064

Compilation of References

Aakula, A., Saini, V., & Ahmad, T. (2024). The Impact of AI on Organizational Change in Digital Transformation. *Internet of Things and Edge Computing Journal*, *4*(1), 75–115.

Aalst, W. (2013). Aalst, W.M.P.: Business process management: A comprehensive survey. *ISRN Software Engineering*, 1–37. https://doi.org/DOI: 10.1155/2013/507984

Aalst, W., Adriansyah, A., & Dongen, B. (2012). Replaying history on process models for conformance checking and performance analysis. *Wiley Interdisciplinary Reviews. Data Mining and Knowledge Discovery*, *2*(2), 182–192. DOI: 10.1002/widm.1045

Aalst, W., & Ter, A. (2004). YAWL: Yet another workflow language. *Information Systems*, *30*(4), 245–275. DOI: 10.1016/j.is.2004.02.002

Aarthi, S., Ravikumar, R. N., & Pardaev, J. (2025). Synergizing multimodal generative AI and blockchain for the future of digital media. In *Advances in Computational Intelligence and Robotics* (pp. 1–34). IGI Global Scientific Publishing. DOI: 10.4018/979-8-3373-1504-1.ch001

Aarthi, S., Ravikumar, R. N., Sobirova, K., & Mamadiyarov, Z. (2025). Empowering a sustainable future through digital learning and the SDGs. In *Harnessing E-Learning to Create a Sustainable Future* (pp. 145–172). IGI Global Scientific Publishing., DOI: 10.4018/979-8-3373-0112-9.ch006

Abbasi, M., Nishat, R. I., Bond, C., Graham-Knight, J. B., Lasserre, P., Lucet, Y., & Najjaran, H. (2025). A review of AI and machine learning contribution in business process management (process enhancement and process improvement approaches). *Business Process Management Journal*, *31*(4), 1414–1452. DOI: 10.1108/BPMJ-07-2024-0555

Abidi, O., Richet, J. L., & Vitari, C. (2025). Digital Transformation and Resilience: Dimensions and Interactions. Https://Services.Igi-Global.Com/Resolvedoi/Resolve.Aspx?Doi=10.4018/JGIM.367873, *33*(1), 1–64. DOI: 10.4018/JGIM.367873

Abourida, M., Short, M., Klymenko, O. V., Khamis, N. M., Mahammedi, C., Al-Mhdawi, M. K. S., & Sakr, A. H. (2026). The evolving landscape of AI-driven risk management in the biogas production: A systematic and bibliometric review. *Waste Management Bulletin*, *4*(1), 100271. Advance online publication. DOI: 10.1016/j.wmb.2025.100271

Abuadbba, A., Sultan, N., Nepal, S., & Jha, S. (2026). *Human Society-Inspired Approaches to Agentic AI Security : The 4C Framework* (arXiv:2602.01942). arXiv. https://doi.org//arXiv.2602.01942DOI: 10.48550

Abushab, A. M., & Hamdan, A. M. (2025). A Model for Integrating Sustainability Metrics into Architectural Project Lifecycle Management. *Engineering And Technology Journal*, *10*(8), 6083–6105. DOI: 10.47191/etj/v10i08.07

Acitelli, G., Alman, A., Maggi, F. M., & Marrella, A. (2025). Achieving framed autonomy in AI-augmented business process management systems through automated planning. *Information Systems*, *133*, 102573. DOI: 10.1016/j.is.2025.102573

Adamakis, M., & Rachiotis, T. (2025). Artificial Intelligence in Higher Education: A State-of-the-Art Overview of Pedagogical Integrity, Artificial Intelligence Literacy, and Policy Integration. *Encyclopedia*, *5*(4), 180. DOI: 10.3390/encyclopedia5040180

Adekunle, B. I., Chukwuma-Eke, E. C., Balogun, E. D., & Ogunsola, K. O. (2021). Machine learning for automation: Developing data-driven solutions for process optimization and accuracy improvement. *International Journal of Multidisciplinary Research and Growth Evaluation*, *3*(1), 800–808. DOI: 10.54660/.IJMRGE.2021.2.1.800-808

Adensamer, A., & Rueckel, D. (2018). Differences Between BPM and ACM Models for Process Execution. *Lecture Notes in Business Information Processing*, *319*, 270–279. DOI: 10.1007/978-3-319-94214-8_18

Adesanya, O. S., Akinola, A. S., & Oyeniyi, L. D. (2021). Robotic Process Automation ensuring regulatory compliance within finance by automating complex reporting and auditing. *International Journal of Multidisciplinary Research and Growth Evaluation*, *2*(4), 1051–1067. DOI: 10.54660/.IJMRGE.2021.2.4.1051-1067

Agostinelli, S., De Luzi, F., di Canito, U., Ferraro, J., Marrella, A., & Mecella, M. (2022). A Data-Centric Approach to Design Resilient-Aware Process Models in BPMN. *Lecture Notes in Business Information Processing, 458 LNBIP*, 38–54. DOI: 10.1007/978-3-031-16171-1_3

Ahmad, T., & Looy, A. Van. (2020). Business Process Management and Digital Innovations: A Systematic Literature Review. *Sustainability 2020, Vol. 12, Page 6827, 12*(17), 6827. DOI: 10.3390/su12176827

Ahmadi, M. M., Hajizadeh, P., & Valmohammadi, C. (2026). Cognitive Management : A Systematic Explication of Applying Cognitive Sciences in Management. *Emerging Technologies and Governance, 1*(1), 66–92. DOI: 10.47176/ETG.2026.1003

Akinbowale, O. E., Zerihun, M. F., & Mashigo, P. (2026). Business process management framework: Systematic review of the trends, potentials and future. *Cogent Business & Management, 13*(1), 2627025. DOI: 10.1080/23311975.2026.2627025

Akinlade, O. F., Filani, O. M., & Nwachukwu, P. S. (2020). Big data approaches enhancing procurement responsiveness through optimized cycle time reduction. *International Journal of Multidisciplinary Futuristic Development, 1*(2), 29–39. DOI: 10.54660/IJMFD.2020.1.2.29-39

Al-Busaidi, A. S., Raman, R., Hughes, L., Albashrawi, M. A., Malik, T., Dwivedi, Y. K., Al- Alawi, T., AlRizeiqi, M., Davies, G., Fenwick, M., Gupta, P., Gurpur, S., Hooda, A., Jurcys, P., Lim, D., Lucchi, N., Misra, T., Raman, R., Shirish, A., & Walton, P. (2024). Redefining boundaries in innovation and knowledge domains: Investigating the impact of generative artificial intelligence on copyright and intellectual property rights. *Journal of Innovation & Knowledge, 9*(4), 100630. DOI: 10.1016/j.jik.2024.100630

Alfaqiyah, E., Alzubi, A., Aljuhmani, H. Y., & Öz, T. (2025). How Industry 4.0 Technologies Enhance Supply Chain Resilience: The Interplay of Agility, Adaptability, and Customer Integration in Manufacturing Firms. *Sustainability 2025, Vol. 17, Page 7922, 17*(17), 7922. DOI: 10.3390/su17177922

Alkhatib, A. W., Zaid, M., & Issa, A. (2025). Antecedents and outcomes of artificial intelligence adoption on the sustainable performance : The TOE framework perspective. *Information Discovery and Delivery*, •••, 1–16. DOI: 10.1108/IDD-04-2025-0076

Alla, P. B. (2026). Augmenting Intelligent Process Automation through Generative AI for Human-in-the-Loop Decision Systems. *Digital Engineering, 8*, 100071. Advance online publication. DOI: 10.1016/j.dte.2025.100071

Al-Mahdawi, E.Dr Emad Al-Mahdawi. (2025). From Generative to Agentic : How AI Shapes the Future of Work. *European Journal of Sports Science Technology, 15*(59), 30–39. DOI: 10.58305/ejsst.v15i59.736

Alshammari, K. H., & Alshammari, A. F. (2026). AI as a Cognitive Partner : Investigating Knowledge Augmentation and Its Role in Digital Transformation Outcomes. *Systems*, *14*(1), 65. DOI: 10.3390/systems14010065

Amer, Y., Soufali, A., & Zaghwan, A. (2026). A digital twin-based framework for predictive quality assurance and supply chain resilience in the automotive industry. *Advanced Engineering Informatics*, *69*, 103969. DOI: 10.1016/j.aei.2025.103969

Anang, A. N., Obidi, P. O., Mesogboriwon, A. O., & Obidi, J. O. (2024). THE role of Artificial Intelligence in industry 5.0: Enhancing human-machine collaboration. *World Journal of Advanced Research and Reviews*, *24*(2), 380–400. DOI: 10.30574/wjarr.2024.24.2.3369

Ananthi, M., Lakshmana Kumar, R., Muthu, B., & Punitha, P. (2026). Adaptive marine intelligence and sensing architecture for autonomous underwater ecosystem monitoring using AI and IoT integration. *Intelligent Data Analysis*, *30*(1), 162–181. DOI: 10.1177/1088467X251339271

Andriyani, Y., Suripto, , Yohanitas, W. A., Kartika, R. S., & Marsono, . (2024). Adaptive innovation model design: Integrating agile and open innovation in regional areas innovation. *Journal of Open Innovation*, *10*(1), 100197. DOI: 10.1016/j.joitmc.2023.100197

Angelis, A., & Kousiouris, G. (2025). An Overview on the Landscape of Self-Adaptive Cloud Design and Operation Patterns: Goals, Strategies, Tooling, Evaluation, and Dataset Perspectives. *Future Internet 2025, Vol. 17, Page 434, 17*(10), 434. DOI: 10.3390/fi17100434

Angulo Martínez, I., Garcia, A., González-Potes, A., Martínez-Castro, D., Paredes, C. M., Ochoa-Brust, A., Mena, L. J., Martínez-Peláez, R., Félix, V. G., & Félix-Cuadras, R. A. (2026). Hybrid AI and LLM-Enabled Agent-Based Real-Time Decision Support Architecture for Industrial Batch Processes: A Clean-in-Place Case Study. *AI 2026, Vol. 7, Page 51, 7*(2), 51. DOI: 10.3390/ai7020051

Antczak, J., Kabus, J., Miciuła, I., Lenort, R., Brzozowska, A., Wolniak, R., & Kana, R. (2025). Artificial Intelligence in business management - SWOT analysis. *Procedia Computer Science*, *270*, 4410–4420. DOI: 10.1016/j.procs.2025.09.566

Antunes, P. (2011). BPM and exception handling: Focus on organizational resilience. *IEEE Transactions on Systems, Man, and Cybernetics. Part C, Applications and Reviews*, *41*(3), 383–392. DOI: 10.1109/TSMCC.2010.2062504

Antunes, P., & Mourão, H. (2011). Resilient Business Process Management: Framework and services. *Expert Systems with Applications, 38*(2), 1241–1254. DOI: 10.1016/j.eswa.2010.05.017

Areta Hiziroglu, O., & Dogan, O. (2025). A Framework for Sustainability Performance Measurement Through Process Mining: Integration of GRI Metrics in Operational Processes. *Systems 2025, Vol. 13, Page 547, 13*(7), 547. DOI: 10.3390/systems13070547

Arno, P. (2025). From Theory to Practice: Empirical Insights on BPM Lifecycle Adoption and Low-Code Utilization in SMEs. *IEEE Access : Practical Innovations, Open Solutions, 13*, 117385–117403. DOI: 10.1109/ACCESS.2025.3585261

Ayoubi, M., Shinwari, A. R., & Khan, M. S. (2019). *An event-driven service oriented architecture approach for E-governance systems*. Kardan Journal of Engineering and Technology., DOI: 10.31841/KJET.2021.1

Aziz, A., & Nuryana, I. K. (2025). Business process analysis and modeling using the business process model notation (BPMN) method (Case study: Simo Jaya Printing). [JEISBI]. *Journal of Emerging Information Systems and Business Intelligence, 6*(3). Advance online publication. DOI: 10.26740/jeisbi.v6i3.70359

Babalola, O., Iluore, E., Bakare, A., & Udechukwu, L. M. (2022). AI-Enabled Business Process Optimization Engine for risk-Aware Cloud Operations. *Journal of Frontiers in Multidisciplinary Research, 3*(2), 211–216. DOI: 10.54660/.IJFMR.2022.3.2.211-216

Baker, J. (2012). The Technology–Organization–Environment Framework. In Dwivedi, Y. K., Wade, M. R., & Schneberger, S. L. (Eds.), *Information Systems Theory* (Vol. 28, pp. 231–245). Springer New York., DOI: 10.1007/978-1-4419-6108-2_12

Baldea, M., Georgiou, A. T., Gopaluni, B., Mercangöz, M., Pantelides, C. C., Sheth, K., Zavala, V. M., & Georgakis, C. (2025). From automated to autonomous process operations. *Computers & Chemical Engineering, 196*, 109064. DOI: 10.1016/j.compchemeng.2025.109064

Bandara, W., Gable, G. G., & Rosemann, M. (2005). Factors and measures of business process modelling: Model building through a multiple case study. *European Journal of Information Systems, 14*(4), 347–360. DOI: 10.1057/palgrave.ejis.3000546

Bandi, A., Kongari, B., Naguru, R., Pasnoor, S., & Vilipala, S. V. (2025). The Rise of Agentic AI : A Review of Definitions, Frameworks, Architectures, Applications, Evaluation Metrics, and Challenges. *Future Internet, 17*(9), 404. DOI: 10.3390/fi17090404

Barata, J., & Kayser, I. (2024). How will the digital twin shape the future of industry 5.0? *Technovation*, *134*, 103025. DOI: 10.1016/j.technovation.2024.103025

Barocas, S., Hardt, M., & Narayanan, A. (2023). *Fairness and machine learning : Limitations and opportunities*. MIT press.

Batini, C., & Scannapieco, M. (2016). *Data and Information Quality : Dimensions, Principles and Techniques*. Springer International Publishing., DOI: 10.1007/978-3-319-24106-7

Bazan-Muñoz, A., Pautasso, C., Ortiz, G., & Garcia-de-Prado, A. (2026). BPSmart-CARE: A framework for managing contextualized actions in IoT systems through the integration of business process modelling and complex event processing. *Internet of Things : Engineering Cyber Physical Human Systems*, *36*, 101887. DOI: 10.1016/j.iot.2026.101887

Beerepoot, I., Di Ciccio, C., Reijers, H. A., Rinderle-Ma, S., Bandara, W., Burattin, A., Calvanese, D., Chen, T., Cohen, I., Depaire, B., Di Federico, G., Dumas, M., van Dun, C., Fehrer, T., Fischer, D. A., Gal, A., Indulska, M., Isahagian, V., Klinkmüller, C., & Zerbato, F. (2023). The biggest business process management problems to solve before we die. *Computers in Industry*, *146*, 103837. DOI: 10.1016/j.compind.2022.103837

Beheshti, A., Yang, J., Sheng, Q. Z., Benatallah, B., Casati, F., Dustdar, S., Motahari Nezhad, H. R., Zhang, X., & Xue, S. (2023). *ProcessGPT: Transforming Business Process Management with Generative Artificial Intelligence*.

Beigi, H., & Franconi, A. (2025). Design for Traceability (DfT). *Proceedings of the 6th Product Lifetimes and the Environment Conference (PLATE2025)*, (6). DOI: 10.54337/plate2025-10322

Bello, M., Bello, R., García, M. M., Nowé, A., Sevillano-García, I., & Herrera, F. (2025). A Three-level Framework for LLM-enhanced Explainable AI: From Technical Explanations to Natural Language. *Information Systems Frontiers*, ●●●, 1–22. DOI: 10.1007/s10796-025-10668-1

Ben Hassen, M., Zahaf, S., & Gargouri, F. (2024). Business View Specification of Enterprise Information Systems Based on Core Ontologies. *Procedia Computer Science*, *237*, 380–388. DOI: 10.1016/j.procs.2024.05.118

Benhmama, A., & Sabiri, B. (2025). Can AI improve individual employee performance? *2025 International Conference on Innovation in Artificial Intelligence and Internet of Things (AIIT)*, 1-6. https://ieeexplore.ieee.org/abstract/document/11082834/

Benjamin, R. (2019). Assessing bias in AI systems: Toward fairer design models. *Science, 366*(6464), 421–423. DOI: 10.1126/science.aaz3873

Bernardi, M. L., Casciani, A., Cimitile, M., & Marrella, A. (2024). Conversing with business process-aware Large Language Models: the BPLLM framework. *Journal of Intelligent Information Systems*. DOI: 10.21203/rs.3.rs-4125790/v1

Berti, A., Wang, X., Kourani, H., & Van der Aalst, W. M. P. (2025). Specializing large language models for process modeling via reinforcement learning with verifiable and universal rewards. *Process Science 2025 2:1, 2*(1), 26-. DOI: 10.1007/s44311-025-00034-4

Berti, A., Kourani, H., & van der Aalst, W. M. P. (2024). PM-LLM-Benchmark: Evaluating Large Language Models on Process Mining Tasks. *Lecture Notes in Business Information Processing, 533*, 610–623. DOI: 10.1007/978-3-031-82225-4_45

Bertrand, Y., Stevens, A., Deforce, B., De Smedt, J., De Weerdt, J., & Serral, E. (2025). Approaches for IoT-enhanced predictive process monitoring. *Process Science 2025 2:1, 2*(1), 7-. DOI: 10.1007/s44311-025-00011-x

Besigomwe, K. (2025). Self-healing digital twins for manufacturing process resilience. *Cognizance Journal of Multidisciplinary Studies, 5*(3), 23–38. DOI: 10.47760/cognizance.2025.v05i03.003

Bhardwaj, A. (2025). Systematic literature review on explainable AI in finance: Methods, applications, and research gaps. *International Journal For Multidisciplinary Research, 7*(5), 58963. Advance online publication. DOI: 10.36948/ijfmr.2025.v07i05.58963

Bidmeshk, O. G., Woo, C., & Mehraeen, M. (2025). Managing operational alignment complexity: A recommender system approach. *Journal of Open Innovation, 11*(4), 100627. DOI: 10.1016/j.joitmc.2025.100627

Bobba, A. R. (2025). Cloud-native AI across industries: Healthcare, finance, and manufacturing. *European Modern Studies Journal, 9*(5), 619–627. DOI: 10.59573/emsj.9(5).2025.56

Bocciarelli, P., & D'Ambrogio, A. (2014). A model-driven method for enacting the design-time QoS analysis of business processes. *Software & Systems Modeling, 13*(2), 573–598. DOI: 10.1007/s10270-013-0345-5

Bocharov, N., Torres, P., & Matos, J. (2025). A practical approach to industrial digitalization through data acquisition and systems integration for predictive maintenance. *EAI Endorsed Transactions on Digital Transformation of Industrial Processes, 1*(2). Advance online publication. DOI: 10.4108/dtip.9700

Bondac, G.-T., Stanescu, S.-G., Ionescu, C. A., Duica, A., & Uzlău, M. C. (2026). Decision-Making in Complex Systems Using AI-Based Decision Support: The Role of Trust, Transparency, and Data Quality. *Electronics 2026, Vol. 15, Page 372, 15*(2), 372. DOI: 10.3390/electronics15020372

Bornay-Barrachina, M., López-Cabrales, Á., & Salas-Vallina, A. (2025). Sensing, seizing, and reconfiguring dynamic capabilities in innovative firms: Why does strategic leadership make a difference? *Business Research Quarterly, 28*(2), 399–420. DOI: 10.1177/23409444231185790;PAGEGROUP:STRING:PUBLICATION

Boulitama, O., Lhamidi, S., Taifi, R., & Brahim, S. (2025). Digital transformation and supply chain capabilities: A sectoral maturity perspective in morocco. *Polish Journal of Management Studies, 32*(1), 61–80. DOI: 10.17512/pjms.2025.32.1.04

Brandao, P. R. (2025). The Impact of Artificial Intelligence on Modern Society. *AI 2025, Vol. 6, Page 190, 6*(8), 190. DOI: 10.3390/ai6080190

Braniff, A., Akundi, S. S., Liu, Y., Dantas, B., Niknezhad, S. S., Khan, F., Pistikopoulos, E. N., & Tian, Y. (2025). Real-time process safety and systems decision-making toward safe and smart chemical manufacturing. *Digital Chemical Engineering, 15*, 100227. DOI: 10.1016/j.dche.2025.100227

Brocke, J., & Rosemann, M. (2015). *Strategic alignment, governance, people and culture*.

Brown, T. (2009). Design thinking. *Harvard Business Review, 86*(6), 84–92.

Brückner, M., Thyssen, C., & Huwer, J. (2026). AI for adaptive science teaching: Strengthening teacher self-efficacy and perceived usefulness. *Frontiers in Education, 11*, 1759274. DOI: 10.3389/feduc.2026.1759274

Brynjolfsson, E., & Mcafee, A. (2017). The business of artificial intelligence. *Harvard Business Review, 7*(1), 1–2.

Burgueño, L., Izquierdo, J. L. C., & Planas, E. (2021). An empirical study on the impact of introducing a modeling tool in a requirement engineering course. In *2021 ACM/IEEE International Conference on Model Driven Engineering Languages and Systems Companion (MODELS-C)* (pp. 712–720). DOI: 10.1109/MODELS-C53483.2021.00115

Cadet, E., Etim, E. D., Essien, I. A., Erigha, E. D., Babatunde, L. A., Ajayi, J. O., & Obuse, E. (2024). Large Language Models for cybersecurity policy compliance and risk mitigation. *International Journal of Scientific Research in Humanities and Social Sciences, 1*(2), 612–643. DOI: 10.32628/IJSRSSH242560

Calabrò, A., Marchetti, E., Skrzypek, P., & Marchel, J. (2025). CONCERN: A model-based monitoring infrastructure. *Internet of Things : Engineering Cyber Physical Human Systems, 31*, 101557. DOI: 10.1016/j.iot.2025.101557

Calvo, R. A., & D'Mello, S. (2010). Affect detection: An interdisciplinary review of models, methods, and their applications. *IEEE Transactions on Affective Computing, 1*(1), 18–37. DOI: 10.1109/T-AFFC.2010.1

Canals, J. (2020). The evolving role of general managers in the age of AI. *The Future of Management in an AI World: Redefining Purpose and Strategy in the Fourth Industrial Revolution*, 37-64.

Cao, Z., Zhao, H., Wang, Y., He, C., Zhou, D., & Han, X. (2025). A Resilience Quantitative Assessment Framework for Cyber–Physical Systems: Mathematical Modeling and Simulation. *Applied Sciences 2025, Vol. 15, Page 8285, 15*(15), 8285. DOI: 10.3390/app15158285

Capolupo, N., Romeo, E., Giampaola, V., & Adinolfi, P. (2025). To AI or not to AI? Unpacking the Hamletian dilemma in public organizations' work processes through a socio-technical perspective. *Business Process Management Journal*, •••, 1–29. DOI: 10.1108/BPMJ-07-2025-1188

Caramihai, S. I., Dumitrache, I., Moisescu, M. A., & Sacala, I. S. (2020). Bio-inspired autonomous enterprise systems. *IFAC-PapersOnLine, 53*(2), 10879–10884. DOI: 10.1016/j.ifacol.2020.12.2820

Casas-Ortiz, A., Jimenez-Tellez, N., Portaz, M., & Santos, O. C. (2026). A holistic framework to create intelligent systems for psychomotor learning: a systematic review and insights from the virtual reality Kenpo learning simulator. *Virtual Reality 2026 30:2, 30*(2), 72-. DOI: 10.1007/s10055-026-01326-1

Casati, F., & Shan, M.-C. (2001). Dynamic and adaptive composition of e-services. *Information Systems, 26*(3), 143–163. DOI: 10.1016/S0306-4379(01)00014-X

Cazorla, A., Flogie, A., Woodruff, E., & Hewitt, J. (2026). Epistemic Agency in the Age of Large Language Models: Design Principles for Knowledge-Building AI. *AI 2026, Vol. 7, Page 99, 7*(3), 99. DOI: 10.3390/ai7030099

Chandrasekaran, S., Edin, P., & Lohr, T. (2025). *The agentic AI advantage : Unlocking the next level of AI value*. KPMG., https://kpmg.com/us/en/articles/2025/the-agentic-ai-advantage.html

Chaturvedi, A., & Dasgupta, M. (2024). Managerial Perception of AI in Strategic Decision-Making. *Journal of Computer Information Systems*, •••, 1–13. DOI: 10.1080/08874417.2024.2418403

Chen, H., Wen, Y., Zhu, M., Huang, Y., Xiao, C., Wei, T., & Hahn, A. (2021). From Automation System to Autonomous System: An Architecture Perspective. *Journal of Marine Science and Engineering 2021, Vol. 9, Page 645, 9*(6), 645. DOI: 10.3390/jmse9060645

Chen, S. C., Chen, H. M., Chen, H. K., & Li, C. L. (2024). Multi-Objective Optimization in Industry 5.0: Human-Centric AI Integration for Sustainable and Intelligent Manufacturing. *Processes 2024, Vol. 12, Page 2723, 12*(12), 2723. DOI: 10.3390/pr12122723

Cheng, C., Ahmad, S. F., Irshad, M., Alsanie, G., Khan, Y., Ahmad, A. Y. A. B., & Aleemi, A. R. (2023). Impact of Green Process Innovation and Productivity on Sustainability: The Moderating Role of Environmental Awareness. *Sustainability 2023, Vol. 15, Page 12945, 15*(17), 12945. DOI: 10.3390/su151712945

Chen, J., & Zhang, Y. (2022). Knowledge graph-driven AI systems for environmental management. *Environmental Science & Technology*, *56*(7), 4410–4417.

Chen, R., Zhang, J., & Wang, H. (2025). Autonomous fleet management system in smart ports: Practical design and analytical considerations. *Multimodal Transportation*, *4*(3), 100211. Advance online publication. DOI: 10.1016/j.multra.2025.100211

Chen, Y.-K. (2012). Challenges and opportunities of internet of things. In *17th Asia and South Pacific Design Automation Conference* (pp. 383–388). DOI: 10.1109/ASPDAC.2012.6164978

Chineme Scholar Okonkwo, C. S. O., Jibril Agbabiaka, J. A., Winner Mayo, W. M., & Obinna ThankGod Okeke, O. T. O. (2024). Supply Chain Automation Framework Using Service Management Platforms. *Shodhshauryam International Scientific Refereed Research*, (SHISRRJ2472157), 157. DOI: 10.32628/SHISRRJ2472157

Choo, C. W. (1996). The knowing organization : How organizations use information to construct meaning, create knowledge and make decisions. *International Journal of Information Management*, *16*(5), 329–340. DOI: 10.1016/0268-4012(96)00020-5

Cohen, W. M., & Levinthal, D. A. (1990). Absorptive capacity : A new perspective on learning and innovation. *Administrative Science Quarterly*, *35*(1), 128–152. DOI: 10.2307/2393553

Collins, C., Dennehy, D., Conboy, K., & Mikalef, P. (2021). Artificial intelligence in information systems research: A systematic literature review and research agenda. *International Journal of Information Management*, *60*, 102383. DOI: 10.1016/j.ijinfomgt.2021.102383

Conz, E., & Magnani, G. (2020). A dynamic perspective on the resilience of firms: A systematic literature review and a framework for future research. *European Management Journal, 38*(3), 400–412. DOI: 10.1016/j.emj.2019.12.004

Corradini, F., Pettinari, S., Re, B., Rossi, L., & Tiezzi, F. (2023). Executable Digital Process Twins: Towards the Enhancement of Process-Driven Systems. *Big Data and Cognitive Computing 2023, Vol. 7, Page 139, 7*(3), 139. DOI: 10.3390/bdcc7030139

Correa-Peralta, M., Vinueza-Martínez, J., & Castillo-Heredia, L. (2025). Evolution, topics and relevant research methodologies in business intelligence and data analysis in the academic management of higher education institutions. A literature review. In *Results in Engineering* (Vol. 25). Elsevier B.V., DOI: 10.1016/j.rineng.2024.103782

Costache, I., Turetken, O., Aysolmaz, B., & Winter, K. (2025). Process Mining Guidelines for Greenhouse Gas Emission Management in Production Processes. *Lecture Notes in Business Information Processing, 533*, 711–724. DOI: 10.1007/978-3-031-82225-4_52

Couckuyt, D., & Looy, A. Van. (2019). Green BPM as a Business-Oriented Discipline: A Systematic Mapping Study and Research Agenda. *Sustainability 2019, Vol. 11, Page 4200, 11*(15), 4200. DOI: 10.3390/su11154200

Couckuyt, D., & Van Looy, A. (2020). A systematic review of Green Business Process Management. *Business Process Management Journal, 26*(2), 421–446. DOI: 10.1108/BPMJ-03-2019-0106

Crawford, K. (2021). *The atlas of AI: Power, politics, and the planetary costs of artificial intelligence*. Yale University Press.

Cunningham, J., Davies, N., Devaney, S., Harding, M., Holm, S., Neumann, V., & Ainsworth, J. (2025). The development of a blockchain-based business model simulator with an application to civic medical data management. *Digital Business, 5*(1), 100112. Advance online publication. DOI: 10.1016/j.digbus.2025.100112

Czvetkó, T., Sebestyén, V., & Abonyi, J. (2025). Key factors of industry 5.0-based organizational sustainability. *Technology in Society, 83*, 102966. DOI: 10.1016/j.techsoc.2025.102966

Dauerer, A. (2025). A systematic literature review of performance measurement systems and the integration of ESG factors. *Environmental and Sustainability Indicators, 27*, 100746. DOI: 10.1016/j.indic.2025.100746

Davenport, T. (1993). *Reengineering work through information technology.*

Davenport, T. H., & Ronanki, R. (2018). Artificial intelligence for the real world. *Harvard Business Review*, *96*(1), 1.

Davenport, T., & Short, J. (1989). The new industrial engineering: Information technology and business process redesign. *Sloan Management Review*, •••, 31.

De Luzi, F., Leotta, F., Marrella, A., & Mecella, M. (2024). On the Interplay Between Business Process Management and Internet-of-Things. *Business & Information Systems Engineering 2024 67:2, 67*(2), 265–288. DOI: 10.1007/s12599-024-00859-6

De Nicola, R., Jähnichen, S., & Wirsing, M. (2020). Rigorous engineering of collective adaptive systems: Special section. *International Journal of Software Tools for Technology Transfer*, *22*(4), 389–397. DOI: 10.1007/s10009-020-00565-0

Debbadi, R. K., & Boateng, O. (2025). Enhancing cognitive automation capabilities with reinforcement learning techniques in robotic process automation using UiPath and automation anywhere. *International Journal of Science and Research Archive*, *14*(2), 733–752. DOI: 10.30574/ijsra.2025.14.2.0450

Denkena, B., Wichmann, M., Kettelmann, S., Matthies, J., & Reuter, L. (2022). Ecological Planning of Manufacturing Process Chains. *Sustainability 2022, Vol. 14, Page 2681, 14*(5), 2681. DOI: 10.3390/su14052681

Devi, A. K. R. S.Arun Kumar Ramachandran Sumangala Devi. (2024). The future of healthcare data governance AI-enabled ETL testing frameworks on data warehousing testing automation using ML, AI and NLP. *International Journal For Multidisciplinary Research*, *6*(5), 28559. Advance online publication. DOI: 10.36948/ijfmr.2024.v06i05.28559

Devlin, J., Chang, M.-W., Lee, K., & Toutanova, K. (2019). BERT: Pre-training of deep bidirectional transformers for language understanding. *Proceedings of NAACL-HLT*, 4171–4186. DOI: 10.48550/arXiv.1810.04805

Dhaigude, A. S., & Kamath, G. B. (2025). Mapping responsible artificial intelligence in business and management: Trends, influence, and emerging research directions. In *Journal of Open Innovation: Technology, Market, and Complexity* (Vol. 11, Number 4). Elsevier B.V. DOI: 10.1016/j.joitmc.2025.100640

Dong, M. M., Stratopoulos, T. C., & Wang, V. X. (2024). A scoping review of ChatGPT research in accounting and finance. *International Journal of Accounting Information Systems*, *55*, 100715. DOI: 10.1016/j.accinf.2024.100715

Dormehl, J. A., Andrews, R., Kratsch, W., Röglinger, M., Wynn, M. T., & Zetzsche, F. (2026). Domain experts in the loop: Leveraging generative artificial intelligence for interactive data validation in process mining. *Information Systems*, *140*, 102715. DOI: 10.1016/j.is.2026.102715

Dorrer, M. G. (2020). The digital twin of the business process model. *Journal of Physics: Conference Series*, *1679*(3), 032096. DOI: 10.1088/1742-6596/1679/3/032096

Dumas, M., Milani, F., & Chapela-Campa, D. (2026). *Agentic Business Process Management Systems* (arXiv:2601.18833). arXiv. /arXiv.2601.18833DOI: 10.1007/978-3-032-13426-4_1

Dumas, M., Fournier, F., Limonad, L., Marrella, A., Montali, M., Rehse, J.-R., Accorsi, R., Calvanese, D., De Giacomo, G., Fahland, D., Gal, A., La Rosa, M., Völzer, H., & Weber, I. (2023). AI-augmented Business Process Management Systems : A Research Manifesto. *ACM Transactions on Management Information Systems*, *14*(1), 1–19. DOI: 10.1145/3576047

Dumas, M., La Rosa, M., Mendling, J., & Reijers, H. (2013). *Fundamentals of business process management*. Springer., DOI: 10.1007/978-3-642-33143-5

Dumas, M., La Rosa, M., Mendling, J., & Reijers, H. A. (2018). Introduction to Business Process Management. In Dumas, M., La Rosa, M., Mendling, J., & Reijers, H. A. (Eds.), *Fundamentals of Business Process Management* (pp. 1–33). Springer., DOI: 10.1007/978-3-662-56509-4_1

Dumitrache, I., Predescu, L. S., Caramihai, S. I., & Moisescu, M. A. (2023). Collaborative Decisions in Knowledge Management for Intelligent Cyber-Enterprises. *Procedia Computer Science*, *221*, 955–962. DOI: 10.1016/j.procs.2023.08.074

Dziembek, D., & Turek, T. (2025). A Model for Integrating Artificial Intelligence with ERP Systems - Towards Autonomous Business Management Systems. *Procedia Computer Science*, *270*, 6260–6269. DOI: 10.1016/j.procs.2025.10.096

Ehrlich-Sommer, F., Eberhard, B., & Holzinger, A. (2025). ForestGPT and Beyond: A Trustworthy Domain-Specific Large Language Model Paving the Way to Forestry 5.0. *Electronics 2025, Vol. 14, Page 3583, 14*(18), 3583. DOI: 10.3390/electronics14183583

Eisenhardt, K. M., & Martin, J. A. (2000). Dynamic capabilities : What are they? *Strategic Management Journal*, *21*(10-11), 1105–1121. DOI: 10.1002/1097-0266(200010/11)21:10/11<1105::AID-SMJ133>3.0.CO;2-E

Emery, F. E., & Trist, E. L. (1965). The causal texture of organizational environments. *Human Relations; Studies Towards the Integration of the Social Sciences*, *18*(1), 21–32. DOI: 10.1177/001872676501800103

Erasmus, J., Vanderfeesten, I., Traganos, K., & Grefen, P. (2020). Using business process models for the specification of manufacturing operations. *Computers in Industry*, *123*, 103297. DOI: 10.1016/j.compind.2020.103297

Eshuis, R., Hull, R., & Yi, M. (2019). Reasoning about property preservation in adaptive Case Management. *ACM Transactions on Internet Technology*, *19*(1), 1–21. DOI: 10.1145/3177778

European Commission. (2025, February 2). *AI Act | Shaping Europe's digital future*. European Commission. https://digital-strategy.ec.europa.eu/en/policies/regulatory-framework-ai

Fahland, D., Fournier, F., Limonad, L., Skarbovsky, I., & Swevels, A. J. E. (2025). How well can a large language model explain business processes as perceived by users? *Data & Knowledge Engineering*, *157*, 102416. DOI: 10.1016/j.datak.2025.102416

Feng, P., Peng, X., Zhang, Z., Lu, Z., Drakopoulos, P., Malousoudis, P., Nousias, N., Tsakalidis, G., & Vergidis, K. (2026). Do LLMs Speak BPMN? An Evaluation of Their Process Modeling Capabilities Based on Quality Measures. *Computation 2026, Vol. 14, Page 10, 14*(1), 10. DOI: 10.3390/computation14010010

Floridi, L. (2019). Establishing the rules for building trustworthy AI. *Nature Machine Intelligence*, *1*(6), 261–262. DOI: 10.1038/s42256-019-0055-y

Floridi, L., & Cowls, J. (2019). A unified framework of five principles for AI in society. *Harvard Data Science Review*, *1*(1). Advance online publication. DOI: 10.1162/99608f92.8cd550d1

Floridi, L., Cowls, J., Beltrametti, M., Chatila, R., Chazerand, P., Dignum, V., Luetge, C., Madelin, R., Pagallo, U., Rossi, F., Schafer, B., Valcke, P., & Vayena, E. (2018). AI4People—An Ethical Framework for a Good AI Society : Opportunities, Risks, Principles, and Recommendations. *Minds and Machines*, *28*(4), 689–707. DOI: 10.1007/s11023-018-9482-5

Fontoura, L., Luiz de Mattos Nascimento, D., Neto, J. V., & Gusmão Caiado, R. G. (2025). Energy Gen-AI technology framework: A perspective of energy efficiency and business ethics in operation management. *Technology in Society*, *81*, 102847. Advance online publication. DOI: 10.1016/j.techsoc.2025.102847

Fornari, F., Compagnucci, I., Callisto De Donato, M., Bertrand, Y., Beyel, H. H., Carrión, E., Franceschetti, M., Groher, W., Grüger, J., Kilic, E., Koschmider, A., Leotta, F., Li, C. Y., Lugaresi, G., Malburg, L., Mangler, J., Mecella, M., Pastor, O., Riss, U., & Valderas, P. (2025). Digital Twins of Business Processes: A Research Manifesto. *Internet of Things : Engineering Cyber Physical Human Systems, 30*, 101477. DOI: 10.1016/j.iot.2024.101477

Fouda, E. (2026). *Data science in action*. DOI: 10.1007/979-8-8688-2013-7_1

Fountaine, T., McCarthy, B., & Saleh, T. (2019). Building the AI-powered organization. *Harvard Business Review*, *97*(4), 62–73.

Gabriel, V. S.Victor Samuel Gabriel. (2025). AI and distributed manufacturing systems: Strengthening healthcare supply chains for national biosecurity. *World Journal of Advanced Research and Reviews*, *26*(2), 4080–4086. DOI: 10.30574/wjarr.2025.26.2.2033

Gaddam, S. G. S., & Raigoni, G. S. (2026). Agentic Artificial Intelligence as a Strategic HR Partner : Redefining Decision-Making Authority and Strategic Roles. *International Journal of Innovative Research in Engineering*, *7*(1), 81–86. https://www.doi.org/10.59256/ijire.20260701010. DOI: 10.59256/ijire.20260701010

Galbraith, J. (1973). Designing complex organizations. *Reading, Mass*. https://search.proquest.com/openview/2124f3c691b3cca453388622c2a20df4/1?pq-origsite=gscholar&cbl=1817083

Gamboa-Cruzado, J., Pineda-Delacruz, V., Salcedo-Mera, H., Alzamora Rivero, C., Coveñas Lalupu, J., & Narro-Andrade, M. (2025). Blockchain and data management security for sustainable digital ecosystems: A systematic literature review. *Sustainability (Basel)*, *18*(1), 185. DOI: 10.3390/su18010185

Garcia Medeiro, J. (2025). Business process governance in the age of digital transformation. *Journal of Interdisciplinary Debates*, *6*(04), 46–57. DOI: 10.51249/jid.v6i04.2688

Garrido-Moreno, A., Martín-Rojas, R., & García-Morales, V. J. (2024). The key role of innovation and organizational resilience in improving business performance: A mixed-methods approach. *International Journal of Information Management*, *77*, 102777. DOI: 10.1016/j.ijinfomgt.2024.102777

Gavetti, G., & Levinthal, D. (2000). Looking Forward and Looking Backward : Cognitive and Experiential Search. *Administrative Science Quarterly*, *45*(1), 113–137. DOI: 10.2307/2666981

Gebreselassie, B. A., Garcia, N. M., & Midekso, D. (2025). A Systematic Review of Self-Adaptive Mobile Applications with Cooperative Dimension. *IoT 2026, Vol. 7, Page 6, 7*(1), 6. DOI: 10.3390/iot7010006

Georgescu, I., Bocean, C. G., Vărzaru, A. A., Rotea, C. C., Mangra, M. G., & Mangra, G. I. (2024). Enhancing Organizational Resilience: The Transformative Influence of Strategic Human Resource Management Practices and Organizational Culture. *Sustainability 2024, Vol. 16, Page 4315, 16*(10), 4315. DOI: 10.3390/su16104315

Geremicca, F., Fascetti, A., Brigham, J. C., & Bilec, M. M. (2026). Digital twins for sustainable buildings: From framework to strategy guidelines and application. *Energy and Building*, *359*, 117332. DOI: 10.1016/j.enbuild.2026.117332

Gero, J. S. (1990). Design prototypes: A knowledge representation schema for design. *AI Magazine*, *11*(4), 26–36. DOI: 10.1609/aimag.v11i4.854

Ghaljahi, M., Omidi, L., & Karimi, A. (2025). Resilience assessment in process industries: A review of literature. *Heliyon*, *11*(4), e42498. DOI: 10.1016/j.heliyon.2025.e42498

Ghobakhloo, M., Iranmanesh, M., Mubarak, M. F., Mubarik, M., Rejeb, A., & Nilashi, M. (2022). Identifying industry 5.0 contributions to sustainable development: A strategy roadmap for delivering sustainability values. *Sustainable Production and Consumption*, *33*, 716–737. DOI: 10.1016/j.spc.2022.08.003

Gignac, G. E., & Szodorai, E. T. (2024). Defining intelligence: Bridging the gap between human and artificial perspectives. *Intelligence*, *104*, 101832. DOI: 10.1016/j.intell.2024.101832

Goodfellow, I., Pouget-Abadie, J., Mirza, M., Xu, B., Warde-Farley, D., Ozair, S., Courville, A., & Bengio, Y. (2014). Generative adversarial nets. *Advances in Neural Information Processing Systems*, *27*, 2672–2680. DOI: 10.48550/arXiv.1406.2661

Graen, G. B., & Uhl-Bien, M. (1995). Relationship-based approach to leadership : Development of leader-member exchange (LMX) theory of leadership over 25 years: Applying a multi-level multi-domain perspective. *The Leadership Quarterly*, *6*(2), 219–247. DOI: 10.1016/1048-9843(95)90036-5

Grohs, M., Abb, L., Elsayed, N., & Rehse, J. R. (2024). Large Language Models Can Accomplish Business Process Management Tasks. *Lecture Notes in Business Information Processing, 492 LNBIP*, 453–465. DOI: 10.1007/978-3-031-50974-2_34

Grønvald, J. M., Nørgaard, M., Christensen, C. K. F., & Mortensen, N. H. (2025). Quantifying the Effects of Modular Product Architectures: A Data-Driven Framework for Evaluating Product Variety and Complexity. *Applied Sciences 2025, Vol. 15, Page 12284, 15*(22), 12284. DOI: 10.3390/app152212284

Grznár, P., Burganová, N., Mozol, Š., & Mozolová, L. (2023). A Comprehensive Digital Model Approach for Adaptive Manufacturing Systems. *Applied Sciences 2023, Vol. 13, Page 10706, 13*(19), 10706. DOI: 10.3390/app131910706

Grznár, P., Papánek, L., Marčan, M., Krajčovič, M., Antoniuk, I., Mozol, Š., & Mozolová, L. (2025). Enhancing Production Efficiency Through Digital Twin Simulation Scheduling. *Applied Sciences 2025, Vol. 15, Page 3637, 15*(7), 3637. DOI: 10.3390/app15073637

Gudipudi, S. (2025). Toward autonomous business intelligence: Research trends in automation and cloud integration. *European Journal of Computer Science and Information Technology, 13*(49), 163–177. DOI: 10.37745/ejcsit.2013/vol13n49163177

Gupta, K., Singh, N., & Kapur, B. (2023). Greenwashed word of mouth (GWWOM): A vibrant influence on customer green behaviour. *Journal of Global Responsibility, 13*(4), 472–487. DOI: 10.1108/JGR-11-2021-0094

Halliday, N. (2023). A conceptual framework for financial network resilience integrating cybersecurity, risk management, and digital infrastructure stability. *International Journal of Advanced Multidisciplinary Research and Studies, 3*(2), 1253–1263. DOI: 10.62225/2583049X.2023.3.2.4887

Hamdani, R., & Chihi, I. (2025). Adaptive human-computer interaction for industry 5.0: A novel concept, with comprehensive review and empirical validation. *Computers in Industry, 168*, 104268. DOI: 10.1016/j.compind.2025.104268

Handler, A., Larsen, K. R., & Hackathorn, R. (2024). Large language models present new questions for decision support. *International Journal of Information Management, 79*, 102811. DOI: 10.1016/j.ijinfomgt.2024.102811

Hao, X., Demir, E., & Eyers, D. (2025). Beyond human-in-the-loop: Sensemaking between artificial intelligence and human intelligence collaboration. *Sustainable Futures : An Applied Journal of Technology, Environment and Society, 10*, 101152. DOI: 10.1016/j.sftr.2025.101152

Haripriya. (2025). Transforming governance: The role of automation in tax, compliance, and public sector efficiency. In *Rethink 2025: Navigating Technology Trends Through Multidisciplinary Insights* (pp. 94–99). Royal Book Publishing. DOI: 10.26524/300.13

Helfat, C. E., & Peteraf, M. A. (2015). Managerial cognitive capabilities and the microfoundations of dynamic capabilities. *Strategic Management Journal, 36*(6), 831–850. DOI: 10.1002/smj.2247

Hetman, K. N. E. U. named after V., Sahaidak, M., Maryniuk, I., & Technology, S. U. of E. and. (2025). Conceptual foundations for the formation and development of organizational compliance management systems under digital transformation. In *Integration of Marketing into the Public Administration System: Ukrainian Experience and Prospects* (pp. 69–102). Scientific Center of Innovative Research. DOI: 10.36690/IM-PAS-69-102

He, X., Zheng, X., & Ding, H. (2023). Existing barriers faced by and future design recommendations for direct-to-consumer health care artificial intelligence apps: Scoping review. *Journal of Medical Internet Research, 25*, e50342. DOI: 10.2196/50342 PMID: 38109173

Hodgkinson, G. P. (2001). Cognitive Processes in Strategic Management : Some Emerging. *Handbook of Industrial, Work & Organizational Psychology: Volume 2: Organizational Psychology*, 416.

Hollan, J., Hutchins, E., & Kirsh, D. (2000). Distributed cognition : Toward a new foundation for human-computer interaction research. *ACM Transactions on Computer-Human Interaction, 7*(2), 174–196. DOI: 10.1145/353485.353487

Hosseini, S., & Seilani, H. (2025). The role of agentic AI in shaping a smart future : A systematic review. *Array (New York, N.Y.), 26*, 100399. DOI: 10.1016/j.array.2025.100399

Hrebiniak, L. G., & Joyce, W. F. (1985). Organizational Adaptation : Strategic Choice and Environmental Determinism. *Administrative Science Quarterly, 30*(3), 336–349. DOI: 10.2307/2392666

Huang, F., Luo, X., Xiao, M., & Dong, W. (2025). Dual performance of business model innovation in emerging market enterprises: A configurational approach. *Journal of Innovation & Knowledge, 10*(4), 100755. Advance online publication. DOI: 10.1016/j.jik.2025.100755

Huang, K. (Ed.). (2025). *Agentic AI : Theories and Practices*. Springer Nature Switzerland., DOI: 10.1007/978-3-031-90026-6

Hughes, L., Dwivedi, Y. K., Li, K., Appanderanda, M., Al-Bashrawi, M. A., & Chae, I. (2025). AI Agents and Agentic Systems : Redefining Global it Management. *Journal of Global Information Technology Management, 28*(3), 175–185. DOI: 10.1080/1097198X.2025.2524286

Hughes, L., Mavi, R. K., Aghajani, M., Fitzpatrick, K., Gunaratnege, S. M., Shekarabi, S. A. H., Hughes, R., Khanfar, A., Khatavakhotan, A., Mavi, N. K., Li, K., Mahmoud, M., Malik, T., Mutasa, S., Nafar, F., Yates, R., Alahmad, R., Jeon, I., & Dwivedi, Y. K. (2025). Impact of artificial intelligence on project management (PM): Multi-expert perspectives on advancing knowledge and driving innovation toward PM2030. *Journal of Innovation & Knowledge*, *10*(5), 100772. DOI: 10.1016/j.jik.2025.100772

Hundogan, O., Lu, X., Du, Y., & Reijers, H. A. (2023). CREATED: Generating viable counterfactual sequences for predictive process analytics. In *International Conference on Advanced Information Systems Engineering*. DOI: 10.1007/978-3-031-34560-9_32

Huser, V. (2012). Process mining: Discovery, conformance and enhancement of business processes. *Journal of Biomedical Informatics*, *42*. Advance online publication. DOI: 10.1016/j.jbi.2012.06.007

Hutchins, E. (1995). *Cognition in the Wild*. MIT press. DOI: 10.7551/mitpress/1881.001.0001

Huy, P. Q., & Phuc, V. K. (2025). Unveiling how business process management capabilities foster dynamic decision-making for effectiveness of sustainable digital transformation. *Business Process Management Journal*, *31*(8), 67–103. DOI: 10.1108/BPMJ-06-2024-0467

Hu, Z. (2023). Application and exploration of artificial intelligence large model in digital culture and tourism industry. *Communication and Information Technology*, (4), 104–107.

Ihlenfeldt, S., Wunderlich, T., Süße, M., Hellmich, A., Schenke, C. C., Wenzel, K., & Mater, S. (2021). Increasing Resilience of Production Systems by Integrated Design. *Applied Sciences 2021, Vol. 11, Page 8457, 11*(18), 8457. DOI: 10.3390/app11188457

Ike, P. N., Ogbuefi, E., Nnabueze, S. B., Olatunde-Thorpe, J., Aifuwa, S. E., Oshoba, T. O., & Akokodaripon, D. (2021). Supplier Relationship Management strategies fostering innovation, collaboration, and resilience in global supply chain ecosystems. *International Journal of Multidisciplinary Evolutionary Research*, *2*(2), 52–62. DOI: 10.54660/IJMER.2021.2.2.52-62

Indulska, M., Recker, J., Rosemann, M., & Green, P. (2009). Business process modeling: Current issues and future challenges. https://doi.org/DOI: 10.1007/978-3-642-02144-2_39

Islam, M. A., Somu, S., & Aldaihani, F. M. F. (2026). The Rise of Agentic AI : Synthesis of Current Knowledge and Future Research Agenda. *Global Business and Organizational Excellence*, *45*(3), 402–416. DOI: 10.1002/joe.70019

Islam, R., Hossain, M. N., & Tusar, M. Z. H. (2021). PREDICTIVE ANALYTICS IN SUPPLY CHAIN MANAGEMENT A REVIEW OF BUSINESS ANALYST-LED OPTIMIZATION TOOLS. *Review of Applied Science and Technology*, *6*(1), 34–73. DOI: 10.63125/5aypx555

Jayaram, Y. (2025). AI-powered ECM automation with agentic AI for adaptive, policy-driven content processing pipelines. *International Journal of Artificial Intelligence, Data Science, and Machine Learning*, *6*, 124–134. DOI: 10.63282/3050-9262.IJAIDSML-V6I3P118

Jin, Z. (2018). The System Dynamic Adaptability Concern. *Environment Modeling-Based Requirements Engineering for Software Intensive Systems*, 217–239. DOI: 10.1016/B978-0-12-801954-2.00012-1

Jorzik, P., Klein, S. P., Kanbach, D. K., & Kraus, S. (2024). AI-driven business model innovation: A systematic review and research agenda. *Journal of Business Research*, *182*, 114764. DOI: 10.1016/j.jbusres.2024.114764

Junjia, Y., Jiawen, L., Alias, A. H., Haron, N. A., & Abu Bakar, N. (2025). Reinforcement learning in risk management for pharmaceutical construction projects: Frontiers, challenges, and improvement strategies. In *Sustainable Futures* (Vol. 10). Elsevier Ltd., DOI: 10.1016/j.sftr.2025.101534

Kadłubek, M. (2021). Expectations for the use of Intelligent Transport Systems applications in the management of freight transport enterprises. *Procedia Computer Science*, *192*, 2318–2329. DOI: 10.1016/j.procs.2021.08.222

Kamal, A., Tanvir Ansari, M., & Chapaneri, K. (2025). *Agentic AI –The new frontier in GenAI: An executive playbook*. PricewaterhouseCoopers., https://www.pwc.ch/en/publications/2025/ch-accelerating-process-automation-playbook.pdf

Kanbach, D. K., Heiduk, L., Blueher, G., Schreiter, M., & Lahmann, A. (2024). The GenAI is out of the bottle: Generative artificial intelligence from a business model innovation perspective. *Review of Managerial Science*, *18*(4), 1189–1220. DOI: 10.1007/s11846-023-00696-z

Kantabutra, S., & Ketprapakorn, N. (2021). Toward an Organizational Theory of Resilience: An Interim Struggle. *Sustainability 2021, Vol. 13, Page 13137*, *13*(23), 13137. DOI: 10.3390/su132313137

Kaplan, A., & Haenlein, M. (2020). Rulers of the world, unite! The challenges and opportunities of artificial intelligence. *Business Horizons*, *63*(1), 37–50. DOI: 10.1016/j.bushor.2019.09.003

Karim, D. N., & Asaduzzaman, M. (2026). Linking talent management practices to autonomous work motivation and proactive customer service behavior: Insights from hospital nurses of Bangladesh. *Social Sciences & Humanities Open*, *13*, 102444. Advance online publication. DOI: 10.1016/j.ssaho.2026.102444

Karras, A., Theodorakopoulos, L., Karras, C., Krimpas, G. A., Giannaros, A., & Bakalis, C.-P. (2025). LLM-driven big data management across digital governance, marketing, and accounting: A spark-orchestrated framework. *Algorithms*, *18*(12), 791. DOI: 10.3390/a18120791

Kerrison, S., Jusak, J., & Huang, T. (2023). Blockchain-enabled IoT for rural healthcare: Hybrid-channel communication with digital twinning. *Electronics (Basel)*, *12*(9), 2128. DOI: 10.3390/electronics12092128

Khan, M. Q., Alvi, M. A. H., Nawaz, H. H., & Umar, M. (2025). Impact of Digital Twins on Real Practices in Manufacturing Industries. *Inventions 2025, Vol. 10, Page 106, 10*(6), 106. DOI: 10.3390/inventions10060106

Khatoon, U. T., Babgi, M., Hadi, N. T., Mir, R. N., & Velidandi, A. (2025). Technology-Driven Change in Human Resource Management: Reshaping Talent Management and Organizational Design. *Administrative Sciences 2025, Vol. 15, Page 452, 15*(11), 452. DOI: 10.3390/admsci15110452

Kim, E. J., & Kim, J. (2025). Privacy-by-design framework for large language model chatbots in urology. *International Neurourology Journal*, *29*(Suppl 2), S65–S72. DOI: 10.5213/inj.2550274.137 PMID: 41355258

Kir, H., & Erdogan, N. (2021). A knowledge-intensive adaptive business process management framework. *Information Systems*, *95*, 101639. DOI: 10.1016/j.is.2020.101639

Klessascheck, F., Weber, I., & Pufahl, L. (2025). SOPA: A framework for sustainability-oriented process analysis and re-design in business process management. *Information Systems and e-Business Management*, *23*(2), 489–537. DOI: 10.1007/s10257-024-00695-x

Klingbeil, A., Grützner, C., & Schreck, P. (2024). Trust and reliance on AI — An experimental study on the extent and costs of overreliance on AI. *Computers in Human Behavior*, *160*, 108352. DOI: 10.1016/j.chb.2024.108352

Klun, M., & Trkman, P. (2018). Business process management – at the crossroads. *Business Process Management Journal, 24*(3), 786–813. DOI: 10.1108/BPMJ-11-2016-0226

Ko, R. K. L., Lee, S. S. G., & Lee, E. W. (2009). Business process management (BPM) standards: A survey. *Business Process Management Journal, 15*(5), 744–791. DOI: 10.1108/14637150910987937

Korzynski, P., Edwards, A., Gupta, M. C., Mazurek, G., & Wirtz, J. (2025). Humanoid robotics and agentic AI : Reframing management theories and future research directions. *European Management Journal, 43*(4), 548–560. DOI: 10.1016/j.emj.2025.06.002

Kourani, H., Berti, A., Schuster, D., & van der Aalst, W. M. P. (2025). Evaluating large language models on business process modeling: Framework, benchmark, and self-improvement analysis. *Software & Systems Modeling*, •••, 1–36. DOI: 10.1007/s10270-025-01318-w

Krogstie, J. (2012). Model-based development and evolution of information systems: A quality approach. *Model-Based Development and Evolution of Information Systems: A Quality Approach*, 1–439. https://doi.org/DOI: 10.1007/978-1-4471-2936-3/COVER

Krogstie, J. (2016). *Business process modeling in practice*.

Krúpová, S., & Koman, G. (2026). The Role of IoT and Big Data in Smart City Transport Management: Lessons Learned from Case Studies in Slovakia and the EU. *Transportation Research Procedia, 93*, 948–959. DOI: 10.1016/j.trpro.2025.12.027

Kumar, R., Pillai, S., O'connell, E., & Denny, P. (2026). Role of Unified Namespace (UNS) and Digital Twins in Predictive and Adaptive Industrial Systems. *Machines 2026, Vol. 14, Page 252, 14*(2), 252. DOI: 10.3390/machines14020252

Kumar, B. R. (2025). *Handbook of brand management*.

Kumar, V., & Kotler, P. (2024). A market management approach to transformative business operations. *Marketing Strategy Journal, 1*, 100005. DOI: 10.1016/j.msj.2025.100005

Kurniati, A. P., Johnson, O., Hogg, D., & Hall, G. (2016). Process mining in oncology: A literature review. In *2016 6th International Conference on Information Communication and Management (ICICM)* (pp. 291–297). DOI: 10.1109/INFOCOMAN.2016.7784260

Lawal, O. A., & Oduleye, T. E. (2020). Process automation and financial reporting integrity: A conceptual governance model. *International Journal of Multidisciplinary Research and Growth Evaluation, 1*(5), 583–594. DOI: 10.54660/.IJMRGE.2020.1.5.583-594

Lawter, L., & Garnjost, P. (2025). Green Human Resource Management and Organizational Performance: A Systematic Review †. *Sustainability (Basel), 17*(7), 3132. DOI: 10.3390/su17073132

Lazaropoulos, A. G. (2026). Simulation-Based Analysis of Performance Role Transformation in SLA-Aware BPMN IT Processes: The Enabling Role of AI in Sustainable Process Governance. *Sustainability 2026, Vol. 18, Page 3369, 18*(7), 3369. DOI: 10.3390/su18073369

LeCun, Y., Bengio, Y., & Hinton, G. (2015). Deep learning. *nature, 521*(7553), 436-444.

Ledro, C., Nosella, A., Vinelli, A., Dalla Pozza, I., & Souverain, T. (2025). Artificial intelligence in customer relationship management: A systematic framework for a successful integration. *Journal of Business Research, 199*, 115531. DOI: 10.1016/j.jbusres.2025.115531

Lee, J. W., & Lee, K. (2025). Building a consensus: Harmonizing AI ethical guidelines and legal frameworks in Korea for enhanced governance. *Government Information Quarterly, 42*(3), 102060. Advance online publication. DOI: 10.1016/j.giq.2025.102060

Leno, V., Dumas, M., Maggi, F. M., La Rosa, M., & Polyvyanyy, A. (2020). Automated discovery of declarative process models with correlated data conditions. *Information Systems, 89*, 101482. DOI: 10.1016/j.is.2019.101482

Lewańska, E., & Stróżyna, M. (2025). Enhancing ESG Reporting Through Process Mining: A Taxonomy of ESG Events. *Procedia Computer Science, 270*, 3450–3458. DOI: 10.1016/j.procs.2025.09.470

Leymann, F., & Roller, D. (2001). *Production workflow – Concepts and techniques.*

Li, B., Mellou, K., Zhang, B., Pathuri, J., & Menache, I. (2023). *Large Language Models for Supply Chain Optimization.*

Li, H., Feng Tian, ·, Tian, F., Li, H., & Tian, F. (2026). Advancing Decision-Making through AI-Human Collaboration: A Systematic Review and Conceptual Framework. *Group Decision and Negotiation 2026 35:2, 35*(2), 26-. DOI: 10.1007/s10726-026-09980-1

Li, H., Mohamed El-Amine, H., & Mohamed, H. (2014). Merging several business process variants. In *The 26th Chinese Control and Decision Conference (2014 CCDC)* (pp. 5218–5223). DOI: 10.1109/CCDC.2014.6853112

Liang, K., Ren, X., Yin, H., Jiang, J., Xiang, L., & Dai, W. (2025). An Adaptive Process Monitoring Method Based on an Improved ADALINE Neural Network. *Applied Sciences 2025, Vol. 15, Page 2090, 15*(4), 2090. DOI: 10.3390/app15042090

Liang, L., & Li, Y. (2024). How does organizational resilience promote firm growth? The mediating role of strategic change and managerial myopia. *Journal of Business Research, 177*, 114636. Advance online publication. DOI: 10.1016/j.jbusres.2024.114636

Liang, R., Deng, K., Mao, Z., & Xu, P. (2019). Research on semantic matching method of combat plan based on military knowledge graph. *Journal of Command and Control, 5*(2), 115–120.

Liao, S. M., Haykel, I., Cheung, K., & Matalon, T. (2025). Navigating the complexities of AI and digital governance: The 5W1H framework. *Journal of Responsible Technology, 23*, 100127. DOI: 10.1016/j.jrt.2025.100127

Licardo, J. T., Tanković, N., & Etinger, D. (2026). BPMN Assistant: An LLM-Based Approach to Business Process Modeling. *Applied Sciences 2026, Vol. 16, Page 2213, 16*(5), 2213. DOI: 10.3390/app16052213

Liedtka, J. (2015). Why design thinking works. *Harvard Business Review, 97*(5), 72–79.

Li, M., Zhu, Y., Chen, F., & Anaba, O. A. (2026). Research on the multiple promotion paths of green innovation performance in Chinese high-end equipment manufacturing enterprises: A configuration study based on TOE framework. *European Research on Management and Business Economics, 32*(1), 100300. DOI: 10.1016/j.iedeen.2026.100300

Lin, K. Y., Li, M. H., Lo, F., Huang, H.-C., Matsuno, K., & Watanabe, R. (2025). Adaptive learning with human factors and Artificial Intelligence: Associations with training effectiveness in programming education. *International Journal of Industrial Ergonomics, 110*, 103834. DOI: 10.1016/j.ergon.2025.103834

Lin, Z. (2026). Large Language Models as Psychological Simulators: A Methodological Guide. *Advances in Methods and Practices in Psychological Science, 9*(1), 25152459251410153. Advance online publication. DOI: 10.1177/25152459251410153

Li, P., Chen, Y., & Guo, X. (2025). Digital transformation and supply chain resilience. *International Review of Economics & Finance, 99*, 104033. DOI: 10.1016/j.iref.2025.104033

Liu, O., Fu, D., Yogatama, D., & Neiswanger, W. (2024). *DeLLMa: Decision Making Under Uncertainty with Large Language Models.*

Liu, S., Zhang, F., & Wang, L. (2021). The application of knowledge graphs in healthcare decision-making. *Healthcare Informatics Research, 27*(4), 239–245.

Liu, X., Vasantha, G., Goh, K., Lin, W., & Marzano, A. (2026). Data-driven smart product design, smart service design, and smart product-service system design – A comprehensive review. *Advanced Engineering Informatics, 73*, 104534. DOI: 10.1016/j.aei.2026.104534

Liu, Z., Hansen, D. W., & Chen, Z. (2023). Leveraging digital twins to support industrial symbiosis networks: A case study in the Norwegian wood supply chain collaboration. *Sustainability (Basel), 15*(3), 2647. DOI: 10.3390/su15032647

Liu, Z., & Yin, X. (2025). A Review of Cognitive Control : Advancement, Definition, Framework, and Prospect. *Actuators, 14*(1), 32. DOI: 10.3390/act14010032

Li, W., & Li, S. (2023). Design of digital twin watershed knowledge platform based on big language model and knowledge graph. *Haihe Water Conservancy*, (12), 75–78.

Li, X., & Li, Y. (2022). Predictive modeling in product prototyping: A machine learning perspective. *Journal of Engineering Design, 33*(5), 431–448. DOI: 10.1080/09544828.2022.2034567

Lübke, D., Ahrens, M., & Schneider, K. (2021). Influence of diagram layout and scrolling on understandability of BPMN processes: An eye tracking experiment with BPMN diagrams. *Information Technology and Management, 22*(2), 1–33. DOI: 10.1007/s10799-021-00327-7

Luthans, F., & Stewart, T. I. (1977). A General Contingency Theory of Management. *Academy of Management Review, 2*(2), 181. DOI: 10.2307/257902

Lyell, D., Coiera, E., Chen, J., Shah, P., & Magrabi, F. (2021). How machine learning is embedded to support clinician decision making: An analysis of FDA-approved medical devices. *BMJ Health & Care Informatics, 28*(1), e100301. DOI: 10.1136/bmjhci-2020-100301 PMID: 33853863

Ma, H., Wang, S., Gu, J., Xie, Q., Wang, W., Dong, S., Sun, C., Li, J., Ma, C., Dong, J., & Liu, H. (2026). IoT-based smart environmental control and air emission management in animal farming: A systematic review. *Information Processing in Agriculture*. Advance online publication. DOI: 10.1016/j.inpa.2026.02.007

Malki, A., Atlam, E. S., & Gad, I. (2022). Machine learning approach of detecting anomalies and forecasting time-series of IoT devices. *Alexandria Engineering Journal*, *61*(11), 8973–8986. DOI: 10.1016/j.aej.2022.02.038

Manning, L., Brewer, S., Craigon, P. J., Frey, J., Gutierrez, A., Jacobs, N., Kanza, S., Munday, S., Sacks, J., & Pearson, S. (2023). Reflexive governance architectures: Considering the ethical implications of autonomous technology adoption in food supply chains. In *Trends in Food Science and Technology* (Vol. 133, pp. 114–126). Elsevier Ltd., DOI: 10.1016/j.tifs.2023.01.015

March, J. G. (1991). Exploration and Exploitation in Organizational Learning. *Organization Science*, *2*(1), 71–87. DOI: 10.1287/orsc.2.1.71

Markaj, A., Mercangöz, M., & Fay, A. (2024). Design and implementation of an Autonomous Systems Training Environment framework for control algorithm evaluation in autonomous plant operation. *Computers & Chemical Engineering*, *189*, 108798. DOI: 10.1016/j.compchemeng.2024.108798

Ma, T., Wang, P., Yang, D., & Wang, Z. (2025). Towards inclusive conservation governance in forest landscape co-management: Insights from the Yarlung Tsangpo river basin. *Trees, Forests and People*, *19*, 100762. Advance online publication. DOI: 10.1016/j.tfp.2024.100762

McCarthy, J., Minsky, M., Rochester, N., & Shannon, C. (1955). *A proposal for the Dartmouth summer research project on artificial intelligence*.

Mendling, J., Reijers, H. A., & Aalst, W. (2010). Seven process modeling guidelines (7PMG). *Information and Software Technology, 52*, 127–136. *Handbook on business process management 1: Introduction, methods, and information systems.* (2010).DOI: 10.1016/j.infsof.2009.08.004

Mendonca, R. S., da Silva, M., Ayres, F. A. C.Jr, Bessa, I. V., Medeiros, R. L. P., & Lucena, V. F.Jr. (2025). Development of a novel retrofit framework considering Industry 4.0 concepts: A case study of a modular production system. *Processes (Basel, Switzerland)*, *13*(1), 136. DOI: 10.3390/pr13010136

Miedema, E., Waschull, S., & Emmanouilidis, C. (2026). Towards trustworthy artificial intelligence for decision-making: A lifecycle perspective on knowledge- and data-driven artificial intelligence systems. *Computers in Industry*, *174*, 104409. DOI: 10.1016/j.compind.2025.104409

Mikalef, P., Islam, N., Parida, V., Singh, H., & Altwaijry, N. (2023). Artificial intelligence (AI) competencies for organizational performance: A B2B marketing capabilities perspective. *Journal of Business Research*, *164*, 113998. DOI: 10.1016/j.jbusres.2023.113998

Miles, R. E., Snow, C. C., & Pfeffer, J. (1974). Organization-Environment [Object Identifier] [DOI]. *Concepts and Issues.*, *13*(3), 244–264. DOI: 10.1111/j.1468-232X.1974.tb00581.xDigital

Mintzberg, H. (2009). *Managing*. Berrett-Koehler Publishers.

Mishra, S. S., Panwar, D., Chouhan, R. P. S., & Shanmugapriya, V. (2024). The Internet of Things and its impact on EV charging stations. In *2024 2nd International Conference on Advancements and Key Challenges in Green Energy and Computing (AKGEC)* (pp. 1–6). IEEE. https://doi.org/DOI: 10.1109/AKGEC62572.2024.10868079

Mishra, S. S., Sharma, S., Sharma, Y., & Singh, M. (2026). Artificial intelligence for climate action and sustainable transitions: Conceptual frameworks, governance models, and responsible innovation pathways. In *Responsible AI for climate action: Governance, justice, and sustainable transitions*. IGI Global., DOI: 10.4018/979-8-2600-2020-3.ch005

Mitchell, T. M. (1997). Does machine learning really work? *AI Magazine*, *18*(3), 11–11.

Mittelstadt, B. D., Allo, P., Taddeo, M., Wachter, S., & Floridi, L. (2016). The ethics of algorithms : Mapping the debate. *Big Data & Society*, *3*(2), 2053951716679679. DOI: 10.1177/2053951716679679

Mohammed, B. (2026). The sentient infrastructure paradigm: A review of novel sensors, digital twins, and AI for civil structural health monitoring. *Sensor Review*, •••, 1–19. DOI: 10.1108/SR-10-2025-0822

Mollahosseini, A., Hasani, B., & Mahoor, M. H. (2019). AffectNet: A database for facial expression, valence, and arousal computing in the wild. *IEEE Transactions on Affective Computing*, *10*(1), 18–31. DOI: 10.1109/TAFFC.2017.2740923

Moreno-Sánchez, P. A., Del Ser, J., van Gils, M., & Hernesniemi, J. (2026). A design framework for operationalizing trustworthy artificial intelligence in healthcare: Requirements, tradeoffs and challenges for its clinical adoption. *Information Fusion, 127*, 103812. DOI: 10.1016/j.inffus.2025.103812

Mosallam, A. S., & Mouhim, S. (2025). Contextual Modeling and Intelligent Decision-Making for IoT Systems: A Combined Ontology and Machine Learning Approach. *Engineering Proceedings 2025, Vol. 112, Page 71, 112*(1), 71. DOI: 10.3390/engproc2025112071

Mossa, Y., Smith, P., & Bland, K. (2025). Reconceptualizing Enterprise Resource Planning (ERP) Systems from a Software Architecture Perspective Using a Framework Based on ERP System Characteristics. *Procedia Computer Science, 256*, 174–189. DOI: 10.1016/j.procs.2025.02.110

Mpanza, S. S. (2025). Revisiting the Technological-Organizational-Environmental (TOE) Framework and Diffusion of Innovation (DOI) : A Theoretical Review for Artificial Intelligence (AI) Adoption. *International Journal of Applied Research in Business and Management, 6*(5). Advance online publication. https://www.wr-publishing.org/index.php/ijarbm/article/view/441. DOI: 10.51137/wrp.ijarbm.441

Mukherjee, A., & Chang, H. H. (2025). *Agentic AI : Autonomy, Accountability, and the Algorithmic Society* (arXiv:2502.00289). arXiv. /arXiv.2502.00289DOI: 10.2139/ssrn.5123621

Munnangi, S. (2018). Seamless Automation: Integrating BPM and RPA with Pega. *Turkish Journal of Computer and Mathematics Education, 9*(3). Advance online publication. DOI: 10.61841/turcomat.v9i3.14971

Murire, O. T. (2024). Artificial Intelligence and Its Role in Shaping Organizational Work Practices and Culture. *Administrative Sciences, 14*(12), 316. DOI: 10.3390/admsci14120316

Mutschler, B., & Reichert, M. (2013). Understanding the costs of business process management technology. In *Studies in Computational Intelligence* (pp. 157–194). Springer Berlin Heidelberg., DOI: 10.1007/978-3-642-28409-0_7

Najana, M., Bhattacharya, S., Kewalramani, C., & Pandiya, D. K. (2024). AI and organizational transformation : Navigating the future. *International Journal of Global Innovations and Solutions (IJGIS)*. https://ijgis.pubpub.org/pub/u5r681dy

Nangi, P. R., & Obannagari, C. K. R. N. (2025). AI-driven security automation for continuous compliance monitoring in regulated cloud environments. *International Journal of Emerging Trends in Computer Science and Information Technology, 6*, 95–105. DOI: 10.63282/3050-9246.IJETCSIT-V6I2P112

Nangi, P. R., Obannagari, C. K. R. N., & Settipi, S. (2022). Self-auditing deep learning pipelines for automated compliance validation with explainability, traceability, and regulatory assurance. *International Journal of Artificial Intelligence, Data Science, and Machine Learning, 3*, 133–142. DOI: 10.63282/3050-9262.IJAIDSML-V3I1P114

Napier, E., Liu, S. Y. H., & Liu, J. (2024). Adaptive strength: Unveiling a multilevel dynamic process model for organizational resilience. *Journal of Business Research, 171*, 114334. DOI: 10.1016/j.jbusres.2023.114334

National Institute of Standards and Technology. (2023). *Artificial Intelligence Risk Management Framework (AI RMF 1.0)*. DOI: 10.6028/NIST.AI.100-1

National Institute of Standards and Technology. (2024). *Artificial Intelligence Risk Management Framework: Generative Artificial Intelligence Profile*. DOI: 10.6028/NIST.AI.600-1

Natta, P. K.BENTONVILLE USA. (2025). Scalable governance frameworks for AI-driven enterprise automation and decision-making. *International Journal of Research Publications in Engineering. Technology and Management, 8*(6). Advance online publication. DOI: 10.15662/IJRPETM.2025.0806022

Naveed, H., Khan, A. U., Qiu, S., Saqib, M., Anwar, S., Usman, M., Akhtar, N., Barnes, N., & Mian, A. (2025). A Comprehensive Overview of Large Language Models. *ACM Transactions on Intelligent Systems and Technology, 16*(5), 1–72. Advance online publication. DOI: 10.1145/3744746

Nayebi, F. (2025). *Foundations of Agentic AI for Retail : Concepts, Technologies, and Architectures for Autonomous Retail Systems (1re* (Divergence, G., Ed.).

Niederman, F. (2021). Process Theory: Background, Opportunity, and Challenges. *Foundations and Trends in Information Systems, 5*(1), 1–230. DOI: 10.1561/2900000017

Niehaves, B., Plattfaut, R., & Becker, J. (2012). Business process governance: A comparative study of Germany and Japan. *Business Process Management Journal, 18*(2), 347–371. DOI: 10.1108/14637151211225234

Nikulina, I. E., & Khomenko, I. V. (2015). Cognitive Management : Theory and Practice in the Organization. *Procedia - Social and Behavioral Sciences, Proceedings of The International Conference on Research Paradigms Transformation in Social Sciences 2014 (RPTSS-2014), 166*, 441-445. DOI: 10.1016/j.sbspro.2014.12.550

Nisa, U., Shirazi, M., Saip, M. A., & Pozi, M. S. M. (2026). Agentic AI : The age of reasoning—A review. *Journal of Automation and Intelligence*, *5*(1), 69–89. DOI: 10.1016/j.jai.2025.08.003

Njima, C., Ghedira, C., Gamha, Y., & Romdhane, L. (2024). Web service composition in mobile environment: A survey of techniques. *IEEE Transactions on Services Computing*, *17*(2), 1–16. DOI: 10.1109/TSC.2024.3361655

Noetzold, D., Leithardt, V. R. Q., Santana, J. F. de P., & Barbosa, J. L. V. (2026). Oraculum: A model for self-adaptive system optimization in smart environments. *Expert Systems with Applications*, *315*, 131705. DOI: 10.1016/j.eswa.2026.131705

Nonaka, I., & Takeuchi, H. (1995). *The Knowledge-Creating Company : How Japanese Companies Create the Dynamics of Innovation*. Oxford University Press., DOI: 10.1093/oso/9780195092691.001.0001

Nosova, S., Norkina, A., Makar, S., Gerasimenko, T., & Medvedeva, O. (2022). Artificial intelligence as a driver of business process transformation. *Procedia Computer Science, 2022 Annual International Conference on Brain-Inspired Cognitive Architectures for Artificial Intelligence: The 13th Annual Meeting of the BICA Society, 213*, 276-284. DOI: 10.1016/j.procs.2022.11.067

Nwachukwu, P. S., Chima, O. K., & Okolo, C. H.Priscilla Samuel NwachukwuOnyeka Kelvin ChimaChinelo Harriet Okolo. (2025). The artificial intelligence governance framework for finance: A control-by-design approach to algorithmic decision-making in accounting. *Finance & Accounting Research Journal*, *7*(8), 350–379. DOI: 10.51594/farj.v7i8.2016

Odogwu, R., Ogeawuchi, J. C., Abayomi, A. A., Agboola, O. A., & Owoade, S. (2022). Optimizing productivity in asynchronous remote project teams through. *International Journal of Multidisciplinary Research and Growth Evaluation*, *3*(4), 628–634. DOI: 10.54660/.IJMRGE.2022.3.4.628-634

OECD. (2019). Artificial Intelligence in Society. In *Artificial Intelligence in Society*. OECD Publishing., DOI: 10.1787/eedfee77-en

Okoruwa, P. O., Babatope, O. M., Akokodaripon, D. A., & Akinleye, O. K. (2024). Developing integrated digital platforms for enhancing transparency in procurement and supply chain management. *International Journal of Multidisciplinary Research and Growth Evaluation, 5*(6), 1719–1729. DOI: 10.54660/.IJMRGE.2024.5.6.1719-1729

Okunlola, F.New Mexico Black Leadership Council United States. (2025). A cyber-physical framework for data assurance and emergency response readiness in critical energy infrastructure. *Journal of Engineering and Applied Sciences (Asian Research Publishing Network), 4*(1), 1–10. DOI: 10.70560/6ag3wg21

Oluwafemi, I. O., Clement, T., Adanigbo, O. S., Gbenle, T. P., & Adekunle, B. I. (2021). Evaluating the efficacy of DID chain-enabled blockchain frameworks for real-time provenance verification and anti-counterfeit control in global pharmaceutical supply chains. *International Journal of Multidisciplinary Research and Growth Evaluation, 2*(2), 436–444. DOI: 10.54660/.IJMRGE.2021.2.2.436-444

Ong, H. J., Lim, B. G., Tan, R., & Ikeda, K. (2025). Causal discovery in additive noise models using beam search. *Artificial Life and Robotics, 31*(1), 317–326. Advance online publication. DOI: 10.1007/s10015-025-01095-1

Oukharijane, J., Chaâbane, M. A., Ben Said, I., Andonoff, E., & Bouaziz, R. (2025). An assessment taxonomy for self-adaptation business process solutions. *Data & Knowledge Engineering, 155*, 102374. DOI: 10.1016/j.datak.2024.102374

Ousghir, S., Benhmama, A., & Sabiri, B. (2024). HR Team Performance in the Realm of AI-HRM : A Case Study of Moroccan Small and Medium Companies. *International Conference on Digital Technologies and Applications*, 202-211. https://link.springer.com/chapter/10.1007/978-3-031-68650-4_20

Pahwa, N., & Gupta, K. (2025). Using artificial intelligence to forecast consumer trends in food consumption. In *Generative AI in food systems: Predictive demand, smart supply chains, and sustainable service futures*. IGI Global., DOI: 10.4018/979-8-3373-5182-7.ch004

Paladugu, N. (2025). Intelligent data governance frameworks for multi-cloud financial environments: An AI-driven approach to compliance automation. *European Modern Studies Journal, 9*(4), 1304–1315. DOI: 10.59573/emsj.9(4).2025.121

Pandit, S., Ravikumar, R. N., Joshi, N., Pandit, S., Kishore, S., & Aarthi, S. (2025). Retrieval-Augmented Generation (RAG) for enterprise intelligence. In *Leveraging LLMs for Business Innovation* (pp. 69–106). IGI Global Scientific Publishing. DOI: 10.4018/979-8-3373-3912-2.ch003

Pandit, S., Ravikumar, R. N., Pandit, S., Joshi, N., Kishore, S., & Aarthi, S. (2025). Explainable Federated Learning for secure and transparent wireless edge networks. In *Advances in Computational Intelligence and Robotics* (pp. 127–168). IGI Global Scientific Publishing. DOI: 10.4018/979-8-3373-3576-6.ch004

Panichella, S., & Zaugg, N. (2020). An Empirical Investigation of Relevant Changes and Automation Needs in Modern Code Review. *Empirical Software Engineering*, *25*(6), 4833–4872. DOI: 10.1007/s10664-020-09870-3

Panwar, D., Mishra, S. S., & Chouhan, R. P. S. (2024). A detailed examination of solar photovoltaic technology, design, efficiency improvement and applications. In *2024 International Conference on Communication, Computing and Energy Efficient Technologies (I3CEET)* (pp. 1709–1714). IEEE. https://doi.org/DOI: 10.1109/I3CEET61722.2024.10994009

Papagiannidis, E., Mikalef, P., & Conboy, K. (2025). Responsible artificial intelligence governance: A review and research framework. *The Journal of Strategic Information Systems*, *34*(2), 101885. DOI: 10.1016/j.jsis.2024.101885

Parasuraman, R., & Riley, V. (1997). Humans and Automation : Use, Misuse, Disuse, Abuse. *Human Factors*, *39*(2), 230–253. DOI: 10.1518/001872097778543886

Park, M. J. (2026). AI as a cognitive collaborator : Assimilation and accommodation in human–machine teaming for innovation. *Journal of Innovation & Knowledge*, *12*, 100892. DOI: 10.1016/j.jik.2025.100892

Parnell, J. A., Wright, P., & Tu, H. S. (1996). Beyond the Strategy-Performance Linkage : The Impact of the Strategy-Organization-Environment Fit on Business Performance. *American Business Review*, *14*(2), 41.

Pasas-Farmer, S., & Jain, R. (2025). From discovery to delivery: Governance of AI in the pharmaceutical industry. *Green Analytical Chemistry*, *13*, 100268. Advance online publication. DOI: 10.1016/j.greeac.2025.100268

Pasi, B. N., Dhamak, P. S., Rane, S. B., Todkari, V. C., & Mishra, A. R. (2026). Dual impact of AI on business processes: Fuzzy TOPSIS prioritization and theoretical framework for strategic adoption. *Business Process Management Journal*, •••, 1–33. DOI: 10.1108/BPMJ-11-2025-1896

Päßler, J., ter Beek, M. H., Damiani, F., Johnsen, E. B., & Tapia Tarifa, S. L. (2025). Analysing Self-Adaptive Systems as Software Product Lines. *Journal of Systems and Software*, *222*, 112324. DOI: 10.1016/j.jss.2024.112324

Pastakia, K., Scotis, S., Scoble-Williams, N., Elliott, S., & Lettau, M. (2025). *Work, reworked : What it takes to win in the age of Agentic AI.* Deloitte., https://www.deloitte.com/cbc/en/services/consulting/perspectives/human-agentic-workforce.html

Patel, J. (2025). Exploring the role of agentic AI in enhancing Human-AI collaboration. *International Journal of Engineering Technology Research & Management, 9*(8), 51–61. DOI: 10.5281/zenodo.16739658

Patel, R., & Mehta, A. (2023). Explainable AI in human-centered design. *Journal of Artificial Intelligence Research, 72*, 345–362.

Pelosi, D., Cacciagrano, D., & Piangerelli, M. (2025). Explainability and Interpretability in Concept and Data Drift: A Systematic Literature Review. *Algorithms 2025, Vol. 18, Page 443, 18*(7), 443. DOI: 10.3390/a18070443

Petropoulos, F., Akkermans, H., Aksin, O. Z., Ali, I., Babai, M. Z., Barbosa-Povoa, A., Battaïa, O., Besiou, M., Boysen, N., Brammer, S., Brandon-Jones, A., Briskorn, D., Browning, T. R., Buijs, P., Centobelli, P., Chiarini, A., Cousins, P., Cudney, E. A., Davies, A., & Zheng, T. (2026). Operations & supply chain management: Principles and practice. *International Journal of Production Research, 64*(1), 330–513. DOI: 10.1080/00207543.2025.2555531

Petrovska, A., Hutzelmann, T., & Kugele, S. (2023). A Theoretical Framework for Self-Adaptive Systems: Specifications, Formalisation, and Architectural Implications. *Proceedings of the ACM Symposium on Applied Computing, 23*, 1440–1449. DOI: 10.1145/3555776.3577665

Peykani, P., Ramezanlou, F., Tanasescu, C., & Ghanidel, S. (2025). Large Language Models: A Structured Taxonomy and Review of Challenges, Limitations, Solutions, and Future Directions. *Applied Sciences 2025, Vol. 15, Page 8103, 15*(14), 8103. DOI: 10.3390/app15148103

Pitelis, C. N., Teece, D. J., & Yang, H. (2024). Dynamic Capabilities and MNE Global Strategy: A Systematic Literature Review-Based Novel Conceptual Framework. *Journal of Management Studies, 61*(7), 3295–3326. DOI: 10.1111/JOMS.13021;-JOURNAL:JOURNAL:14676486;WGROUP:STRING:PUBLICATION

Plattfaut, R., & Grisold, T. (2025a). Business process management as a continuous enabler for digital innovation – insights from top executives. *Business Process Management Journal, 31*(8), 244–266. DOI: 10.1108/BPMJ-05-2025-0759

Porter, M. E., & Heppelmann, J. E. (2017). Why every organization needs an augmented reality strategy. *Harvard Business Review, 95*(6), 46–57.

Pramod, D., Patil, K. P., Bharathi, S. V., Vijaykumar, B. V., Birabar, S., & Sahoo, B. (2024). Artificial intelligence and machine learning adoption in automotive organizations using technology-organization-environment framework. *Journal of Information and Optimization Sciences*, *45*(7), 1963–1976. DOI: 10.47974/JIOS-1748

Qin, S., Zheng, Z., Gu, Y., & Lu, X. (2023). Application test and discussion of large language model in construction engineering. *Industrial Architecture*, *53*(9), 162–169.

Raina, K., Sharma, G. D., Taheri, B., Dev, D., & Chavriya, S. (2026). Artificial intelligence-driven management: Bridging innovation, knowledge creation, and sustainable business practices. *Journal of Innovation & Knowledge*, *11*, 100860. DOI: 10.1016/j.jik.2025.100860

Raisch, S., & Krakowski, S. (2021). Artificial Intelligence and Management : The Automation–Augmentation Paradox. *Academy of Management Review*, *46*(1), 192–210. DOI: 10.5465/amr.2018.0072

Ramalingam, R., Arthi, K., Monica Bhavani, M., & Sunitha, T. (2025). AI-Enhanced Security Information and Event Management (SIEM) System. In Kumar, R., Peng, S., Jain, P., & Elngar, A. (Eds.), *Deep Learning Innovations for Securing Critical Infrastructures* (pp. 75–94). IGI Global Scientific Publishing. [REMOVED HYPERLINK FIELD] DOI: 10.4018/979-8-3373-0563-9.ch005

Raman, R., Mandal, S., Gunasekaran, A., Papadopoulos, T., & Nedungadi, P. (2025). Transforming business management practices through metaverse technologies: A Machine Learning approach. In *International Journal of Information Management Data Insights* (Vol. 5, Number 1). Elsevier B.V. DOI: 10.1016/j.jjimei.2025.100335

Rame, R., Purwanto, P., & Sudarno, S. (2024). Industry 5.0 and sustainability: An overview of emerging trends and challenges for a green future. *Innovation and Green Development*, *3*(4), 100173. DOI: 10.1016/j.igd.2024.100173

Ramu, D. S., & Bansal, P. (2025). A study on AI's Transformative Impact on Strategic Decision-Making. *International Journal of Science & Technoledge*, *16*(2), 6335. Advance online publication. DOI: 10.71097/IJSAT.v16.i2.6335

Ransbotham, S., Kiron, D., Khodabandeh, S., Iyer, S., & Das, A. (2025). *The Emerging Agentic Enterprise : How Leaders Must Navigate a New Age of AI*. MIT Sloan Management Review & Boston Consulting Group., DOI: 10.63383/jAXE2583

Rathinavel, P., Ravikumar, R. N., Aarthi, S., Habib, M., Salaeva, N., & Pardaev, J. (2025). Augmenting human capabilities. In *Advances in Computational Intelligence and Robotics* (pp. 1–30). IGI Global Scientific Publishing. DOI: 10.4018/979-8-3373-2372-5.ch001

Rauch, H., Henry, C., Mühlhaus, M., Kurz, F., Hellekes, J., & Merkle, N. (2025). Large-scale mapping of urban parking from aerial images: A case study in Berlin, Germany. *International Archives of the Photogrammetry, Remote Sensing and Spatial Information Sciences*, XLVIII-4/W16-2025, 83–89. DOI: 10.5194/isprs-archives-XLVIII-4-W16-2025-83-2025

Ravikumar, & Aarthi, S. (2025a). Conscious AI Integration in business education. In *Advances in Computational Intelligence and Robotics* (pp. 73–104). IGI Global Scientific Publishing. DOI: 10.4018/979-8-3373-4377-8.ch003

Ravikumar, R. N., & Aarthi, S. (2025a). Federated Learning for cross-enterprise supply chain collaboration. In *Advances in Computational Intelligence and Robotics* (pp. 47–84). IGI Global Scientific Publishing. DOI: 10.4018/979-8-3373-2833-1.ch002

Ravikumar, R. N., & Aarthi, S. (2025b). AI-driven digital twins and autonomous operations for sustainable logistics and procurement. In *Advances in Computational Intelligence and Robotics* (pp. 65–100). IGI Global Scientific Publishing. DOI: 10.4018/979-8-3373-7006-4.ch003

Ravikumar, R. N., & Aarthi, S. (2025c). Human-AI synergy for organizational resilience. In *Advances in Computational Intelligence and Robotics* (pp. 287–322). IGI Global Scientific Publishing. DOI: 10.4018/979-8-3373-7416-1.ch011

Ravikumar, R. N., & Aarthi, S. (2025d). Real-time IoT, quantum advances, and ethical implications in autonomous AI systems. In *Advances in Computational Intelligence and Robotics* (pp. 185–216). IGI Global Scientific Publishing. DOI: 10.4018/979-8-3373-3146-1.ch007

Ravikumar, R. N., & Aarthi, S. (2025d). Towards collaborative intelligence. In *Advances in Computational Intelligence and Robotics* (pp. 1–32). IGI Global Scientific Publishing. DOI: 10.4018/979-8-3373-5747-8.ch001

Ravikumar, R. N., Aarthi, S., & Mamadiyarov, Z. (2026). Building the foundations of augmented organizations through AI empowerment. In *Advances in Computational Intelligence and Robotics* (pp. 101–130). IGI Global Scientific Publishing. DOI: 10.4018/979-8-3373-8607-2.ch006

Ravikumar, R. N., & Aarthi, S. (2025b). Fine-tuning large Language Models for domain-specific business applications. In *Leveraging LLMs for Business Innovation* (pp. 107–136). IGI Global Scientific Publishing., DOI: 10.4018/979-8-3373-3912-2.ch004

Ravikumar, R. N., & Aarthi, S. (2025c). Digital Twin applications in urban planning. In *Building Information Modeling and Geographic Information Systems for Next-Generation Spatial Intelligence* (pp. 29–58). IGI Global Scientific Publishing., DOI: 10.4018/979-8-3373-4058-6.ch002

Ravikumar, R. N., Aarthi, S., Dilshodbek, A., Satheesh Kumar, A., Eshkaraev, S., & Shanmugam, K. B. (2025). Integrating blockchain, smart contracts, and AI for decentralized healthcare in the metaverse. In *Blockchain-Based Healthcare Management in the Metaverse* (pp. 261–294). IGI Global Scientific Publishing., DOI: 10.4018/979-8-3373-1225-5.ch010

Ravikumar, R. N., Aarthi, S., Pardaev, J., & Shukla, P. (2025). Quantum computing for financial modelling. In *Quantum-Driven Financial Intelligence* (pp. 221–252). IGI Global Scientific Publishing., DOI: 10.4018/979-8-3373-2893-5.ch010

Ray, P. P., & Pradhan, M. P. (2025). An evaluation framework for measuring prompt wise metrics for large language models in resource-constrained edge. *BenchCouncil Transactions on Benchmarks. Standards and Evaluations*, *5*(4), 100249. DOI: 10.1016/j.tbench.2025.100249

Reichert, M., & Weber, B. (2012). Enabling flexibility in process-aware information systems. DOI: 10.1007/978-3-642-30409-5_11

Reicherts, L., Zhang, Z. T., Von Oswald, E., Liu, Y., Rogers, Y., & Hassib, M. (2025). AI, Help Me Think—but for Myself : Assisting People in Complex Decision-Making by Providing Different Kinds of Cognitive Support. *Proceedings of the 2025 CHI Conference on Human Factors in Computing Systems*, 1-19. DOI: 10.1145/3706598.3713295

Reijers, H., & Aalst, W. (2005). The effectiveness of workflow management systems: Predictions and lessons learned. *International Journal of Information Management*, *25*(5), 458–472. DOI: 10.1016/j.ijinfomgt.2005.06.008

Relich, M. (2023). Predictive and Prescriptive Analytics in Identifying Opportunities for Improving Sustainable Manufacturing. *Sustainability 2023, Vol. 15, Page 7667*, *15*(9), 7667. DOI: 10.3390/su15097667

Renugadevi, R. (2025). An Optimal Method of Assessing the Quality of Water Using Integrated AI and Blockchain Methodology. In Ruiz-Vanoye, J., & Díaz-Parra, O. (Eds.), *Smart Water Technology for Sustainable Management in Modern Cities* (pp. 261–284). IGI Global Scientific Publishing. DOI: 10.4018/979-8-3693-8074-1.ch011

Renugadevi, R. (2025). The AI Evolution in Nuclear Medicine by Enhancing Precision and Personalization. In Satishkumar, D., & Sivaraja, M. (Eds.), *AI Insights on Nuclear Medicine* (pp. 355–376). IGI Global Scientific Publishing. DOI: 10.4018/979-8-3373-1275-0.ch017

Renugadevi, R., Muthumeenakshi, R., & Karthika, G. (2025). Empowering Cybersecurity with LLMs: Overcoming Automation Challenges in Threat Intelligence and Detection Systems. In Zangana, H., Al-Karaki, J., & Omar, M. (Eds.), *Revolutionizing Cybersecurity With Deep Learning and Large Language Models* (pp. 237–270). IGI Global Scientific Publishing. [REMOVED HYPERLINK FIELD] DOI: 10.4018/979-8-3373-3296-3.ch008

Reuter, E. (2024). *Cognitive Capacity in the Management and Organization Research : A Review and Agenda for Future Research* (SSRN Scholarly Paper 5040625). Social Science Research Network. DOI: 10.4018/979-8-3693-5777-4.ch001

Ritala, P., Aaltonen, P., Ruokonen, M., & Nemeh, A. (2024). Developing industrial AI capabilities: An organisational learning perspective. *Technovation, 138*, 103120. DOI: 10.1016/j.technovation.2024.103120

Röglinger, M., Plattfaut, R., Borghoff, V., Kerpedzhiev, G., Becker, J., Beverungen, D., vom Brocke, J., Van Looy, A., del-Río-Ortega, A., Rinderle-Ma, S., Rosemann, M., Santoro, F. M., & Trkman, P. (2022). Exogenous Shocks and Business Process Management: A Scholars' Perspective on Challenges and Opportunities. *Business & Information Systems Engineering*, *64*(5), 669–687. DOI: 10.1007/s12599-021-00740-w

Romeo, G., & Conti, D. (2025). Exploring automation bias in human–AI collaboration : A review and implications for explainable AI. *AI & Society*. Advance online publication. DOI: 10.1007/s00146-025-02422-7

Rosemann, M., Brocke, J. vom, Van Looy, A., & Santoro, F. (2024). Business process management in the age of AI – three essential drifts. *Information Systems and E-Business Management 2024 22:3*, *22*(3), 415–429. DOI: 10.1007/s10257-024-00689-9

Rouzky, R., Sarrafzadeh, A., Sowells-Boone, E., Pasandi, H. B., & Goins, G. (2025). Data governance in smart farming: A systematic review of challenges and gaps. In *Information Processing in Agriculture*. China Agricultural University., DOI: 10.1016/j.inpa.2025.11.006

Rozinat, A., & Aalst, W. (2008). Conformance checking of processes based on monitoring real behavior. *Information Systems*, *33*(1), 64–95. DOI: 10.1016/j.is.2007.07.001

Rožman, N., Corn, M., Kozjek, D., Vrabič, R., & Podržaj, P. (2025). Autonomous production unit: An architecture for blockchain-based shared manufacturing. *Robotics and Computer-integrated Manufacturing, 96*, 103035. Advance online publication. DOI: 10.1016/j.rcim.2025.103035

Rudin, C. (2019). Stop explaining black box machine learning models for high stakes decisions and use interpretable models instead. *Nature Machine Intelligence, 1*(5), 206–215. DOI: 10.1038/s42256-019-0048-x

Rudin, C., & Radin, J. (2019). Why are we using black box models in AI when we don't need to? A lesson from an explainable AI competition. *Harvard Data Science Review, 1*(2), 1–9. DOI: 10.1162/99608f92.5a8a3a3d

Ruotsalainen, P., & Blobel, B. (2025). A system model and requirements for transformation to human-centric digital health. *Journal of Medical Internet Research, 27*, e68661. DOI: 10.2196/68661 PMID: 40294402

Russell, S., Norvig, P., Popineau, F., Miclet, L., & Cadet, C. (2021). *Intelligence artificielle : Une approche moderne (4e édition)*. Pearson France. https://hal.science/hal-04245057

Russell, S. (2019). *Human compatible : AI and the problem of control*. Penguin Uk.

Russell, S., & Norvig, P. (2021). *Artificial intelligence: A modern approach* (4th ed.). Pearson.

Saeed Malik, F., Terzidis, O., Malik, F. S., & Terzidis, O. (2026). The metamorphosis of organizational resilience: A strategic roadmap for SMEs in the digital age. *Review of Managerial Science, 2026*, 1–35. DOI: 10.1007/s11846-026-01002-3

Saemaldaher, K., & Emeagwali, O. L. (2025). The Role of Open Innovation in Enhancing Organizational Resilience and Sustainability Performance Through Organizational Adaptability. *Sustainability 2025, Vol. 17, Page 5846, 17*(13), 5846. DOI: 10.3390/su17135846

Sajjanapu, M. (2025). Cloud-native government: Digital infrastructure transforming public service delivery. *Journal of Computer Science and Technology Studies, 7*(7), 82–90. DOI: 10.32996/jcsts.2025.7.7.6

Salameh, A., & Bass, J. M. (2022). An architecture governance approach for Agile development by tailoring the Spotify model. *AI & Society, 37*(2), 761–780. DOI: 10.1007/s00146-021-01240-x

Salehie, M., & Tahvildari, L. (2009). Self-adaptive software. [TAAS]. *ACM Transactions on Autonomous and Adaptive Systems*, *4*(2), 1–42. Advance online publication. DOI: 10.1145/1516533.1516538

Salvaña, M. L. O., Bolingot, H. J. M., & Tangonan, G. L. (2026). A self-organized criticality model of extreme events and cascading disasters of hub-and-spoke air traffic networks. *International Journal of Disaster Risk Reduction*, *133*, 106009. DOI: 10.1016/j.ijdrr.2026.106009

Sang, L. T., Ravikumar, R. N., Aarthi, S., & Ben, T. L. (2025). Optimizing sales and inventory performance using AI and business intelligence in POS environments. In *Advances in Computational Intelligence and Robotics* (pp. 259–290). IGI Global Scientific Publishing. DOI: 10.4018/979-8-3373-4392-1.ch011

Sapkota, R., Roumeliotis, K. I., & Karkee, M. (2026). AI Agents vs. Agentic AI : A Conceptual taxonomy, applications and challenges. *Information Fusion*, *126*, 103599. DOI: 10.1016/j.inffus.2025.103599

Sarrasin, N., & Ramangalahy, C. (2007). La gestion cognitive des connaissances dans les organisations. *Documentation et Bibliothèques*, *53*(1), 43–51. DOI: 10.7202/1029216ar

Sarvate, D., Mishra, S. S., Shanmugapriya, V., & Panwar, D. (2025). Integrating blockchain and federated learning for enhanced security and privacy in smart cities. In Singh, K. K., Singh, A., & Mahesh, T. R. (Eds.), *Integration of federated learning and blockchain for smart cities*. Wiley., DOI: 10.1002/9781394167760.ch10

Schoenegger, S.. (2024). *Large Language Models: Revolutionizing Strategic Business Decision-Making*. Technology & Operations. [REMOVED HYPERLINK FIELD]

Schuller, B.. (2018). Speech emotion recognition: Two decades in a nutshell, benchmarks, and ongoing trends. *Communications of the ACM*, *61*(5), 90–99. DOI: 10.1145/3129340

Sebova, L., Svec, A., & Chovanova, K. (2025). The Role of Business Process Management in Hotel Innovation and Competitiveness. *Tourism and Hospitality 2025, Vol. 6, Page 122, 6*(3), 122. DOI: 10.3390/tourhosp6030122

Seidel, S., Recker, J., & Vom Brocke, J. (2012). Green Business Process Management. *Green Business Process Management: Towards the Sustainable Enterprise, 9783642274886*, 3–13. DOI: 10.1007/978-3-642-27488-6_1

Semenenko, Y. (2025). Modeling digital twins of business processes as a tool for managing company efficiency. *Scientific Notes of Ostroh Academy National University Series Economics, 1*(38(66)), 99–106. DOI: 10.25264/2311-5149-2025-38(66)-99-106

Senge, P. M. (2006). *The fifth discipline : The art and practice of the learning organization*. Broadway Business.

Shahin, M., Maghanaki, M., & Chen, F. F. (2026). Integration of Lean Analytics and Industry 6.0: A Novel Meta-Theoretical Framework for Antifragile, Generative AI-Orchestrated, Circular–Regenerative, and Hyper-Connected Manufacturing Ecosystems. *Big Data and Cognitive Computing 2026, Vol. 10, Page 65, 10*(2), 65. DOI: 10.3390/bdcc10020065

Sharma, D. T. (2024). Embracing artificial intelligence in management: Navigating the futuristic landscape. In *Futuristic Trends in Management Volume 3 Book 13* (pp. 101–116). Iterative International Publisher, Selfypage Developers Pvt Ltd. DOI: 10.58532/V3BFMA13P3CH1

Shneiderman, B. (2020). Human-centered artificial intelligence: Reliable, safe & trustworthy. *International Journal of Human-Computer Interaction, 36*(6), 495–504. DOI: 10.1080/10447318.2020.1741118

Shoham, Y., & Leyton-Brown, K. (2008). *Multiagent systems : Algorithmic, game-theoretic, and logical foundations*. Cambridge University Press. DOI: 10.1017/CBO9780511811654

Siino, M. (2025). Exploring the use of LLMs in the Italian legal domain: A survey on recent applications. *Computer Law & Security Review, 58*, 106164. DOI: 10.1016/j.clsr.2025.106164

Sincorá, L. A., de Oliveira, M. P. V., Zanquetto-Filho, H., & Alvarenga, M. Z. (2023). Developing organizational resilience from business process management maturity. *Innovation & Management Review, 20*(2), 147–161. DOI: 10.1108/INMR-11-2021-0219

Singh, K. S., Shanmugapriya, V., Mishra, S. S., & Singh, M. (2026). Artificial intelligence-driven flow optimization for enhanced efficiency in renewable energy systems. In Awasthi, M. K., & Gupta, R. (Eds.), *Artificial intelligence and computational modeling in heat transfer and fluid dynamics*. Wiley., DOI: 10.1002/9781394433605.ch10

Singh, P. (2024). Using AI to navigate cross-border regulatory complexities in Fin-Tech. *International Journal of Leading Research Publication*, *5*(1), 1596. Advance online publication. DOI: 10.70528/IJLRP.v5.i1.1596

Smallman, H. S., & Cook, M. B. (2013). Proactive supervisory decision support from trend-based monitoring of autonomous and automated systems: A tale of two domains. In *Virtual, Augmented and Mixed Reality. Systems and Applications* (pp. 320–329). Springer Berlin Heidelberg. DOI: 10.1007/978-3-642-39420-1_34

Smirnov, M. (2012). *Business process model abstraction.*

Sohns, T. M., Aysolmaz, B., Figge, L., & Joshi, A. (2023). Green business process management for business sustainability: A case study of manufacturing small and medium-sized enterprises (SMEs) from Germany. *Journal of Cleaner Production*, *401*, 136667. DOI: 10.1016/j.jclepro.2023.136667

Stary, C. (2021). Digital Twin Generation: Re-Conceptualizing Agent Systems for Behavior-Centered Cyber-Physical System Development. *Sensors 2021, Vol. 21, Page 1096*, *21*(4), 1096. DOI: 10.3390/s21041096

Sukharevsky, A., Kerr, D., Hjartar, K., Hämäläinen, L., Bout, S., Di Leo, V., & Dagorret, G. (2025). Seizing the agentic AI advantage. *McKinsey & Company*. https://www.fredericsanz.com/s/Seizing-the-Agentic-AI-Advantage.pdf

Sukharevsky, A., Krivkovich, A., Gast, A., Storozhev, A., Maor, D., Mahadevan, D., Hämäläinen, L., & Durth, S. (2025). *The agentic organization : Contours of the next paradigm for the AI era*. McKinsey & Company.

Sutar, D., Metri, R. A., Naik, R., Mishra, S., Thakkar, J., & Kolhe, M. (2026). Energy-efficient automation of industrial surface pre-treatment processes using PLC and SCADA. In *Proceedings of the Second International Conference on New Power System and Power Electronics (NPSPE 2025)* (Vol. 14129, Article 141292C). SPIE. https://doi.org/DOI: 10.1117/12.3110247

Szczepanek, M., Rudnik, K., Deptuła, A., & Estrada, Q. (2025). Development of approaches to business processes improvements in a digital transformation. *IFAC-PapersOnLine*, *59*(24), 257–262. DOI: 10.1016/j.ifacol.2025.11.874

Szelagowski, M., & Berniak-Woźny, J. (2020). The adaptation of business process management maturity models to the context of the knowledge economy. *Business Process Management Journal*, *26*(1), 212–238. DOI: 10.1108/BPMJ-11-2018-0328

Szelągowski, M., & Lupeikiene, A. (2020). Business process management systems: Evolution and development trends. *Lithuanian Academy of Sciences. Informatica (Vilnius)*, •••, 579–595. DOI: 10.15388/20-INFOR429

Taeihagh, A. (2025). Governance of Generative AI. *Policy and Society*, *44*(1), 1–22. DOI: 10.1093/polsoc/puaf001

Tarasiuk, M. V. (2025). Genesis and evolutionary factors concepts of process-oriented management. *Вісник Київського Національного Лінгвістичного Університету. Серія Історія, Економіка. Філософія*, (30), 166–179. DOI: 10.32589/2412-9321.30.2025.331979

Teece, D. J. (2007). Explicating dynamic capabilities : The nature and microfoundations of (sustainable) enterprise performance. *Strategic Management Journal*, *28*(13), 1319–1350. DOI: 10.1002/smj.640

Teece, D. J. (2018). Business models and dynamic capabilities. *Long Range Planning*, *51*(1), 40–49. DOI: 10.1016/j.lrp.2017.06.007

Teece, D., & Pisano, G. (1994). The dynamic capabilities of firms : An introduction. *Industrial and Corporate Change*, *3*(3), 537–556. DOI: 10.1093/icc/3.3.537-a

Teece, D., Pisano, G., & Shuen, A. (1997). Dynamic capabilities and strategic management. *Strategic Management Journal*, *18*(7), 509–533. DOI: 10.1002/(SICI)1097-0266(199708)18:7<509::AID-SMJ882>3.0.CO;2-Z

Teixeira, A. R., Ferreira, J. V., & Ramos, A. L. (2024). Optimization of Business Processes Through BPM Methodology: A Case Study on Data Analysis and Performance Improvement. *Information 2024, Vol. 15, Page 724, 15*(11), 724. DOI: 10.3390/info15110724

Ter, A., Aalst, W., Adams, M., & Russell, N. (2010). *Modern business process automation: YAWL and its support environment*. DOI: 10.1007/978-3-642-03121-2

Terreberry, S. (1968). The Evolution of Organizational Environments. *Administrative Science Quarterly*, *12*(4), 590–613. DOI: 10.2307/2391535

Tetteh, F. K., & Owusu Kwateng, K. (2025). Digital pathways to supply chain viability: Exploring nice and must-to have practices in dynamic environment –insight from SEM and NCA. *Journal of Enterprise Information Management*, •••, 1–38. DOI: 10.1108/JEIM-05-2025-0367/1308502

Thakur, K., & Gupta, K. (2025). Artificial intelligence in social media advertising: Raising engagement and ethical concerns. In *Improving consumer engagement in digital marketing through cognitive AI*. IGI Global. https://doi.org/ of FormDOI: 10.4018/979-8-3373-3775-3.ch011Bottom

Theodorakopoulos, L., Theodoropoulou, A., & Halkiopoulos, C. (2025). Cognitive Bias Mitigation in Executive Decision-Making: A Data-Driven Approach Integrating Big Data Analytics, AI, and Explainable Systems. *Electronics 2025, Vol. 14, Page 3930, 14*(19), 3930. DOI: 10.3390/electronics14193930

Thoenig, J.-C., & Michaud, C. (2007). *Le management cognitif* [Article]. HAL open access archive. https://shs.hal.science/halshs-00142994/document

Thomas Kohlborn, D. O. M. P. J. P. (2014). Event-Driven Business Process Management: Where are we now? A comprehensive synthesis and analysis of literature. *Business Process Management Journal, 20*(4), 615–633. DOI: 10.1108/BPMJ-07-2013-0092

Tirulo, A., Yadav, M., Lolamo, M., Chauhan, S., Siano, P., & Shafie-khah, M. (2026). Beyond automation: Unveiling the potential of agentic intelligence. *Renewable & Sustainable Energy Reviews, 226*, 116218. DOI: 10.1016/j.rser.2025.116218

Tornatzky, L. G., & Fleischer, M. (1990). *The processes of technological innovation*. Heath & Company.

Troshani, I., & Rowbottom, N. (2024). Corporate sustainability reporting and information infrastructure. *Accounting, Auditing & Accountability Journal, 37*(4), 1209–1237. DOI: 10.1108/AAAJ-01-2023-6244

Tsakalidis, G., & Vergidis, K. (2026). Systematicity and generalizability in business process redesign methodologies: A systematic literature review. *Science of Computer Programming, 248*, 103392. DOI: 10.1016/j.scico.2025.103392

Tumpa, R. J., & Naeni, L. (2025). Improving decision-making and stakeholder engagement at project governance using digital technology for sustainable infrastructure projects. *Smart and Sustainable Built Environment, 14*(4), 1292–1329. DOI: 10.1108/SASBE-10-2024-0451

Tung Khuat, T., Jacob Kedziora, D., & Gabrys, B. (2023). The Roles and Modes of Human Interactions with Automated Machine Learning Systems: A Critical Review and Perspectives. *Foundations and Trends in Human-Computer Interaction, 17*(3–4), 195–387. DOI: 10.1561/1100000091

Turienzo, J., Cabanelas, P., & Lampón, J. F. (2023). Business models in times of disruption: The connected and autonomous vehicles (uncertain) domino effect. *Journal of Business Research, 156*, 113481. Advance online publication. DOI: 10.1016/j.jbusres.2022.113481

Tushman, M. L., & Nadler, D. A. (1978). Information Processing as an Integrating Concept in Organizational Design. *Academy of Management Review*, *3*(3), 613. DOI: 10.2307/257550

Ullagaddi, P. (2024). Digital transformation in the pharmaceutical industry: Ensuring data integrity and regulatory compliance. *International Journal of Business and Management*. Advance online publication. DOI: 10.5539/ijbm.v19n5p16

Ullah, I., Arishi, A., Singh, S. K., Alharbi, F., Ibrahim, A. H., Islam, M., Daradkeh, Y. I., & Choi, C. (2025). Autonomous network management for 6G communication: A comprehensive survey. *Digital Communications and Networks*, *11*(6), 1917–1940. DOI: 10.1016/j.dcan.2025.07.001

Uzoamaka, N.-O. P., Amujiogu, A. A., & Blessing, E. O. (2025). Governance and ethical oversight of agentic artificial intelligence in business process management: Balancing human-autonomy, organizational accountability and algorithmic transparency. *Journal of Basic and Applied Research International*, *31*(6), 121–133. DOI: 10.56557/jobari/2025/v31i610028

Vaishnavi, S., Shobana, J., Renugadevi, R., Ramanathan, S., Arthi, K., & A. V., K. (2025). Revolutionizing Finance and Accounting Through AI-Driven Decision Making. In M. Tariq & R. Sergio (Eds.), *Convergence of AI, Education, and Business for Sustainability* (pp. 261-284). IGI Global Scientific Publishing.

Valmohammadi, C., & Mortaz Hejri, F. (2023). Designing a conceptual green process model in software development: A mixed method approach. *International Journal of Information Management Data Insights*, *3*(2), 100204. DOI: 10.1016/j.jjimei.2023.100204

van der Aalst, W. M. (2016). *Data science in action.*

van Esch, P. (2026). From agentic AI to AI-orchestrated organizations : Understanding the next surge in artificial intelligence. *Business Horizons*. Advance online publication. DOI: 10.1016/j.bushor.2026.03.003

Veeravalli, P. K.Pradeep Kiran Veeravalli. (2025). The synergistic integration of artificial intelligence and cloud orchestration in process automation. *Journal of Computer Science and Technology Studies*, *7*(2), 316–322. DOI: 10.32996/jcsts.2025.7.2.32

Verma, L. (2025). The Transformative Impact of Artificial Intelligence on Business Process Management. *European Journal of Computer Science and Information Technology*, *13*(48), 14–25. Advance online publication. DOI: 10.37745/ejcsit.2013/vol13n481425

Vidgen, R., & Wang, X. (2006). From business process management to business process ecosystem. *Journal of Information Technology, 21*(4), 262–271. DOI: 10.1057/palgrave.jit.2000076

Villamil, R. M., Restrepo-Carmona, J. A., Escobar, A., Aponte-Moreno, A., Herrera, J. A., Gutiérrez-Betancur, S. A., & Fletscher, L. (2025). An Enterprise Architecture-driven service integration model for enhancing fiscal oversight in supreme audit institutions. *Applied System Innovation, 9*(1), 16. DOI: 10.3390/asi9010016

Vincent, N. (2020). Emotion recognition and AI: Emerging tools for design empathy. *International Journal of Human-Computer Studies, 142*, 102–113.

Visagamurthy, G. (2025). Cross-sector digital transformation: Lessons from insurance industry automation and interoperability models. *European Journal of Computer Science and Information Technology, 13*(41), 131–145. DOI: 10.37745/ejcsit.2013/vol13n41131145

Viteri-Sánchez, C., & Novillo-Villegas, S. (2025). A Framework for a Sustainable Adoption of Business Process Management. *Sustainability 2025, Vol. 17, Page 9827, 17*(21), 9827. DOI: 10.3390/su17219827

Vom Brocke, J., Seidel, S., & Recker, J. (2013). Green business process management: Towards the sustainable enterprise. *Green Business Process Management: Towards the Sustainable Enterprise, 9783642274886*, 1–251. DOI: 10.1007/978-3-642-27488-6/COVER

Vrchota, J., Pech, M., Rolínek, L., & Bednář, J. (2020). Sustainability Outcomes of Green Processes in Relation to Industry 4.0 in Manufacturing: Systematic Review. *Sustainability 2020, Vol. 12, Page 5968, 12*(15), 5968. DOI: 10.3390/su12155968

Walmsley, T. G., Patros, P., Yu, W., Young, B. R., Burroughs, S., Apperley, M., Carson, J. K., Udugama, I. A., Aeowjaroenlap, H., Atkins, M. J., & Walmsley, M. R. W. (2024). Adaptive digital twins for energy-intensive industries and their local communities. *Digital Chemical Engineering, 10*, 100139. DOI: 10.1016/j.dche.2024.100139

Wang, C., & Chen, Y. (2026). Evaluating large language models for evidence-based clinical question answering. *Patterns (New York, N.Y.), 101519*(5), 101519. Advance online publication. DOI: 10.1016/j.patter.2026.101519 PMID: 42130942

Wang, N. (2023). Design of an Intelligent Processing System for Business Data Analysis Based on Improved Clustering Algorithm. *Procedia Computer Science, 228*, 1215–1224. DOI: 10.1016/j.procs.2023.11.105

Wang, X., Shi, X., Chen, J., Guo, X., & Li, D. (2024). Exploring optimal pathways for enterprise procurement management systems based on fast neural modeling and semantic segmentation. *Heliyon, 10*(7), e26474. Advance online publication. DOI: 10.1016/j.heliyon.2024.e26474 PMID: 38689967

Wang, Y., Bai, H., & Meng, X. (2023). Intelligent practice of big language model in library reference service. *Information Theory and Practice*, *46*(8), 96–103.

Weick, K. E. (1995). *Sensemaking in organizations* (Vol. 3). Sage publications Thousand Oaks, CA. https://www.torrossa.com/gs/resourceProxy?an=5564159&publisher=FZO137#page=545

Weinzierl, S., Zilker, S., Dunzer, S., & Matzner, M. (2024). Machine learning in business process management: A systematic literature review. *Expert Systems with Applications*, *253*, 124181. DOI: 10.1016/j.eswa.2024.124181

Weyns, D. (2021). *An introduction to self-adaptive systems : a contemporary software engineering perspective*. https://www.wiley.com/en-ie/An+Introduction+to+Self-adaptive+Systems%3A+A+Contemporary+Software+Engineering+Perspective-p-9781119574910

Widad, J. (2025). The Evolution of Autonomy : From Reactive AI to Agentic AI. In Reddy, C. K. K., Joseph, S., Joshi, H., Doss, S., & Ouaissa, M. (Eds.), *The Power of Agentic AI: Redefining Human Life and Decision-Making* (pp. 1–9). Springer Nature Switzerland., DOI: 10.1007/978-3-031-89424-4_1

Wilson, H. J., & Daugherty, P. R. (2018). Collaborative intelligence : Humans and AI are joining forces. *Harvard Business Review*, *96*(4), 114–123.

Winter, S. G. (2003). Understanding dynamic capabilities. *Strategic Management Journal*, *24*(10), 991–995. DOI: 10.1002/smj.318

Wu, Z., Pan, S., Chen, F., Long, G., Zhang, C., & Yu, P. S. (2019). A comprehensive survey on graph neural networks. *IEEE Transactions on Neural Networks and Learning Systems*, *32*(1), 4–24. DOI: 10.1109/TNNLS.2020.2978386 PMID: 32217482

Xiao, L., & Cao, H. (2017). Organizational Resilience: The Theoretical Model and Research Implication. *ITM Web of Conferences, 12*, 04021. DOI: 10.1051/itmconf/20171204021

Xu, W. (2020). The application of AI in creative industries. *Journal of Creativity*, *12*(4), 311–325.

Xu, W., Huang, D., Chen, Z., & Lin, J. (2023). Research and design of AI communication robot based on special equipment knowledge base. *China Special Equipment Safety, 39*(2), 26–29.

Yanamadala, P. (2025). Strategic pathways for AI modernization: A framework for enterprise evolution. *Journal of Computer Science and Technology Studies, 7*(10), 680–688. DOI: 10.32996/jcsts.2025.7.10.68

Yang, W., Some, L., Bain, M., & Kang, B. (2025). A comprehensive survey on integrating large language models with knowledge-based methods. *Knowledge-Based Systems, 318*, 113503. DOI: 10.1016/j.knosys.2025.113503

Yang, Y., & Xue, G. (2026). How does environmental regulation reshape global value chain resilience? *Cleaner Environmental Systems, 20*, 100395. DOI: 10.1016/j.cesys.2026.100395

Yeung, T., Martinez, J. G., Schlenger, J., Borrmann, A., & Sacks, R. (2025). Integrating digital twin and agent-based simulation to support adaptive production system design in building projects. *Automation in Construction, 180*, 106550. DOI: 10.1016/j.autcon.2025.106550

Zahra, S. A., & George, G. (2002). Absorptive Capacity : A Review, Reconceptualization, and Extension. *Academy of Management Review, 27*(2), 185. DOI: 10.2307/4134351

Zelt, S., Recker, J., Schmiedel, T., & vom Brocke, J. (2019). A theory of contingent business process management. *Business Process Management Journal, 25*(6), 1291–1316. DOI: 10.1108/BPMJ-05-2018-0129

Zetta Software Tlc Shpk, A., Benelli, F., Maciariello, F., New Generations Sensors Italy, 56024, Marketing Area) Italy, S. M. la F. (CE, Marku, R., & Department of Economics and Business Albania, T. B. U. C. (2025). Towards a human-centric Physical Internet: The next digital transition. *Journal of Robotics and Automation Research, 6*(4), 01–09. DOI: 10.33140/JRAR.06.04.10

Zhang, Z., Kang, Y., Lu, Y., & Li, P. (2025). The Role of Artificial Intelligence in Business Model Innovation of Digital Platform Enterprises. *Systems 2025, Vol. 13, Page 507, 13*(7), 507. DOI: 10.3390/systems13070507

Zhang, D., Frei, R., Wills, G., Gerding, E., Bayer, S., & Senyo, P. K. (2023). Strategies and practices to reduce the ecological impact of product returns: An environmental sustainability framework for multichannel retail. *Business Strategy and the Environment, 32*(7), 4636–4661. DOI: 10.1002/bse.3385

Zhang, H., Wang, X., Han, L., Li, Z., Chen, Z., & Chen, Z. (2023). Question and answer system research on knowledge graph fusion of large language models. *Computer Science and Exploration*, *17*(10), 2377–2388.

Zhang, J.. (2022). AI-assisted design ideation: Diversity, novelty, and evaluation. *Design Studies*, *75*, 101–118. DOI: 10.1016/j.destud.2021.101030

Zhang, Y.. (2021). Human-AI collaboration in creativity: A case study of co-creation with AI. *Proceedings of the CHI Conference on Human Factors in Computing Systems*, 1–13. DOI: 10.1145/3411764.3445218

Zhang, Y., Wang, P., & Xu, H. (2021). Ethical challenges in artificial intelligence applications: A human-in-the-loop approach. *AI & Society*, *36*, 1115–1128. DOI: 10.1007/s00146-020-01062-0

Zhu, Z., Song, Z., & Palazoglu, A. (2011). Transition process modeling and monitoring based on dynamic ensemble clustering and multiclass support vector data description. *Industrial & Engineering Chemistry Research*, *50*(24), 13969–13983. DOI: 10.1021/ie201792r

Zimek, M., & Baumgartner, R. J. (2024). Systemic sustainability assessment: Analyzing environmental and social impacts of actions on sustainable development. *Cleaner Production Letters*, *7*, 100064. DOI: 10.1016/j.clpl.2024.100064

Ziwoni, C. (2025). Strategic oversight and governance framework for Business Process Management at Old Mutual Namibia. *International Journal of Innovative Science and Research Technology*, *2081–2085*, 2081–2085. Advance online publication. DOI: 10.38124/ijisrt/25jun1428

Zur Muehlen, M., & Indulska, M. (2009). Modeling languages for business processes and business rules: A representational analysis. *Information Systems*, *35*(4), 379–390. DOI: 10.1016/j.is.2009.02.006

About the Contributors

Houssem Chemingui is an Assistant Professor of Information Systems at Brest Business School, France. He earned his PhD in Computer Science from Université Paris 1 Panthéon-Sorbonne, where his doctoral research focused on advanced approaches to software configuration and process-oriented decision support systems. His research lies at the intersection of artificial intelligence, data analytics, and software engineering, with a particular emphasis on guided product configuration and decision support for complex organizational processes. His work addresses several application domains, including business process management, logistics, and higher education, exploring how intelligent systems can enhance decision-making and organizational performance. He has published and presented his research in international conferences and scientific venues, and maintains active collaborations with both academic institutions and industry partners. He also serves as a reviewer for academic journals published by Elsevier and Springer. Through his research and teaching activities, he aims to bridge theoretical contributions and their practical implementation in real-world contexts.

***

Jumayev Akbar is affiliated with the Economic Statistics Department at Tashkent State University of Economics, Tashkent, Uzbekistan. He is engaged in academic and research activities in the field of economic statistics, with a focus on the application of statistical methods to economic analysis, policy evaluation, and data-driven decision-making. His work supports the advancement of quantitative approaches in higher education and contributes to the understanding of economic trends and development indicators. He is committed to promoting analytical thinking, research excellence, and the effective use of statistical tools in academic and practical contexts. Dr. Jumayev contributes to teaching, research, and scholarly discussions that strengthen the role of statistics in economics and public policy. His academic interests include economic measurement, applied statistics, econometric analysis, and the interpretation of socio-economic data for strategic planning and institutional development.

Gafurov Anvar Bazarbayevich is the Head of the Department of Economics at Alfraganus University, Tashkent, Uzbekistan. His academic and professional interests focus on economic development, economic policy, and the analysis of contemporary economic systems. He is actively involved in academic leadership, teaching, and research activities that contribute to strengthening economics education and advancing scholarly research in the field. His work emphasizes the application of economic theories and analytical methods to address challenges related to economic growth, public policy, and sustainable development. As a departmental leader, he also plays an important role in curriculum development, academic collaboration, and research coordination. Through his academic and administrative engagement, Gafurov Anvar Bazarbayevich aims to promote high-quality economic education and support the preparation of professionals capable of addressing modern economic and policy challenges.

Ismailov Azizbek is affiliated with the Faculty of Economics at Alfraganus University, Tashkent, Uzbekistan. His academic interests focus on economic theory, economic development, and the analysis of contemporary economic issues in both national and global contexts. He is actively engaged in teaching and academic activities that promote a deeper understanding of economic principles, policy analysis, and sustainable economic growth. His work emphasizes the application of economic models, quantitative methods, and analytical approaches to address challenges related to market development, financial systems, and public policy. Ismailov Azizbek also contributes to academic discussions on economic planning, development strategies, and economic policy formulation. Through his scholarly engagement and educational activities, he aims to strengthen economics education and support the development of students capable of analyzing and addressing complex economic challenges.

Othman Boulitama, who holds a PhD in Management Sciences, has received advanced academic training and built a distinguished career marked by successful teaching experience at different educational levels. His expertise notably encompasses the teaching of finance and logistics, fields in which he has consistently combined scientific rigor with an innovative pedagogical approach. He also serves as a reviewer for the Journal of Information and Knowledge Management (JIKM) as well as for books and journals published by IGI Global, while concurrently holding the position of Associate Editor of the International Journal of Innovation, Digital Economy and Management Sciences (IJIDEMS). Recognized for his ability to adopt and integrate new teaching methods, he is committed to creating a stimulating learning environment that fosters both knowledge acquisition and the personal development of students. His patience, rigor, and strong organizational skills constitute the foundations of his professional commitment. Driven by a sustained passion for education and research, he continues his academic work with the aim of exploring new perspectives and actively contributing to the advancement of pedagogical and scientific practices in the field of management sciences.

Kritika Gupta is an experienced academician with over 13 years in the education sector. She holds a Ph.D. in Business Management and is currently an Assistant Professor at the University Business School, Chandigarh University. Her research contributions include publications in reputed Q1 journals indexed in SCOPUS and Web of Science. Dr. Gupta is passionate about delivering quality education and advancing academic research, consistently striving to enhance learning outcomes and contribute to the development of knowledge in the field of business management.

Mavlanov Normumin serves as the Director of the Department of Scientific Research and Innovation at Tashkent State University of Economics, Tashkent, Uzbekistan. He is actively involved in strengthening the university's research culture and promoting innovation-driven academic development. His professional work focuses on advancing scientific research activities, supporting interdisciplinary collaboration, and encouraging the practical application of knowledge to address contemporary academic and societal challenges. He contributes to the design and implementation of strategies that enhance research productivity, innovation capacity, and institutional academic excellence. Dr. Mavlanov plays an important role in fostering a dynamic environment for scholars, researchers, and students by promoting quality research initiatives and innovation-oriented programs. His interests include research management, innovation systems, higher education development, and the integration of scientific inquiry with institutional growth. Through his leadership and commitment, he continues to support the expansion of impactful research and innovation practices that strengthen the university's national and international academic standing.

Nidhi Oswal is currently working as Head of the Human Resource Management department, Faculty of Business at Liwa College AbuDhabi, UAE. Program Leader and Assistant Professor for the Business Administration department at Khawarizmi International College, UAE. I worked as a lecturer at Punjab Institute of Management and Technology, India. In addition, she has been a recruitment consultant for companies in UAE and India. She has earned the SHRM Certified Professional® (SHRM-CP) credential. Nidhi has attended various conferences within and outside the UAE. She holds an ABA level 1 certification through KHDA. She specializes in recruitment, selection, and training & development. The main research areas are recruitment, technology implications in HR, human behaviour, and Sustainable HR development. Furthermore, she serves as a reviewer for a range of reputable journals.

Ravikumar R. N. is currently a Postdoctoral Researcher at Universiti Malaysia Perlis, Malaysia, and also serves as an Associate Professor in the Department of Computer Science and Engineering at Marwadi University, Rajkot, India. He has more than 14 years of national and international teaching experience. He completed his Ph.D. from Amity University, Rajasthan, M.E. from VMRF Deemed University, and B.E. from Anna University, Chennai. He actively trains and mentors students in research activities, contributing to various scholarly publications. He has published numerous research articles in Scopus and SCI-indexed journals, book chapters, and conference proceedings. He also holds several patents and has authored books. His areas of specialization include Machine Learning, Deep Learning, Federated Learning, Recommendation Systems, and Optimization Techniques.

R. Renugadevi completed her Ph.D. in 2022 from Anna University, Chennai, and currently serves as an Associate Professor at Saveetha Engineering College. With over two decades of teaching experience in esteemed engineering institutions, she made significant contributions to the academic and research domains. She has published more than 45 research papers in renowned international journals and conferences, and holds 10 patents, reflecting her innovative approach to technology and engineering. She holds the distinction of being an approved research supervisor at Anna University, with six scholars under her supervision. Her research interests span across cutting-edge fields such as Machine Learning, Internet of Things (IoT), Wireless Sensor Networks, and Cloud Computing. As a passionate academician, she is a life member of professional organizations, including the Indian Society for Technical Education (ISTE), Internet Society and the International Association of Engineers (IAENG).

Aarthi S is a dedicated Assistant Professor in Computer Science with over six years of teaching experience. She is currently pursuing her Ph.D. at Marwadi University, Gujarat, and holds an M.Phil. and MCA from Bharathiar University, Tamil Nadu. Throughout her academic career, Aarthi has specialized in programming languages such as C++, Java, and C, and contributed to research in recommender systems and AI-related data privacy. She has published journal articles, contributed to book chapters, and actively participated in technical workshops. Known for her student mentorship and engagement in academic events, Aarthi remains committed to fostering innovation and excellence in computer science education.

Brahim Sabiri is Assistant Professor at the Moroccan School of Engineering Sciences and PhD in Economics from Hassan II University of Casablanca. His research focuses on individual and organizational behavior, behavioral and experimental approaches, as well as the use of new technologies in the service of management.

Karimov Rashid Salayevich is affiliated with the Department of Accounting at Mamun University, Khiva, Uzbekistan. His academic interests focus on accounting systems, financial reporting, auditing, and financial management. He is actively involved in teaching and academic activities aimed at strengthening students' understanding of accounting principles and modern financial practices. His work emphasizes transparent financial reporting, the application of international accounting standards, and the use of contemporary analytical tools in accounting and finance. Karimov Rashid Salayevich also contributes to academic discussions related to corporate finance, business management, and financial regulation. Through his teaching and scholarly engagement, he aims to enhance accounting education and support the development of competent professionals capable of addressing modern financial and organizational challenges.

Gautam Shivaraj, with a comprehensive academic background including a Bachelor's in Mechanical Engineering, Master of Engineering Management (Australia), MBA and also pursuing Ph.D., brings an impressive 07 years of teaching experience alongside a brief stint in the industry of 5 years. Currently serving as an Assistant Professor at the Adarsh Institute of Management and Information Technology in Bangalore, Karnataka, India, he has distinguished himself in the academic and research fields. Prof.Gautam Shivaraj professional journey is a testament to his dedication to academia, research, in the evolving world.

Yadavalli Srinivasa Rao is an Assistant Professor in the Department of Management Studies at Vignan's Deemed to be University, Guntur, with over 14 years of teaching and research experience. He holds a Ph.D. in Financial Management from JNTU Anantapur, focusing on SME financial performance. His expertise includes financial management, investment analysis, and AI-based financial research. He has published research papers in reputed journals and contributed book chapters with international publishers like Springer. He also holds patents in AI-based financial systems. Dr. Rao actively participates in conferences, workshops, and academic development programs, contributing to innovative and industry-oriented management education.

Dilmurod Turaev is affiliated with the Department of Information Technology and Exact Sciences at Termez University of Economics and Service, Termez, Uzbekistan. His academic interests focus on information technology, computer science, and the application of computational methods in scientific and educational environments. He is actively involved in teaching and academic activities that aim to strengthen students' knowledge in programming, data analysis, and modern computing technologies. His work emphasizes the integration of digital technologies, problem-solving skills, and analytical approaches to address real-world technological challenges. Dilmurod Turaev also contributes to academic discussions related to emerging technologies, information systems, and the advancement of IT education. Through his teaching and scholarly engagement, he seeks to promote innovation in technology education and support the development of skilled professionals capable of contributing to modern digital and scientific advancements.

Amit Verma is an Assistant Professor in the department of Commerce, Kalicharan PG College Lucknow, Uttar Pradesh, India. He had done LLB, MBA, Ph.D. in Commerce & qualified UGC NET Two Times. He was also awarded with Honorary Ph.D. in Management. Recently, he worked as ICSSR Project Director for Short-term Empirical Research Project (Collaborative) entitled- Digital Transformation of Health Services- A Comparative Study of Bhopal and Lucknow. He was also invited as Resource Person and Keynote Speaker in various National Seminars sponsored by ICSSR. He has been nominated as Academic Member of Communication Institute of Greece, Athens, Greece. He was also invited by the Saint Augustine University of Tanzania (SAUT) in Mwanza, Tanzania, East Africa as a Chief speaker in Media Management Seminar. He served in various capacities in different reputed Colleges and a Private University of Delhi NCR, Uttar Pradesh & Uttarakhand and provided in-depth academic knowledge in various subjects. He has extensive hands-on experience in conceptualizing strategies, building and knitting confidence level, measuring and evaluating efficiency of performance of young learners. He participated in various development programs, workshops, conferences, seminars and FDP'S. His strong point does not stop at creating an interest for the concept and communicating the required knowledge, counselling (personal & professional), planning a specific career, shifting attitude towards solving problems, nurturing basic values which added worth to the workplace. He has strong belief in self-motivation, an urge for excellence and the feeling of independence which helped in ensuring updated knowledge and a structured reorientation of the subject matter. He is very much inspired by Swami Vivekananda and his philosophy and he willingly follows the Swami Vivekananda philosophy of life. He is Secretary of Commerce and Management Association, India. He is life member of Lucknow Management Association, Lucknow and President of Swatantrata Sangram Senani

Shrikrishna Verma Charitable trust, Lucknow. He is also member of Editorial Board in different International and National Journals such as International Journal of Multi-disciplinary Research and Technology, KAVYA Publication, ACME, KVIM Journal of Management & many others. He is the author of Research Papers in peer-reviewed journals, including in ABDC, SSCI, and SCOPUS-indexed journals. He has also reviewed a large number of research papers. His publications encompass edited volumes that bridge the gap between academics and industry. He wrote Books on the subjects Essentials of Management, Organisational Behaviour and Research Methodology. He also edited books Entrepreneurship Amidst Pandemic COVID19, Emerging Business Practices and Trends during COVID-19, Innovative Business Practices in the Present Digital Era and Future of Marketing: Trends and Changes, The Vision of Viksit Bharat: Holistic Growth and Development of All Sectors, Catalysts of Change Interdisciplinary Research on India's Growth Story etc.

Index

A

B

C

D

E

F

G

I

www.ingramcontent.com/pod-product-compliance
Lightning Source LLC
LaVergne TN
LVHW060006031026
849161LV00025B/1500